RESEARCH METHODS IN BEHAVIORAL SCIENCES

Fred Leavitt
California State University—Hayward

Wm. C. Brown Publishers

Book Team

Editor *Michael Lange*
Developmental Editor *Carla J. Aspelmeier*
Production Editor *Anne E. Gardiner*
Designer *Laurie J. Entringer*
Art Editor *Donna Slade*
Permissions Editor *Karen L. Storlie*
Visuals Processor *Kenneth E. Ley*

Wm. C. Brown Publishers

President *G. Franklin Lewis*
Vice President, Publisher *George Wm. Bergquist*
Vice President, Publisher *Thomas E. Doran*
Vice President, Operations and Production *Beverly Kolz*
National Sales Manager *Virginia S. Moffat*
Advertising Manager *Ann M. Knepper*
Marketing Manager *Kathy Law Laube*
Production Editorial Manager *Colleen A. Yonda*
Production Editorial Manager *Julie A. Kennedy*
Publishing Services Manager *Karen J. Slaght*
Manager of Visuals and Design *Faye M. Schilling*

Cover illustration: © Karen Watson

Photo credit: David Caras

Copyright © 1991 by Wm. C. Brown Publishers. All rights reserved

Library of Congress Catalog Card Number: 89-081362

ISBN 0-697-11260-8

No part of this publication may be reproduced, stored in a retrieval system, or transmitted, in any form or by any means, electronic, mechanical, photocopying, recording, or otherwise, without the prior written permission of the publisher.

Printed in the United States of America by Wm. C. Brown Publishers, 2460 Kerper Boulevard, Dubuque, IA 52001

10 9 8 7 6 5 4 3 2 1

I dedicate this book to Diane, Jess, Mel,
and the memory of my mother.

CONTENTS

Preface x

PART I Introduction 1

1 Introduction and Overview 3

The Rewards of Science 5
What Is Science? 8
    Scientists Organize Facts 8
    Scientists Create Falsifiable Theories 9
    A Few Noncritical Features 10
Limitations of Science 10
    Scientists Can Answer Only Certain Kinds of Questions 10
    Scientists Cannot Establish Absolute Truth 11
    The Fundamental Uncertainty 14
We Are All Scientists 15
What Makes a Person a Good Scientist 17
    Scientists Versus Data Collectors 18
Objectives for This Book 20
    To Show the Relevance of Social Science Research 20
    To Show the Importance of Diverse Approaches 21
    To Show the Unity of Scientific Method 21
    To Introduce Controversial Issues 21
    To Show How Exciting Science Can Be 22
Summary 22
Key Terms 23

PART II Preparing for Research 25

2 Finding an Interesting Problem 27

Be Attentive to Both Routine and Unusual Occurrences 28
Become a Careful Observer 28
Read-I 28
Read-II 29
Seek Analogies 29
Examples 30

3 Studying the Problem Area in Depth 33

Abstracts 36
Citation Indexes 36
Index Medicus 40
Reviews of Research 41
Current Contents 41
Other Strategies 41
How Should You Read? 42
Summary 43
Key Terms 43

4 Evaluating What You Read: Single Studies 44

Problems of Internal Validity 46
Problems of Construct Validity 47
Problems of External Validity 48
    Scorer 49
    Items 49
    Time 49
    Setting 49
    Method 50
    Dimensions 50
Statistical Validity 51
The Relationship Between Data and Theory 54
Cheating, Distortions, and Related Problems 54

Contents   v

Questions for Consideration *58*
Summary *59*
Key Terms *59*
Answers *60*

5   Asking Valuable Questions *62*

Features of Good Questions *63*
    Good Scientific Questions Must Be Answerable Empirically *64*
    Good Scientific Questions Indicate the Nature of Acceptable Answers *64*
    Good Questions are Worth Answering *64*
Types of Questions *65*
    Questions That Lead Directly to Research Projects *65*
    Questions That Organize Research Programs *68*
Summary *74*
Key Terms *74*
Answers *74*

6   Selecting and Measuring Variables *76*

Scales of Measurement *79*
Principles of Classifying *80*
Properties of Good Measures *81*
    Measures Must Be Reliable *82*
    Measures Must Be Sensitive *84*
    Measures Must Be Valid *86*
Summary *94*
Key Terms *94*

PART III   Developing a Strategy for Answering the Questions *95*

7   Developing a Strategy for Testing the Speculations *97*

Surveys *101*
Case Studies *101*
Correlational Studies *102*
Secondary Analysis *104*
Experiments *104*
Why Choose One Method Over Another? *105*
Additional Comments *105*
Summary *106*
Key Terms *107*
Answers *108*

8   Developing an Ethical Strategy *109*

Key Ethical Issues Facing Scientists Who Use Human Subjects *111*
    Is It Ever Appropriate to Harm Anybody for the Advancement of Science? *112*
    Is It Ever Proper to Deceive Anybody In the Interests of Research? *115*
    Do Researchers Have the Right to Invade the Privacy of Others? *115*
    Must Informed Consent of Subjects Be a Prerequisite to Research? *116*
Methods for Reducing Ethical Problems *116*
An Ethical Dilemma *118*
Should Ethical Concerns Ever Lead to the Suppression of Research? *119*
Ethics of Research with Animals *120*
Summary *124*
Answers *124*

9   A Few Suggestions for Creative Problem Solving *126*

Redefining and Analyzing Problems *127*
    Boundary Examinations *127*
    Five W's and H *128*
    Considering Related Problems *128*
    Working Backwards *129*
Generating Ideas *129*
    Seek Analogies *129*
    Cliches, Proverbs, and Maxims *130*
    Draw *131*
Evaluating and Selecting Ideas *131*
    Highlighting *131*
Implementing Ideas *131*
    Potential Problem Analysis *131*
Work *132*
Relax *132*
Pretend *133*
Be Happy *133*
Don't Be Smug *133*
Summary *134*
Answers *134*

PART IV The Different Strategies 135

10 Observing 137

Reasons for Doing Observational Studies 138
    To Describe Nature 138
    To Generate and Test Hypotheses 138
    To Assess Behavior 139
    Because Other Methods are Impractical 140
Becoming a Good Observer 140
    Decide on Type of Observation 140
    Develop a List of Behaviors 142
    Decide How to Record the Behaviors 144
    Use Technology 144
    Be Attentive to Setting 145
    Use Unobtrusive Measurements 145
    Practice 145
    Measure the Reliability of Observations 147
Participant Observation 149
Some Problems to Be Aware Of 150
    Confounding 150
    Computing Relative Frequencies 150
    Feedback May Lead to Bias 150
    Sloppy Observing Can Be Dangerous 151
Summary 151
Key Terms 151

11 Surveying 152

Selecting the Sample 154
    Size of Sample 154
    Choosing Individual Subjects 157
    Nonresponse 159
Choosing the Method of Data Collection 160
Making Up the Questionnaire 162
    Writing Questions About Behavior 163
    Writing Questions About Knowledge 165
    Writing Questions About Attitudes 165
    Putting the Questions in Order 166
    Assembling the Questionnaire 167
Why Survey Houses Disagree 167
    Sampling Error 167
    Variability in Sample Composition 167
    Different Interviewers 168
    Wording, Sequence, and Context of Questions 168
Other Sources of Inaccuracy in Surveys 169
The Kinsey Report: A Case Study 171
Summary 172
Key Terms 173

12 Single Subject Research 174

After-the-Fact Case Studies 175
Complex Analytic After-the-Fact Case Studies 176
Single Case Experimental Designs 177
Summary 177
Key Terms 177

13 Using Correlation Strategies to Predict and Assess Relationships 178

Interpreting Correlation Coefficients: Part I 179
When Should a Correlational Study Be Considered? 181
Predicting by Means of Regression Equations 183
The Correlation Coefficient for Variables Measured with an Interval Scale—The Pearson Product-Moment Correlation 185
Interpreting Correlation Coefficients: Part II 186
Three Potential Problems 187
When There Are Only Two Categories for Each Variable—Phi Coefficient 189
Partial Correlation 190
Correlation and Causation—Additional Comments 191
    Lazarsfeld's Approach 199
    Cross-Lagged Panel Correlation 198
Summary 199
Key Terms 200
Answers 200

14 Experimenting: Two Groups 201

Defining Characteristics of Experiments 202
Reasons for Experimenting 202
    Seeking Answers to "What If" Questions 202
    Testing Logically Deduced Consequences from Theories 202

Sample Size *203*
    Types of Errors *203*
    Power *205*
How to Randomly Assign Subjects *208*
Variability in Experiments *209*
    Implications for Experimenters of the Analysis of Variance Approach *212*
    Is It Bad to Decrease Within-Group Variability? *216*
Experiments Do Not Eliminate the Need for Judgment *218*
    Plausible Alternatives May Remain *218*
    Experimenters Must Balance the Costs and Benefits of Research *218*
    Results are Interpreted Within the Framework of Existing Knowledge *219*
    Experimenters Must Be Sensitive to the Complexity of Variables *219*
    Generalizations Cannot Be Made Routinely *220*
Additional Statistics for Analyzing Results from Two-Group Experiments *222*
    When the Dependent Variable is Measured Nominally *222*
    Quick and Dirty Methods *223*
Summary *224*
Key Terms *224*
Answers *225*

15 Experimenting—Variations on the Two-Group Design *228*

Studying an IV at More Than Two Levels *229*
Factorial Designs *232*
    Eight Possible Outcomes of a 2 × 2 Factorial Study *233*
    Interpretation of Interactions *235*
    An Example *236*
    More on Interactions *237*
    Subject Variables *240*
    Blocking *241*
    The Experimenter as a Variable *241*
Within-Subject Designs *242*
    Single Case Experimental Designs *242*
    Procedures for Conducting Single Case Experimental Designs *244*
    Data Analysis *244*
    Experimenter as Subject *245*
Summary *247*
Key Terms *247*
Answers *248*

16 Experimenting—Sequential Sampling *250*

Overview of Procedures *252*
    Steps in Sequential Analysis *252*
    For Ordinal Data *252*
    For Interval Data *255*
Summary *258*

17 Analyzing Existing Data— Secondary Analysis *259*

Reasons for Using Existing Data Sources *260*
Locating the Appropriate Materials *261*
Types of Secondary Analyses *261*
    Direct Use of Statistical Material *262*
    Documents Used to Develop and Support a Position *262*
    Experiments and Surveys *264*
    Meta-Analysis *265*
    Content Analysis *265*
    Weaknesses of Secondary Analysis *268*
Summary *270*
Key Terms *270*

18 Designing, Implementing, and Evaluating Social and Medical Programs *271*

Evaluating Program Design *273*
    Is a Problem Documented? *273*
    Have the Program Goals and Objectives for Accomplishing Them Been Stated Clearly? *273*
    Have the Activities for Meeting the Objectives Been Clearly Stated? *274*
    Have Criteria Been Established for Deciding If the Objectives Have Been Met? *275*
Evaluating How Well a Program Has Been Implemented *276*
Evaluating Outcomes *277*
    Outcome Criteria *277*
    Cost Effectiveness Analysis *280*
    Randomization *281*
Using Evaluations *287*
    Generalizing *287*
    For Policymakers *287*
    Rules for Evaluators *287*
Summary *288*
Key Terms *289*
Answers *289*

19 Studying the Development of Behavior 290

Cross-Sectional Designs *291*
Longitudinal Designs *292*
    Further Comments on Cross-Sectional and Longitudinal Designs *293*
Advanced Approaches *294*
Miscellaneous Techniques *295*
Summary *295*
Key Terms *296*
Answer *296*

PART V Controversies 297

20 Using Nonhuman Animals in Research 299

Why Use Animals in Research? *300*
    Using Animals to Test How Independent Variables Affect or Are Likely to Affect Humans *300*
    Creating Animal Models of Human Conditions *301*
    Using Animals to Learn About Human Systems *303*
    Studying a Species Because It Has an Interesting Property *305*
    Studying Animals from an Evolutionary Perspective *307*
Summary *314*
Answers *314*

21 Using Private Data 315

Some Introspective Reports Are Verifiable *318*
    The Test of Self-Reports Is Objective and Verifiable *318*
    The Correspondence Between Descriptions of Inner Experiences and Other Data Can Often Be Evaluated *318*
Hypotheses Generated from Introspective Data Can Be Falsified *320*
Some Uses of Self-Report Methods *320*

Methods of Collecting Self-Report Data *323*
    The Focused Interview Method *323*
    The Association Methods *323*
    The Phenomenological Method *324*
    The Critical Incident Method *325*
Summary *325*
Key Terms *326*

22 Testing Data for Statistical Significance 327

The Effects of Statistical Significance Testing on Theory *329*
Solutions *331*
    Measure Effect Size in Addition to Significance Level *331*
    Find Confidence Limits *331*
    Seek Functional Relationships *333*
Summary *335*
Answers *336*

PART VI Drawing Conclusions 337

23 Writing Research Reports 339

All Good Writing Shows Certain Characteristics *340*
    Checklist for Writing *340*
Qualities Unique to Science Writing *342*
Writing for Scientific Journals *344*
    The Parts of a Manuscript *344*
Summary *347*

24 Making Sense of Groups of Studies 348

Organize a Reviewing Strategy *350*
    Formulate a Precise Question *350*
    Decide Whether the Review is to Test a Specific Hypothesis or is Exploratory *350*
    Decide How Studies Should Be Selected *350*
Express Information Quantitatively *351*
    Are the Effect Sizes Similar? *351*
    What is the Combined Effect Size? *352*

Use Visual Displays *353*
  Evaluate the Findings *354*
    Determine the Population to Which Results Can Be Generalized *354*
    Relate Outcomes to Study Characteristics *354*
  Summary *355*
  Key Terms *355*

Appendixes *357*

A   Biographic Retrieval for the Social and Behavioral Scientist *357*
B   Effects of Knowledge of Results and Delay of Knowledge on the Judgement of the Muller-Lyer Illusion *374*
C   F Distribution *381*
D   Table of Critical Values of D (or C) in the Fisher Test *385*
E   Table of Probabilities Associated with Values as Extreme as Observed Values of Z in the Normal Distribution *400*
F   Transformation of r to z *401*
G   Chi-Square *403*
H   Interval Data Master Sheets *404*

References *413*
Index *426*

PREFACE

I've tried to write an introductory research methods textbook that will be referred to and read even after the final exam. The book covers the standard topics plus several additional ideas for collecting and analyzing data. Included are controversial issues not usually discussed in introductory books, such as the use of private data in psychology, the use of animals in research, and the place of statistical significance testing. I discuss and challenge certain basic assumptions, reflecting my belief that science training should include more than dogmatic presentation of methodological principles. Important as those are, they should be supplemented by both critical and creative thinking. Advanced as well as beginning students may find the discussions useful.

Because of the additional topics, I worried about making the size unwieldy. So I wrote tersely and economized wherever possible (except for occasional inability to resist cracking bad jokes). For example, the statistics sections contain no detailed explanations or derivations—only the circumstances of use and computational steps.

Most of the examples are from psychology, my field of training. But the same principles apply in many fields. I hope that students of nursing, sociology, political science, business, and public health, among others, will find the material helpful.

A one-semester course may not be long enough to cover the entire book. Chapters 1 through 8 plus 20 and parts of 14 would probably suffice for a nonstatistical introduction to research methods. Chapters 10 through 13 and 17 through 19 could be omitted from a course on straight experimental psychology. Many instructors may prefer to omit chapters 16, 21, and 22.

Adopters of *Research Methods for Behavioral Scientists* will receive an instructor's manual and test item file. When writing those materials, as with the text itself, I emphasized concepts as much as possible.

I'd like to thank the administration at Cal State, Hayward for awarding me a sabbatical leave during the 1987–88 academic year; I couldn't have completed the book in such a short time without it. Thanks, too, to Carolyn Lee; she took my vague ideas for cartoons and developed them beautifully. I'd like to pay special tribute to seven former teachers: Simon Certner; Sidney Bloomgarden; Robert Steige; Darold Shutt; Steve Glickman; and the memories of Frank Beach and Leo Zeff.

I would also like to thank the following instructors for their ideas and insights as they reviewed the manuscript:

Kevin S. Seybold,
Grove City College

Donald A. Smith,
Northern Illinois University

Seth C. Kalichman,
University of South Carolina

Clint D. Anderson,
Providence College

Susan E. Dutch,
Westfield State College

Laurence T. White,
Beloit College

Ellen R. Girden,
Nova University

PART 1

Introduction

CHAPTER 1

Introduction and Overview

By the time you finish reading this chapter, you should be able to answer the following questions:

What is the key characteristic of science?

Why is it important how science is defined?

How have scientific theories changed the way people view the world?

What characteristics distinguish good scientists from others?

Should scientists collect data as an end in itself?

What types of questions will scientists never be able to answer?

What are other limitations of the scientific method?

A former colleague of mine, eminent in his field, once related his despair over the state of psychology. He had gone to the annual meeting of the American Psychological Association and sat through an entire day of presentations of new research findings. Having heard virtually nothing that interested or seemed significant to him, he conducted a study of his own. He informally polled the 30 or so speakers about their reasons for doing research. Only one person said it was because of fascination with a problem. The rest talked of status, requirements for doctoral dissertations, or publication pressures from their universities.

One interpretation of my colleague's finding is the recognition of a principle: An uninterested researcher will probably not produce interesting research. Yet this principle is widely ignored by writers of methodology textbooks. They tell how to analyze data statistically and they provide detailed discussions of esoteric subjects like operational definitions and scales of measurement. But knowledge of such materials no more qualifies people as scientists than ability to paint by numbers makes them artists. Scientists must know methods, facts, and formulae, but they should also be able to think creatively and critically about problems. Students of science would do well to learn the history of their disciplines and hear great scientists speak:

Albert Einstein, 1930: . . . I maintain that cosmic religiousness is the strongest and most noble driving force of scientific research. Only the man who can conceive the gigantic effort and above all the devotion, without which original scientific thought cannot succeed, can measure the strength of the feeling from which alone such work . . . can grow. What a deep belief in the intelligence of Creation and what longing for understanding, even if only a meager reflection in the revealed intelligence of this world, must have flourished in Kepler and Newton, enabling them as lonely men to unravel over years of work the mechanism of celestial mechanics. . . . Only the man who devotes his life to such goals has a living conception of what inspired these men and gave them strength to remain steadfast in their aims in spite of countless failures. It is cosmic religiousness that bestows such strength. A contemporary has said, not unrightly, that the serious research scholar in our generally materialistic age is the only deeply religious human being.

Koestler (1964) and Baker (1966) wrote about others who approached their work with religious fervor and a never-ending sense of awe. A few quotes should dispel the notion that scientists are a dispassionate, bloodless lot:

Russian biologist Pyotr Kropotkin: "There are not many joys in human life equal to the joy of the sudden birth of generalization. . . . He who has once in his life experienced this joy of scientific creation will never forget it."

Louis Pasteur, upon making some surprising observations: "All is discovered." He then rushed from the laboratory to embrace an assistant and give him an exuberant account of what had happened.

A. R. Wallace, co-formulator with Charles Darwin of the theory of evolution, after catching a new species of butterfly:

Cited in Beveridge, 1957: None but a naturalist can understand the intense excitement experienced when I at length captured it. On taking it out of my net and opening the glorious wings, my heart began to beat violently, the blood rushed to my head, and I felt much more like fainting than I have done in apprehension of immediate death.

Ptolemy: I know that I am a mortal, a creature of a day; but when I search into the multitudinous revolving spirals of the stars, my feet no longer rest on the earth, but, standing by Zeus himself, I take my fill of ambrosia, the food of the gods.

I don't expect to transport readers to a state of religious ecstasy but I would like to convince you that the study of science can be enjoyable as well as valuable. If you come to appreciate the attitudes of Pasteur and others, you'll have more in common with them than do many people who've learned to mechanically set up flawless and complex experimental designs.

The Rewards of Science

Science is hard work. The hours are long and often irregular, and the pay is usually much less than that of lawyers, doctors, and business executives. Why then do people become scientists? An obvious reason is that they wish to solve practical problems. The field of psychology is especially seductive in this respect because researchers are able to work on some of the most important concerns of humankind: love and hate, creativity, racism, greed, learning and memory, depression, psychosis, and self-actualization.

Many people become scientists because they crave intellectual challenge. Curiosity inspires and drives them on. To go from the sublime of Einstein's feelings to the ridiculous of some contemporary research, I'll describe an experiment of mine, motivated by curiosity. While taking a course at the University of Michigan, I learned of two studies on the sexual behavior of rats. In the first, a pair of rats was allowed to copulate undisturbed, and the males averaged seven ejaculations before they stopped. In the second, the female was replaced by a new one each time the pair went for a short time without copulating. Under this regime the males had almost twice as many ejaculations. The researchers concluded that the first males had stopped not because they had been sexually exhausted but because they had habituated to the once-arousing stimuli of their original partners.

That week I attended a pharmacology lecture in which evidence was presented that the drug scopolamine blocks the habituation process. Because of my simultaneous exposure to the two sets of findings, I conceived the notion of administering scopolamine to a male rat and then allowing it to copulate with a single female. My guess was that the male would never habituate, so would copulate until he or the female contentedly dropped from exhaustion. Unfortunately for the hypothesis and the many people of both sexes on campus who followed the experiment with great interest, scopolamine-injected rats didn't copulate at all. Still, because my curiosity had been aroused, because I wanted to know "What would happen if. . . ?" I worked long hours to find the answer.

Scientists are puzzle solvers, detectives on the spoor of profound mysteries. In addition to "What would happen if. . . ?" they also ask "Why?" Niko Tinbergen won the Nobel Prize in biology for asking and answering why questions. For example, he observed that when black-headed gulls hatch, their parents invariably remove the egg shells from the nests. As a result of their parents' behavior young black-headed gulls are left unprotected for brief periods of time. By conducting a series of experiments (Tinbergen, Brockhuysen, Feekes, Houghton, Kruuk, and Szuk, 1962), some of which are described in Box 1.1, Tinbergen found answers. Before reading what he did, imagine that you had made the initial observations of egg shell removal; can you develop a strategy for answering the question?

Scientific journals, serving professional audiences, require that writers use a technical vocabulary and a terse, impersonal, style. This has the unfortunate consequence of making original research appear dull to students. Unfortunate because people motivated by curiosity frequently develop highly creative strategies for answering questions. Their research, elegantly conceived and executed, can give esthetic and intellectual sensations comparable to those elicited by masterworks of music and art. In the words of Nobel laureate Steven Weinberg: "The effort to understand the universe is one of the very few things that lifts human life a little above the level of farce and gives it some of the grace of tragedy."

BOX 1.1

Egg Shell Removal by Gulls

Black-headed gulls, like birds of many species, remove egg shells from the nest. The gulls fly with the empty shells in their bills and drop them far away. Niko Tinbergen, seeking an explanation, made use of prior observations: (a) Kittiwake gulls do not remove egg shells and (b) black-headed gulls remove most foreign objects that fall into the nest. These suggested to him that nest-cleaning functions to maximize camouflage against predators. (Kittiwakes don't need camouflage because they nest on cliffs and are largely predator-free, but many animals prey on black-headed gull chicks.) He did many experiments to test his ideas.

1. He laid out eggs over a wide area near a gull colony, some by themselves and some with a broken egg shell nearby. More of the latter were discovered and eaten by predators.

2. Empty egg shells did not attract crows; but when fresh eggs were placed near the shells, the crows quickly discovered the eggs; then, when empty shells were scattered, the crows searched in their vicinity. They had learned to associate egg shells with eggs.

3. Eggs painted white were taken more readily than naturally colored ones.

I've known artists and musicians to proudly proclaim their ignorance of science; they are as lacking in culture as scientists who know little of Beethoven, Shakespeare, and Picasso. Psychologist Gustave Le Bon wrote:

The sudden political revolutions which strike the historian most forcibly are often the least important. The great revolutions are those of manners and thought. The true revolutions, those which transform the destinies of people, are most frequently accomplished so slowly that the historians can hardly point to their beginnings. Scientific revolutions are by far the most important.

Read Box 1.2 to get an idea of the extent to which scientific research and theory has influenced twentieth century thought.

The Influence of Science on Modern Thought

Olson (1971) called attention to the influence of science on social, esthetic, and religious beliefs. Its effects have been far more pervasive than generally realized. Prior to Galileo, people believed that angels, heavenly bodies, beasts, and elements all had their place in a fixed hierarchy. But when scientific evidence disputed the ordering of the heavenly bodies and indicated that the earth is not the center of the universe, the structure crumbled. Galileo proved that the universe is much greater than had previously been supposed, and, with the discovery of a new star, that it changes continuously. This was terrifying; the poet John Milton called it a "wide gaping void that threatened even the angels with loss of being." Philosophy was profoundly changed, and even the rules of evidence of law were reevaluated.

In the seventeenth century, Galileo, Descartes, Robert Boyle, and Kepler interpreted all natural phenomena in terms of mechanical motions. Their ideas shaped metaphysics, political theory, and theology. Thomas Hobbes used mechanistic reasoning to show that complex social problems should be broken down into isolated units and considered one at a time.

Newton's gravitational theory provided a unified explanation of phenomena previously thought unrelated; this was a triumph and encouraged the belief that humanity could solve its moral and political problems. Social theorist Saint Simon urged that traditional Christianity be replaced by a Religion of Newton.

Darwin's evolutionary theory was psychologically devastating, as it repudiated the descent of humans from angels; and it was horribly misused to justify slavery, master race theory, and other tyrannies of the well-placed over the needy.

As much as Darwin changed people's self-image, Freud changed it even more. He introduced terms like wish-fulfillment, frustration, regression, repression, unconscious impulses, sublimation, anxieties, neuroses, and defense mechanisms; all are now widely used in character descriptions. He gave people freedom to express interest in sexuality and to question Puritan ethics and he influenced the focus of modern literature and philosophy.

More recently, philosophers have reanalyzed the ancient free-will versus determinism debate in the light of quantum theory. Scientific findings indeed have many ramifications.

BOX 1.3

Some Views on Philosophy of Science

Monod (1975) showed that scientists do not invariably reject theories refuted by experiment. A direct experiment (measurements of heat received by the earth, of the dimensions of the sun, and of fuel caloricity) seemed incompatible with the theory of evolution but did not persuade Darwin to revise it. Chalmers (1976) argued that several theories generally held in high esteem today would have been rejected in their infancy if scientists adhered strictly to falsification. His examples included Newton's gravitational theory (at the time, falsified by observations of the moon's orbit), Bohr's theory of the atom (falsified by the observation that some atomic particles are stable for longer periods than predicted), and the kinetic theory of gases (falsified from the beginning by measurements on the specific heats of gases).

Einstein, told that certain facts refuted his theory, said: "The facts are wrong."

Feyerabend (1975) claimed that not a single interesting theory agrees with all the known facts in its domain. His book, brilliantly argued, is a nightmare for methodologists. He said that, given any rule for science, there will always be circumstances when it is advisable not only to ignore the rule but to adopt its opposite. Every rule is associated with cosmological assumptions, and the assumptions may be wrong. Conducting science according to fixed rules reduces its adaptability. Hypotheses should be developed inconsistent with highly confirmed theories because the evidence that might refute a theory can often be unearthed only with the help of an incompatible alternative.

Medawar (1969) wrote that active scientists rarely pay attention to the advice of philosophers of science, and the scientists give distorted accounts when they describe their own methodological practices for the public.

Toulmin (1961) contrasted the ancient Babylonians and Greeks of Ionia. The Babylonians, by recording past occurrences of the new moon and then applying arithmetic techniques, made accurate predictions of the first visibility of the new moon and lunar eclipses. The Ionians developed many theories but never made accurate forecasts. But we should not conclude that the Babylonians were better scientists; they applied the same methods to predicting earthquakes, plagues of locusts, and political disasters. Modern astronomy owes them a lot, but is also greatly indebted to the Ionians for teaching how to use speculative imagination to try to make sense of observed regularities.

What Is Science?

Karl Popper, perhaps the foremost philosopher of science of the twentieth century, asked how scientific and nonscientific activities can be distinguished. He identified two criteria that, although not accepted by all philosophers (see Box 1.3 for some dissenting opinions) are satisfactory for the purposes of this book.

Scientists Organize Facts

Scientists do not merely catalog facts in actuarial tables and almanacs. They formulate laws and theories to organize the facts, unify diverse phenomena, and make sense of the world. As Poincare put it, "Science is built up of facts, as a house is with stones. But a collection of facts is no more science than a heap of stones is a house." Popper wrote, "Only if it is an answer to a problem—a difficult, a fertile problem, a problem of some depth—does a truth, or a conjecture about the truth, become relevant to science."

Scientists Create Falsifiable Theories

Philosophers and theologians also search for meaning, but unlike scientists do not put their conclusions to empirical tests. To Popper, the empirical tests are what distinguish science from other activities. He wrote (1968): ". . . a system is to be considered as science only if it makes assertions which may clash with observations." In other words, scientists make predictions; and only if they risk being wrong by accepting that failure of the predictions would prove them wrong, are they behaving scientifically.

Popper encouraged scientists to be bold in constructing theories but ruthless in trying to falsify them. Other people, including many who claim to be scientists (and wear white lab coats to prove it), also make predictions but then attempt to salvage unsatisfactory outcomes by any means possible. In doing so they act much like Stephen Leacock's fictional detective, who may have been charming and made lots of money, but was no scientist.

Stephen Leacock

Stories Shorter Still
An Irreducible Detective Story
Hanged by a Hair or A Murder Mystery Minimized

The mystery had now reached its climax. First, the man had been undoubtedly murdered. Secondly, it was absolutely certain that no conceivable person had done it.

It was therefore time to call in the great detective. He gave one searching glance at the corpse. In a moment he whipped out a microscope. "Ha! Ha!" he said, as he picked a hair off the lapel of the dead man's coat. "The mystery is now solved." He held up the hair. "Listen," he said, "we have only to find the man who lost this hair and the criminal is in our hands." The inexorable chain of logic was complete. The detective set himself to the search. For four days and nights he moved, unobserved, through the streets of New York scanning closely every face he passed, looking for a man who had lost a hair. On the fifth day he discovered a man, disguised as a tourist, his head enveloped in a steamer cap that reached below his ears. The man was about to go on board the *Gloritania*. The detective followed him on board. "Arrest him!" he said, and then drawing himself to his full height, he brandished aloft the hair. "This is his," said the great detective. "It proves his guilt." "Remove his hat," said the ship's captain sternly. They did so. The man was entirely bald. "Ha!" said the great detective, without a moment of hesitation. "He has committed not one murder but about a million."

For a nonfictional example of the same attitude, consider Sigmund Freud's daughter Anna's description of a psychoanalytic patient. (Both Freuds believed that penis envy was a universal condition of young girls.) The patient, a young girl, ". . . had succeeded in so completely repressing her envy of her little brother's penis—an affect by which her life was entirely dominated—that even in analysis it was exceptionally difficult to detect any traces of it." (1948)

Popper's definition has implications: accept it and no particular subject matter is included in nor excluded from the domain of science. No sharp line demarcates so-called scientific and nonscientific fields. The criteria involve attitudes and methods, not subject matter. Historical research as conducted by some historians is similar in important respects to paleontological research (observe, guess at relationships, collect data, make predictions, collect new data against which to test the predictions). Ziman (1968) noted that the German word for science, *Wissenschaft,* includes all the branches of scholarship including literary and historical studies. People have often been discouraged from working in certain areas of psychology stigmatized as being "unscientific." Without denigrating the more traditional areas of research, I assert that questions about the nature of love, the achievement of peace and happiness, and full realization of the human potential can be posed and studied scientifically.

A Few Noncritical Features

Nothing has been said about control groups, statistics, replicability, intersubjective agreement, or even experimentation. All are valuable and help scientists make predictions and explain events. Properly used, they are similar to technological advances like the microscope and computer in reducing the likelihood of error. *But they do not define science.* The most important works of Darwin and Einstein among others did not depend on formal experimentation, computers, microscopes, statistical reasoning, or control groups. Even today, some of those are luxuries rarely available to naturalists, astronomers, and paleontologists. (Einstein and Darwin, those poor souls, were doubly handicapped; they had to muddle along without a single course on research methods.)

Limitations of Science

Scientists Can Answer Only Certain Kinds of Questions

The distinction made between philosophers and theologians on the one hand, scientists on the other, points to an important limitation of science. Although it is the most powerful method known for finding answers to certain kinds of questions, it is incapable of answering others. Questions about ethics, the existence of God, and the possibility of an afterlife are not within its realm. To understand why, consider types of statements made in ordinary communication. For convenience, they may be classified as follows:

1. Analytic Statements (Tautologies): These are true by virtue of their structure. "Either it is raining or it is not raining." "If all men are mortal and Socrates is a man, then Socrates is mortal." Tautological statements can be much more complex than these, as in mathematics, and can contribute to scientific advancement.
2. Definitions: These say what terms mean. They are arbitrary, so neither true nor false, but they can be more or less useful. "I will define psychology as the science of behavior." "I will define intelligence as score on a standardized IQ test."
3. Observational Statements: These make assertions about the world that can, at least in principle, be shown to be true or false. "Water, when heated to 100 degrees Celsius, will boil." "The earth is flat." "Frustration during the anal phase of development leads to miserliness." This type of statement constitutes the core of science.
4. Value Statements: These express judgments and preferences. "The weather in Oakland, California is wonderful." "King Kong had a higher set of ethical principles than many government officials."
5. Miscellaneous Statements: These are commands, greetings, exclamations, etc. Exclamations, if results don't turn out as hoped, may come in a rush; they are otherwise irrelevant to science.

As shown in the next section, science is not value-free and research findings may influence decisions on moral issues such as abortion, treatment of criminals, and genetic counseling. But research reports rarely contain explicit statements about values, depending instead on tautologies, definitions, and observational statements. Moreover, questions of definition—What is truth? What is beauty?—are not part of science. No observations can answer them and they cannot be falsified.

Scientists Cannot Establish Absolute Truth

Scientists often derive such accurate predictions from their theories that they seem to have penetrated absolute reality. But modern philosophers have concluded that theories can never be proven, and absolute truth is unattainable. New theories constantly replace old ones and inviolable laws are superseded. Although modern physics and chemistry owe much to their seventeenth century predecessors, they are not mere updated versions but vastly different conceptions.

Scientists often disagree. One reason, which partially explains their inability to establish absolute truth, is that every important phase of their professional lives is influenced by their intellectual histories and values.

Figure 1.1
This classic drawing can be viewed as either a young woman or an old woman.

Intellectual Values Influence Choice of Discipline   Problems can be approached from many perspectives. For example, the tragedy of mental illness has engaged personality theorists, social and environmental psychologists, anthropologists, geneticists, physiologists, neuroanatomists, neurochemists, and pharmacologists. The choice of discipline is influenced by values.

Intellectual Values Influence the Questions Scientists Ask   Hrdy (1986) used the field of sociobiology to show that intellectual values affect the questions scientists ask even within a single discipline. Sociobiologist Robert Trivers (1972) published an important paper in 1972. He argued that, because egg cells are much larger and more physiologically costly than sperm, offspring represent a greater investment for females than males. As a result, females are more selective than males in choosing mates. In addition, females gain no reproductive advantage by copulating while pregnant, whereas males that frequently copulate increase their reproductive success.

Several consequences seemed to follow from the intellectual framework provided by Trivers: females are more coy than males; males are more variable than females; competition between males is much greater than that between females; and males help minimally with infants.

Given the framework, researchers asked questions about male but not female aggression and about female but not male involvement with infants. They asked about the relationship between male dominance and frequency of mating, but not about multiple matings by females. Only in the 1970s, when several women scientists entered the field, did the questions begin to change.

Intellectual History Influences Observational Data   Hanson (1958) argued that two observers with normal eyesight, gazing at the same object, may not see the same thing. He supported his position with many examples from Gestalt psychology. Look at reversible perspective Figures 1.1 and 1.2. At different times, people will see Figure 1.1 as an old woman and as a young woman; and Figure 1.2 as opposing faces and a goblet. Hanson wrote that two people asked to draw what they saw might produce identical results. But if one saw an old woman and the other a young one, they would in an important sense be observing differently.

The history of N-rays (Broad & Wade, 1982) offers additional evidence for the view that observations can be influenced by expectations. In 1903, a distinguished French physicist announced his discovery of a new kind of ray, which he called an N-ray, that emanated from X-ray sources. Soon afterwards, his colleague observed N-rays from the nervous system of the human body. By 1906 at least 40 people including several eminent scientists had observed N-rays, and their properties had been discussed in more than 300 scientific papers. But N-rays do not exist; the preconceptions of the observers had influenced what they saw.

Intellectual Values Influence What Scientists Read Payer (1988) compared medical practices in four countries—the United States, France, Germany, and England—that have roughly equivalent life expectancies. She found enormous differences in diagnosis and treatment. One reason is that physicians read the medical literature selectively. One interviewer found that British doctors were unable to name a single French medical journal, and the French named an average of slightly over one British journal. Payer cited a letter in 1981 in the journal *Arthritis and Rheumatism* crediting Churchill with being the first to document disc-space infections in bacterial endocarditis; she noted that at least 10 articles on the subject had previously been published in the French literature, the first in 1965.

Figure 1.2
Do you see opposing faces or a goblet? Both images are possible.

From Benjamin B. Lahey, *Psychology: An Introduction*, 3d ed. Copyright © 1989 Wm. C. Brown Publishers, Dubuque, Iowa. All Rights Reserved. Reprinted by permission.

Intellectual Values Influence How Scientists Interpret Data Scientists do not have a uniform standard of proof. For example, Hansel (1966) showed how parapsychology researchers could have cheated in their most famous experiments. The parapsychologists responded that most scientists have opportunities to cheat and that Hansel should have either shown that they did cheat or remained silent. I suspect that most readers of Hansel's criticisms accepted or rejected them according to their prior beliefs about the relative probabilities of cheating and extrasensory powers.

Glaser (1977) compared conclusions from three papers within a single issue of the *British Medical Journal*. One reported a near death from eating licorice but drew no conclusions about future safety. A second indicated a 4 percent mortality rate in patients who had undergone intestinal bypass operations for obesity and statistically significant liver damage in all the patients; the conclusion was that "further careful evaluation of the effects of intestinal bypass operations is required." A third paper reported that one of 1,000 patients given an intravenous anesthetic had an adverse reaction, and concluded that this rate made the intravenous injection unacceptable.

Intellectual Values Influence the Theoretical Models Scientists Create
Cell biologists of the 1930s viewed the relationship between nucleus and cytoplasm as similar to that between husband and wife. The Biology and Gender Study Group (1988) described competing views of eminent scientists on the exact nature of the relationship. From autocratic Germany came a model in which the nucleus contained all the executive functions and the cytoplasm acted upon its commands. The leading American geneticist, T. H. Morgan, modeled the cell after his family, with nucleus and cytoplasm exchanging information and the nucleus making the final decision. British socialist C. H. Waddington, who viewed his marriage as a partnership, tried to show the equality of nucleus and cytoplasm. And in the model of American black embryologist E. E. Just, the cytoplasm dominated over the nucleus.

Intellectual Values Influence the Methodologies That Scientists Consider Valid Scientists often resolve disagreements by presenting relevant evidence to a kind of invisible "science court," practitioners within their specialty, who judge the evidence impartially according to rules upon which all agree (Laudan, 1984). But Cochrane (1971) reported that rules differ greatly from country to country. For example,

> . . . If the number of randomized controlled trials were worked out and a map of the world shaded accordingly (black signifying the highest percentage of such trials), one would see the UK in black and scattered black patches in Scandinavia, the USA and a few other countries: the rest would be nearly white.

One scientist may believe that a theory must make surprising predictions that turn out to be correct before accepting it. Another may accept a theory that explains a broad range of phenomena even if it has not made startling predictions. A third may require that a theory be tested against a wide variety of supporting instances. A fourth may believe that a large number of confirming instances is sufficient, even if they have little variety. A fifth may demand both direct and indirect evidence for the existence of entities postulated by the theory. Laudan (1984) noted that prominent advocates in recent science and philosophy support each of the guidelines.

The Fundamental Uncertainty

Stripped to the bone, science is a search for relationships between variables. Scientists observe and experiment to determine how one variable changes with another. Then they try to find equations that relate the variables mathematically and they create theories to explain the equations. For example, an industrial psychologist might vary the number of motivational lectures given to different groups of salespeople, then record their average number of sales per week. She might then record her data in a table like Table 1.1. The equation $Y = X + 1$ fits the data and leads to the prediction that the missing value of Y is 5.0000.

But other equations also fit, such as

$$Y = \frac{X^5 - 17X^4 + 107X^3 - 307X^2 + 576X}{180}$$

and

$$Y = \frac{(X^7 - 1582X^4 + 14651X^3 - 49322X^2 + 101772X)}{32760}$$

When $X = 4$, Y is 5.0667 in the first equation and 5.0974 in the second. The difference between the three Y values may seem trivial; after all, the data still show that number of sales increases with increased frequency of motivational lectures. But now look at Table 1.2. Again, the obvious solution is that $Y = X + 1$ and the missing $Y = 4$. But another equation that fits the data is

$$Y = \frac{(55X^4 - 382X^3 + 749X^2 - 374X)}{24}$$

Table 1.1

X (number of motivational lectures)	Y (number of sales per week)
1	2.0000
2	3.0000
3	4.0000
4	?
5	6.0000
6	7.0000

Table 1.2

X	Y
1	2
2	3
3	?
4	5

For the new equation, when $X = 3$, $Y = -10$. In fact, an infinite number of equations yielding an infinite number of predictions can be found to perfectly fit the data of Table 1.2—and *any* other set of data.

Comments on the Limitations of Science Philosophers are better suited than psychologists for dealing with the implications of the limitations of science. But two points can be safely made. First, the limitations apply to all attempts to understand the world, not just to the scientific approach. Second, the limitations do not lessen the intellectual challenge and excitement of solving scientific problems. Although an infinite number of equations fit the data relating mass, energy, and the speed of light, it took the genius of Einstein to find even one.

We Are All Scientists

Although scientists must be willing to test and falsify their predictions, correct predictions are both more useful and more pleasant. For accuracy of prediction to be a rational goal, scientists must believe that the past in some way foretells the future. They must do research under the assumption that, if all essential aspects of a study are repeated, the outcome will be the same. In short, they must believe in a deterministic universe. (Quantum physicists do not hold this belief, but it is essential for scientists who work with objects larger than atomic particles.) Most philosophers acknowledge that belief in determinism, that every act is the inevitable consequence of prior ones, cannot be logically justified. It is a matter of faith. However, in light of what seems to be a widespread disenchantment with science and a conviction that deterministic beliefs make scientists irrevocable adversaries of artists (cf. Roszak, 1972), I must note that all living creatures, even lowly amoeba, act as though the world is deterministic. Otherwise we wouldn't eat, for food would as likely burn as nourish us; we wouldn't drink, make love, or even move.

Many years ago Thomas Huxley wrote a short essay called "We are All Scientists." Huxley said, "The method of scientific investigation is nothing but the expression of the necessary mode of working of the human mind. It is simply the mode at which all phenomena are reasoned about, rendered precise and exact. There is no more difference, but there is just the same kind of difference, between the mental operations of a man of science and those of an ordinary person, as there is between the operations and methods of a baker or of a butcher weighing out his goods in common scales, and the operations of a chemist in performing a difficult and complex analysis by means of his balance and finely graduated weights. It is not that the action of the scales in the one case, and the balance in the other, differ in their principles of construction or manner of working; but the beam of one is set on an infinitely finer axis than the other, and of course turns by the addition of a much smaller weight."

Huxley went on to describe how a person, having tried two hard, green apples and found them sour, would probably reject a third on the grounds that it was also likely to be sour. This type of induction by observation is exactly what scientists do. Moreover, the strength of belief in the "Law of the Hard, Green Apples" would depend on the extensiveness of the verifications and increase with repetition of the basic finding, especially when conditions of testing were modified. Scientists too seek reliable data, and they have the greatest confidence in generalizations based on diverse types of data.

Huxley's analysis suggests that everybody can profit from studying scientific methodology. Even if your apple-judging skills are already at peak capacity, critical reasoning may help you make sounder decisions about many practical affairs. But I shouldn't go overboard. Scientific controversies rarely revolve around simple issues like lack of a control group. Even experienced methodologists may not be sure if extrasensory perception is genuine; if massive doses of vitamin C cure colds, cancer, and schizophrenia; and if Velikofsky's cosmology is accurate. Still, even people who do not pursue science careers may derive considerable satisfaction from being able to follow the arguments on both sides of a controversy.

They can do more. They can design and carry out simple procedures for data collection and analysis; this will enable them to get sound answers to questions of great personal significance. Here are a few examples:

- A small businessman might want to compare the best ways for advertising and displaying his products.
- A person fascinated by a Sunday supplement article on the effects of color of a room on mood might design a study to test the ideas.
- A psychotherapist might compare different techniques to see which work best with specific clients.
- A gambler might compare the performance of various teams or individuals under different conditions, such as home versus away, after a win versus after a loss, and so forth.

Many years ago, Roger Williams wrote an important book, *Biochemical Individuality* (1956), in which he established that people are enormously variable in anatomical features, physiological functioning, and nutritional requirements. A person's specific needs may depart substantially from what is generally recommended. Readers might wish to determine scientifically what is optimal for them.

My interest in methodological issues started years ago, upon reading several popular accounts of the effects of certain drugs on creativity. I used to enjoy pretending to be a poet (none of my audiences was ever fooled), and was intrigued by the possibility that LSD and similar substances might be beneficial. Statements by experts did not help me decide. For example:

Alan Watts: . . . I think that LSD has been very beneficial. LSD-inspired works have returned the glory to Western art. It's difficult to estimate its value in literature. I can only say from my own point of view that I have derived all kinds of ideas for lectures and writings from it.
Donald Louria: "LSD-induced creativity is a myth."

So I eagerly reviewed every available study in the scientific literature. The result can be summed up by an exam question to my psychopharmacology class (the correct answer is d):

The number of methodologically sound studies on the effects of drugs on creativity is:

   a. between 25 and 50
   b. between 5 and 25
   c. between 1 and 5
   d. fewer than the number of partridges on a pear tree

What Makes a Person a Good Scientist

Good scientists differ from lesser lights in at least three respects. Perhaps the most important is in the types of questions they ask. The following criteria distinguish good questions from bad ones:

1. Is the question answerable? This is fundamental. Many questions are not answerable because:

    a. They are too vague, for example, what are the effects of stress on personality?
    b. No possible observations could have a bearing upon them. For example, is the music of Beethoven more or less pleasurable than that of the Rolling Stones? (The question could be answered if "pleasurable" were defined according to specific criteria. For example, if a Beethoven and a Rolling Stones concert were offered on the same night, the more pleasurable might be defined as the one that more people attended.)
    c. Present technology is incapable of answering them, for example, is there intelligent life on Alpha Centaurus?

2. Does the question apply to a broad or restricted range of phenomena? An answerable question about cures for experimentally-induced neuroses in cats would be less valuable than one about cures for all emotional problems in all animals including humans.

In addition to asking the right questions, good scientists are efficient. They design their studies to answer the research question as thoroughly, inexpensively, and unambiguously as possible. This requires knowledge of apparatus, experimental design, sampling, test administration, and analysis of data. They are aware of known pitfalls in their fields, which is usually accomplished through reviews of the literature.

Another aspect of efficiency is ability to develop a methodology for answering a question. This may take creativity of a high order, as in Tinbergen's studies of black-headed gulls.

Good scientists make fewer errors than poor ones. According to some popular accounts, the good ones are cautious to the point of timidity when interpreting data. "Nothing but the facts, ma'am." The stereotype is exemplified by a story about Cordell Hull, an American statesman, whose friend pointed to some sheep grazing in a field. "Look," he said, "those sheep have just been sheared." Hull answered, "Sheared on this side, anyway." Though cute, the story grossly distorts how scientists think; they willingly risk errors whenever they draw inferences, make generalizations, or construct a theory from data. To insist on freedom from such error is to ensure dull and plodding science. But certain kinds of errors must be minimized: in collecting, tabulating, and interpreting data; and in sampling.

The distinction between the two types of errors can be sharpened by an analogy. Scientists are like crossword puzzle solvers. The puzzler who fills in blank spaces even when in doubt about the correctness of his choices—but always aware that he may be mistaken, so always with eraser in hand—is likely to do better than one who writes in indelible ink only what he's sure of. But the puzzler who puts his answer in three down instead of three across, or who spells "kat" for "a small furry mammal," is headed for trouble.

Scientists Versus Data Collectors

People who feel they must do research need never fear for lack of ideas. It's easy to pick completely "original" research problems. Simply choose a popular independent variable (a variable manipulated by the experimenter) and a sensitive dependent variable (a behavior recorded and measured by the experimenter—see page 67 for further discussion) and study the effects of one on the other. For example, the independent variable can be dose of drug and the dependent variable can be amount eaten. In fact, every drug is a potential independent variable, every type of behavior a dependent variable. If there were only 100 of each kind, there would be 10,000 possible pairs. If two of the independent variables were studied simultaneously, more than 20,000,000 distinct experiments could be performed. And there are not 100, but an infinite number of variables.

Some experimenters have taken the above reasoning to heart. Many drug researchers, for example, find a suitable dependent variable and then routinely test each new drug to which they gain access for its effects on that variable.

But research done this way barely qualifies as science—it is not part of a thought-out program to answer questions of the "how" or "why" sort. Such research can be called "mere data collecting." Don't misunderstand: Data collection is important. But scientists must do more.

Collecting data as an end in itself is akin to doing research for almanacs or sport record books: find the number of physicians for each of the 100 largest cities in the United States, or compile the batting average of each of the players in the major leagues for the past decade. Do people do this with a straight face? You will be surprised to find that this sort of stuff gets published in books, journals, reports, and bulletins for the federal government and private business corporations. Psychologists contribute their share. Studies designed to explore the validity of widely accepted beliefs sometimes yield exciting, counterintuitive findings; these are often cited in introductory textbooks as proof that science is superior to common sense; but an unintended and unfortunate side effect of the citations is to encourage further such studies. Science becomes trivialized, and critics like Jose Ortega y Gasset attack mercilessly:

Experimental science has progressed thanks in great part to the work of men astoundingly mediocre, and even less than mediocre. That is to say, modern science . . . finds a place for the intellectually commonplace man and allows him to work therein with success. The reason for this lies in what is at the same time the great advantage and the gravest peril of the new science, and of the civilization directed and represented by it, namely, mechanisation. A fair amount of the things that have to be done in physics or in biology is mechanical work of the mind which can be done by anyone, or almost anyone. (1951)

Scientists sometimes do research that appears on the surface to be mere data collecting but is actually much more. For example, Donald Dewsbury spent several years studying sexual patterns of rodent species native to the southern United States. His procedure, repeated many times, was to trap animals of a particular species and then observe them during routine tests of sexual activity. Later, having amassed the data (and Dewsbury had this in mind from the start), he organized and classified it. Then he theorized about the probable origins and functions of various patterns of sexual behavior, which generated testable predictions about behaviors of previously untested species. Construction of a good theory is the pinnacle of scientific achievement. Dewsbury's (1975) theorizing led to an important paper on the selection pressures that facilitate development of different behavioral patterns.

Unlike Dewsbury, many scientists collect data without any underlying plan or long-range perspective. Such people not only fail to advance psychological science, they actually hinder it. The reasons are threefold: first, they fight for a share of the research budget with scientists interested in solving genuine, pertinent problems; second, they clutter up journal space so readers must wade through volumes of chaff to find grains of wheat; and third, editors and reviewers must spend inordinate amounts of time reading submissions for publication.

Some years ago, the psychology department of a first-rank academic institution was evaluating candidates for a staff vacancy. The chairman, a renowned scientist and scholar, was unenthusiastic about one person who had greatly impressed the members of the hiring committee. The chairman said that he had never heard of the person. A shocked committee member told the chairman that the candidate had published over 200 papers. The chairman, so the story goes, said: "Two hundred publications and I never heard of him. That's all the more reason not to hire him."

Objectives for This Book

To Show the Relevance of Social Science Research

The United States has a large human potential movement. Many books and seminars dealing with ways of altering consciousness attract large audiences. Esalen, rolfing, aikido, est, bioenergetics, Gestalt therapy, dream analysis, Feldenkrais method, polarity therapy, and scores of others have hordes of adherents; old standbys like astrology and the occult have many more. Some of these fields may ultimately help people achieve harmony with nature, inner peace and contentment, and full expression of creative talents. Many others will serve the intended function of making their promoters rich. The scientific outlook should be applied in such cases, to identify the useful fields and hasten the others to their ends. I hope too that readers will come to understand why many scientists devote their energies to areas that might seem unworthy of their talents.

Experimental psychology courses often lack relevance for students. Chapanis (1968) offered one reason: Laboratory findings often cannot be applied anywhere else (see p. 216 for a discussion of why this is so). A related drawback is that courses fail to prepare students to study phenomena outside the laboratory. No guidance is given for doing research with variables that can't be carefully controlled or subjects that have not been randomly assigned to groups. In physics labs, by contrast, simple preparations like inclined planes and pendulums are used precisely because they demonstrate "real world" principles. Perhaps only in psychology do prominent spokespeople make statements such as the following:

Strupp, 1960: . . . I believe that, up to the present, research contributions have had exceedingly little influence on the practical procedures of psychotherapy. I raise these questions not as a therapeutic nihilist but as a researcher who feels that important answers should come from research. But, I must confess that I am not fully convinced that they will.

Luborsky, 1969, referring to Carl Rogers: I recall that at a conference about 10 years ago, one of the most productive psychotherapy researchers remarked to a group of other psychotherapy researchers that his own therapeutic practice had hardly been influenced in any way by his own psychotherapy research. Most of the psychotherapy researchers present agreed with him.

Researchers should not be required to have practical applications as their sole or even primary criterion. Historically, basic research has led to major advances; even if that were not so, the power of scientific findings to transform our conceptions of the world assures their relevance. Still, results that don't generalize to the world outside the laboratory are unlikely to have much scientific value. And as a corollary, scientists should be able to use their methodological expertise even in situations where data gathering has been faulty—for example, to evaluate social programs implemented (unfortunately) without proper controls.

To Show the Importance of Diverse Approaches

No single research design is invariably superior to others. Different methods have different weaknesses, and various methods complement each other. No single measure reflects the richness of abstract concepts; multiple measures tell a great deal more. Scientists should use many methods, many measures.

To Show the Unity of Scientific Method

Scientific research rarely proceeds as smoothly as sanitized descriptions in journals seem to indicate. Medawar (1969) wrote that ". . . nearly all scientific research leads nowhere—or, if it does lead somewhere, then not in the direction it started off with." Though the various phases of a research project may follow a tortuous path, they should always be guided by an underlying purpose. A literature review supplemented by personal observations should enable the scientist to ask certain kinds of questions. The nature of the questions and the availability of resources should dictate the research design. The setting, subjects, independent and dependent variables, choice of statistics—all should form a coherent whole.

Techniques for analyzing data must be integrated with research designs. Unless data analysis and research design are considered as a unit, the research question will probably not be answered efficiently. Suppose an investigator wants to learn if 15- and 16-year-old girls differ in height (they do, by about a half inch). If she selects a representative sample of 50 15-year-olds and 50 16-year-olds and measures them, she is likely to find a difference at the conventional level of statistical significance (.05 level, see p. 51) only 26 percent of the time. With 20 girls per group, significance would be reached in only 15 percent of cases. But if she uses a power table (see p. 207) to determine sample size, she is much less likely to conduct an uninterpretable experiment.

To Introduce Controversial Issues

In my first science course, the teacher contrasted pre-scientific attitudes with those of our own age of enlightenment. According to her account, the ancient Greeks resolved debates about various matters by consulting authorities. Rather than counting the number of teeth a horse has, they referred to Aristotle. She said that modern scientists constantly question and let the facts decide, which is a superior method for finding answers. But Feyerabend (1975) insists that the unrelenting skepticism of scientists is a myth, and in any case reserved for the ideas of the opposition. In this book, I raise questions and discuss controversial issues. Not about the foundations of scientific inference, a topic best left to philosophers like Feyerabend. And nothing original (though several are original for introductory books on research methods). Still, I hope that students will be stimulated by discussions of such topics as the role of statistical significance testing, the place of private data in psychology, the value of small sample studies, and the ethics of animal experimentation.

To Show How Exciting Science Can Be

Finally, but most importantly, science should be fun. Several aspects of research are often drudgery, but others offer an intellectual challenge and a chance for creativity matched by few other forms of human endeavor. Science should be learned not as a set of rules and formulae to be memorized, but as a method for solving problems.

Summary

Many scientists derive enormous satisfaction from their work. They exercise their creative faculties while trying to solve problems about deep mysteries. The most successful ones have changed how people perceive themselves and their places in the universe.

Scientists collect data and then formulate laws and theories to organize and explain the data. They test their theories by making falsifiable predictions. Scientists use many tools in their work, but the only essential feature (so far as this book is concerned) is that they construct and test falsifiable theories.

Science is limited in several ways. First, the scientific method can answer questions about observational statements only—not about definitions or values. Second, the values of scientists affect virtually all aspects of their work. Third, an infinite number of equations (and hence an infinite number of theories) fit any set of data. Despite the limitations, all people are scientists to some extent, and people can use scientific principles to improve their decision-making ability.

Good scientists ask answerable questions that apply to a broad range of phenomena. They develop efficient methods for answering their questions. They gather data carefully and use proper sampling procedures and research designs; but they are not afraid to risk errors by generalizing beyond their data and constructing falsifiable theories. They use a variety of methods and measures, as individual projects dictate.

Key Terms

Dependent Variable   The behavior measured by a researcher at the end of a study to see how it has been affected by administration of the independent variable. If a researcher studies the effects of alcohol ingestion on sexual activity, sexual activity is the dependent variable. If she studies the effects of sexual activity on amount of alcohol drunk, alcohol is the dependent variable.

Falsifiable   A theory is falsifiable if it can in principle be refuted by data.

Independent Variable   The variable that a researcher manipulates. In the examples given for dependent variables, the independent variable would be alcohol in the first case and sexual activity in the second.

Observational Statement   A statement whose truth or falseness can be established only by empirical means.

PART II

Preparing for Research

CHAPTER 2

Finding an Interesting Problem

By the time you finish reading this chapter, you should be able to answer the following questions:

How do scientists get their ideas for research projects?

How have people used their personal experiences as a source for research ideas?

Should science students read popular magazines?

Of what value are anomalies to scientists?

Be Attentive to Both Routine and Unusual Occurrences

Much of the joy of research comes from intense involvement with a problem. Crossword puzzlers have their problems printed daily in newspapers and detectives have theirs handed to them, but scientists must find their own. Beginning researchers often problem-solve in collaboration with their graduate advisors, then follow up exciting findings with clarifications, extensions, or tests of competing explanations. Choice of a problem area may arise from personal experiences, as when a victim of chronic illness reads the medical literature for possible insights into cures, or a marathoner becomes interested in speculations about "runner's high." Other sources include television documentaries, discussions with friends, and attendance at engrossing lectures. Accidents often stimulate interest. Several active research areas began with observations of people with unusual problems: an obese person with a tumor near the hypothalamus; an aphasic with a lesion in Broca's area; a child who ate handfuls of salt with each meal and was found to have a defective adrenal gland; excessive sleeping in people with lesions or tumors of the reticular formation. The event doesn't have to be life-threatening; perhaps someone you care about reacts strongly to a particular food or color or piece of music, and you want to find out why. The important thing is that you get in the habit of asking questions.

> Why is my cat neurotic?
> How can I help him stop gambling?
> How can I conduct a fair test of astrology?

Become a Careful Observer

Observing should almost always precede experimentation. Somebody describing fellow psychologists once said, "We try to find causes to explain certain observations but no longer bother making the observations." Careful observing requires skills comparable to those needed in other areas of science. For now, recognize that even casual observations can lead to ideas for research.

Read-I

Reading stimulates ideas. Many biographies and autobiographies of great scientists convey the excitement of research and indicate promising areas for future study. Science columns of daily newspapers, and magazines such as *Scientific American, Psychology Today, Science News,* and *Discovery,* offer well-written articles about important topical research. The better introductory textbooks provide valuable overviews of different fields. If a particular area of investigation seems interesting, check *Readers Guide to Periodical Literature* for references to relevant magazine articles. This type of reading is to whet the appetite and suggest potential areas for exploration.

Read-II

Scientists who have already carved out a problem area spend much of their professional time reading original research articles in technical journals. The journals cover an enormous range of specialty areas. Try the following strategies for developing a research project in an area that interests you:

- Search for flaws in published papers. If you find any, do a similar study without the flaws.
- Two flawless projects may nevertheless give rise to conflicting data or interpretations. Examine the procedures to figure out why, then design a new study to resolve the conflict and pit alternative predictions against each other.
- Interesting theories have testable consequences. Deduce a consequence and test it.
- Discussion sections of research articles often suggest still-unanswered questions and possible follow-up studies.
- Look for a practical problem to solve, as do many industrial and human engineering psychologists.
- Consider modifications of a study that would test its generality. Ask yourself if additional work with different kinds of subjects, settings, or ways of measuring variables would be profitable.
- Pay attention to results that don't seem to make sense in light of current beliefs. Scrutinize individual research reports for outliers (see p. 354) and entire fields for aberrant individual findings. These may suggest new directions for investigation. Humphreys (1968) wrote that "both the logical structure of scientific theories and their historical evolution are organized around the identification, clarification, and explanation of anomalies."

Seek Analogies

Koestler (1964) gave several illustrations of the power of analogies—both recognizing and creating them—for generating research ideas. For example, Metchnikoff threw some rose thorns to starfish larvae and observed that the larvae surrounded and dissolved the thorns. From this, he developed the idea that the body defends itself against invading microbes by secreting pus that surrounds and tries to digest them. Kepler drew an analogy between the role of the Father in the Trinity and the sun in the universe. He said, "And I cherish more than anything else the Analogies, my most trustworthy masters. They know all the secrets of Nature. . . ." Kelvin conceptualized the mirror galvanometer as an analogy to the way his monocle reflected light.

The development of analogies presupposes knowledge of something else. So researchers should try to familiarize themselves with past and current work in sciences other than their own.

Examples

Psychology Today magazine publishes occasional interviews with leading scientists in which they discuss, among other topics, how their particular research interests developed. Cohen (1977) collected a series of additional interviews in a book. I've summarized statements by the scientists on how they got started.

1. Benjamin Bloom (Chance, 1987): He was a University Examiner at the University of Chicago and observed that some students were very poor problem solvers. So he sought to find out why.
2. Donald Broadbent (Cohen, 1977): He had two sources of ideas: (a) his advisor gave him a problem, (b) he tried to reconcile a paradox: most laboratory experiments had failed to show any effects of noise on subjects, but industrial experience suggested that noise reduction is good for people.
3. Rosalind Cartwright (Lamberg, 1988): A colleague asked her to help him study the effects of a hallucinogenic drug he had synthesized. The drug-induced fantasies of their subjects sounded so much like dreams that she decided to find out if the drug eliminated the need to dream during sleep. She then began dream research.
4. Noam Chomsky (Cohen, 1977): He was involved in linguistics and failed in conventional linguistic approaches to Hebrew. So he worked on developing a new approach—construction of a system of rules that would give a base for interpreting languages.
5. H. J. Eysenck (Cohen, 1977): He had studied esthetics, hypnosis, and humor but couldn't get a job in any of those fields. So he took what was available—research psychologist in a hospital. He found many things wrong with diagnosis and treatment, so began checking on their validity.
6. Leon Festinger (Cohen, 1977): He read a study by a man who had collected thousands of rumors that were circulating widely after disasters. It puzzled him that people would circulate rumors that increased their anxiety, and he sought for reasons.
7. Liam Hudson (Cohen, 1977): He was doing dull research for no special reason and by accident found something exciting. The results received widespread coverage, and he enjoyed being a celebrity. So he began pursuing research diligently.
8. Michel Jouvet (Cohen, 1977): He confronted conflicting theories: Either Pavlov had been correct and the cerebral cortex controls sleep, waking, and learning; or Magoun was correct and the reticular formation is controller. He tried to design a study to see who was right.

PSYCHOLOGY NEEDS ITS NEWTON.

9. R. D. Laing (Cohen, 1977): He worked in a mental hospital and spent a day on a very difficult ward. He noticed that the most withdrawn patients—12 women—were the ones with whom the nurses interacted least and felt most hopeless about. So he arranged to have two nurses spend about 40 hours a week with them, and within a year all 12 were discharged.
10. David McClelland (Cohen, 1977): He was asked to teach two courses for which he had no training. He became intensely interested in the fields and was additionally fortunate in having a very able student to collaborate with.
11. Ronald Melzack (Warga, 1987): He was an aimless undergraduate student who came into contact with an inspirational teacher, Donald Hebb. Hebb started him on a problem and the results were exciting to Melzack.
12. Neal Miller (Cohen, 1977): He was invited to the Institute of Human Relations at Yale to try to bring together economics, sociology, anthropology, psychiatry, psychology, and law. He did research in several different areas and followed what was working.

13. Joseph Pleck (Kimmel, 1987): He was a frail boy who didn't conform to the traditional masculine sex role—he was a poor athlete and excellent student who studied piano. He read Margaret Mead and realized that sex roles differ in different cultures. Today, he works on the nature of masculinity and the social psychology of sex roles.
14. Martin Seligman (Trotter, 1987): When a graduate student, he saw a group of dogs that had failed a learning experiment, and he found out that they had been exposed to inescapable shock. He wondered if people exposed to inescapable, aversive circumstances also behave in a helpless manner.
15. B. F. Skinner (Cohen, 1977): His behavioristic philosophy and research program were shaped by his readings of Watson and Pavlov and his discussions with Keller, a fellow student.
16. Niko Tinbergen (Cohen, 1977): His studies of animal behavior stemmed from an early and lifelong love of animals. He was stimulated by reading Darwin and corresponding with Konrad Lorenz.

CHAPTER 3

Studying the Problem Area in Depth

By the time you finish reading this chapter, you should be able to answer the following questions:

How much of the scientific literature in their fields should prospective researchers read?

Are scientists most efficient when they read only in their narrow specialty or when they sacrifice depth of reading for breadth?

How important to scientists are sources of information other than written materials?

Where would you look to find summaries of recently published articles in a particular field?

If the only article you know of in a field was published several years ago, how can you use it to learn of more recent developments?

Where can you read reviews of psychology books?

"Knowledge is of two kinds. We know a subject ourselves, or we know where we can find information upon it."

Samuel Johnson, English writer (1709–1784)

Once you've become interested in a particular field, you should turn to the scientific literature. Reading recent original research articles will help you refine the problem and perhaps reformulate the research question. You will also reduce your chances of repeating work that has already been done and of making mistakes. You will be able to keep abreast of the most sophisticated techniques.

Philosopher of science Carl Hempel (1966) quoted economist A.B. Wolfe on how an ideal scientist would work:

If we try to imagine how a mind of superhuman power and reach, but normal so far as the logical processes of its thought are concerned, would use the scientific method, the process would be as follows: First, all facts would be observed and recorded, without selection or a priori guess as to their relative importance. Secondly, the observed and recorded facts would be analyzed, compared, and classified, without hypothesis or postulates other than those necessarily involved in the logic of thought. Third, from this analysis of the facts generalizations would be inductively drawn as to the relations, classificatory or causal, between them. Fourth, further research would be deductive as well as inductive, employing inferences from previously established generalizations.

But Hempel quoted Wolfe in order to emphatically disagree with the advice. The scientist who followed Wolfe's methodology would be paralyzed. She would never get past the first phase, since collection of all the facts relevant to a problem, let alone "all facts," is impossible. The facts of interest depend on the hypotheses of the researcher—what is irrelevant for one may be crucial to another. Hempel gave the example of Ignaz Semmelweiss, who was concerned with the high incidence of childbed fever in one of the maternity wards where he practiced medicine from 1844–1848; about 10 percent of the mothers there died in childbirth. One belief was that a priest, preceded by an attendant ringing a bell, was responsible. The pair periodically walked through the wards to administer the last sacrament to dying women. This was assumed to scare and debilitate the patients, thus increasing their likelihood of contracting and dying from childbed fever. Semmelweiss tested the hypothesis, plausible for that time period, by having the priest change his routine. But the deaths continued. Then he hypothesized that the disease was spread by infectious microorganisms carried by medical students; they frequently examined the women in that ward immediately after dissecting cadavers and without first disinfecting their hands. When Semmelweiss required the students to wash before examining their patients, deaths from childbed fever were greatly reduced. Wolfe's ideal scientist, with no hypothesis to guide him, would still be collecting data.

How then should prospective researchers review their fields? No consensus exists, and the different points of view reduce to matters of temperament. Some methodologists advise readers to saturate themselves in their subject. The advantages are clear, but there is a major drawback: the scientific literature is overwhelming. In addition to books, monographs, symposiums, reports, and newsletters, more than half a million articles are published each year. Ziman (1980) estimated that the scientific literature doubled in bulk from 1960 to 1975, and this expansion continues though at a somewhat lower rate. On the assumption that people read about the same amount each year, a scientist who kept up with three-fourths of the literature in a particular area in 1973 would keep up with only three-eighths in 1988. (To get a feel for the proliferation of studies, go to the reference section of a university library and compare the sizes of Index Medicus (see p. 40) for 1950 and 1989.)

Selectivity of reading is clearly required. It is physically impossible to read, let alone assimilate, the mass of literature; in addition, oversaturation with the works in a particular field may reduce the possibility of developing new perspectives. Claude Bernard wrote, "It is that which we do know which is the great hindrance to learning, not that which we do not know." Two types of evidence support his view. First, many great revolutionary ideas in science have been greeted with scepticism by the experts; and second, many outstanding scientists, Pasteur, Metchnikoff, and Galvani among them, were not trained in the fields in which they made their most important discoveries.

Ziman (1980) urged scientists to read in fields other than their own, partly because the classification of scientific fields is somewhat arbitrary. Scientists may group themselves according to the types of organisms they study, the instruments they use, the theories that guide them, or particular practical problems they are trying to solve. If they read only in journals covering their narrow specialty, they may miss many important papers. In Ziman's opinion, they should focus their reading on five or six core journals with high editorial standards that publish relevant articles.

Enrico Fermi said that scientists should spend two-thirds of their working time on their own problems and one-third on other ideas. He added that the other ideas should come less from reading than from attendance at lectures, since lectures afford the opportunity to ask questions. Along those lines, Kasperson (1978) used various criteria to classify scientists as either creative and productive, productive but not creative, or non-creative and non-productive. He found that the more creative ones placed great store in information gathered at conventions, conferences, and meetings.

Maini & Nordbeck (1973) questioned scientists about events the scientists considered crucial in the development of their research. Interpersonal informal situations were mentioned more often than lectures and literature reviews. And Olby (1970), discussing the working habits of Nobel Prize winner Francis Crick and his colleagues, said: "One hour's discussion of their problems per day is the rule. On such occasions they say 'anything without having to justify it up to the hilt'. . . . They interrupt each other, to continue either in 'dialogue' or 'duologue', the ideas tumbling helter-skelter from Crick to be met by a relentless questioning from Brenner."

Hans Selye said, "Just as in eating, the feeling of satiety should be our guardian against excess in reading; different people have different capacities and should gorge themselves with literature accordingly."

Hempel's description of Semmelweiss' methods shows the impossibility of knowing in advance all that is relevant for solving a particular problem. When problems are conceived narrowly, potentially relevant data and concepts from other disciplines may be overlooked. In addition, as noted in the previous chapter, one of the most fruitful methods for advancing science is to seek analogies between different areas. This presupposes familiarity with more than one area. So don't stop reading popular magazines like *Scientific American.*

Scientists who wish to read in depth should start in the reference section of a good library. Several excellent sources can simplify the task.

Abstracts

Psychological Abstracts is a collection of abstracts (summaries, usually about 100 to 200 words) of research articles and some Ph.D. dissertations and books on psychological topics. Each monthly issue covers more than 850 journals. Information printed with the abstract tells readers where to find the unabridged articles. Abstracts with both author and detailed subject indexes are published in many scientific fields. With a bit of practice it becomes easy to go quickly from the indexes to the abstracts; after you read the abstract, you'll have a pretty good idea if the unabridged article is appropriate for your needs. Some important abstracts for psychologists are *Biological Abstracts, Child Development Abstracts, Psychopharmacology Abstracts, Sociological Abstracts,* and *Women Studies Abstracts.* Figure 3.1 shows in greater detail how to use *Psychological Abstracts.*

Citation Indexes

The citation indexes (*Science Citation Index* for chemistry, physics, biology, and the like; *Social Science Citation Index* for sociology, anthropology, and related; both indexes for psychology) enable readers to find additional articles once a single key article has been identified. The value of the indexes stems from the fact that reference sections of scientific articles cite other articles dealing with similar subject matter. (Most scientific publications are cited most frequently shortly after being printed, and about half as often by the fifth year (Margolis, 1967), but important articles may continue to be cited for many years.) The citation indexes list alphabetically by author all articles cited during the time covered by the index. If Einstein's 1905 paper on relativity is cited in 1988, then the issue of the Index covering 1988 will have a listing for Einstein, Albert. "On the electrodynamics of moving bodies." Underneath that will be a list of all the articles published in 1988 that refer to his paper. Thus, with the citation indexes, readers can follow the most recent developments in a field. Figure 3.2 shows how to use the *Social Science Citation Index.*

BRIEF SUBJECT INDEX

Consumer Behavior 17672, 20931, 20932, 20936, 20938, 20942, 20943, 20944, 20945, 20946, 20948, 20950, 20952, 20955, 20963
Consumer Research 20931, 20935, 20944, 20948, 20954, 20955
Contagion 18870, 20391
Content Analysis 18477, 18539, 19257, 19675, 20153, 20155, 20961, 20967, 20969
Content Analysis (Test) 20685
Contextual Associations 17539, 17811, 17812, 17815, 17821, 17822, 17857, 17862, 17878, 17881, 17888, 17901, 17908, 17909, 17911, 17958, 18386, 18391, 18405, 18410, 18478, 18483, 18594, 18866, 18872, 18887, 18979, 19444, 20134, 20518, 20521
Contingency Management [See Also Token Economy Programs] 17918, 19912
Continuous Reinforcement [See Reinforcement Schedules]
Contour [See Form and Shape Perception]
Controls (Instrument) [See Instrument Controls]
Conversation 17640, 17812, 18488, 18602, 18731, 18795, 18873, 18884, 18890, 18896, 18905, 18907, 18919, 18919, 19907
Conversion Hysteria [See Conversion Neurosis]
Conversion Neurosis 19659
Convulsions 18290, 18328, 18406, 19503, 19585
Cooperation 18052, 18882, 18892, 20376, 20822, 20851, 20878
Cooperative Therapy [See Cotherapy]
Coordination (Motor) [See Motor Coordination]
Coordination (Perceptual Motor) [See Perceptual Motor Coordination]
Coping Behavior 17561, 17570, 17575, 17591, 17639, 18167, 18664, 18749, 18777, 18787, 19141, 19199, 19273, 19291, 19415, 19420, 19480, 19559, 19651, 19664, 19883, 19904, 20029, 20072, 20099, 20217, 20219, 20340, 20383, 20444, 20629, 20710
Copulation (Animal) [See Animal Mating Behavior]
Coronary Disorders [See Cardiovascular Disorders]
Coronary Prone Behavior 17542, 18172, 18180, 18185, 18186, 18193, 18657, 18905, 18952, 19510, 19528, 19549, 19569, 19649, 19653, 19910
Corporal Punishment [See Punishment]
Corporations [See Business Organizations]
Corpus Callosum 18081, 18089, 18099, 18276, 19208, 19209, 19548, 19554
Corpus Striatum [See Basal Ganglia]
Correctional Institutions 19884
Correlation (Statistical) [See Statistical Correlation]
Cortex (Cerebral) [See Cerebral Cortex]
Cortex (Motor) [See Motor Cortex]
Cortex (Somatosensory) [See Somatosensory Cortex]
Cortex (Visual) [See Visual Cortex]
Cortical Evoked Potentials 18093, 18100, 18117, 18123, 18183, 18184, 18199, 18563, 19349, 19622
Corticoids [See Corticosteroids]
Corticosteroids [See Also Corticosterone, Hydrocortisone, Prednisolone] 17963, 19961
Corticosterone 18240
Corticotropin 18354, 18374, 18423, 18437, 19212, 19522, 19974, 19989, 19997
Cortisol [See Hydrocortisone]
Cost Effectiveness [See Costs and Cost Analysis]
Costa Rica 18734
Costs and Cost Analysis 18835, 20067, 20114, 20859, 20936, 20949
Cotherapy 19717, 20342

Counselees [See Clients]
Counseling (Group) [See Group Counseling]
Counseling [See Also Related Terms] 18525, 18868, 20076, 20153, 20155, 20156, 20158, 20162, 20316, 20820
Counselor Attitudes 18850, 19801, 20389, 20662, 20737
Counselor Characteristics [See Also Counselor Attitudes] 17649, 19737, 19790, 19809, 20154
Counselor Client Interaction [See Psychotherapeutic Processes]
Counselor Education 20343, 20350, 20358, 20410, 20411
Counselor Effectiveness [See Counselor Characteristics]
Counselor Personality [See Counselor Characteristics]
Counselor Role 20412, 20739
Counselor Trainees 20377, 20395
Counselors [See Also Rehabilitation Counselors, School Counselors] 20157, 20384, 20402, 20414, 20972
Countertransference 19765, 19768, 19778, 19822, 19826
Countries [See Developing Countries]
Couples 18774
Course Objectives [See Educational Objectives]
Courts [See Adjudication]
Courtship (Animal) [See Animal Courtship Behavior]
Covert Sensitization 20992
Craving [See Appetite]
Creative Writing [See Literature]
Creativity (17502) 17868, 18081, 18959, 18968, 18969, 18975, 19886, 19001, 19147, 19225, 20451, 20798
Credibility 18813, 18819, 18876, 19282, 20946
Crime [See Also Child Abuse, Driving Under The Influence, Homicide, Infanticide, Rape, Sex Offenses, Shoplifting, Theft, Vandalism] 19148, 19266, 19280, 19294, 19312, 19338, 20163, 20305, 20324
Crime Prevention 18830
Crime Victims 19338
Criminally Insane [See Mentally Ill Offenders]
Criminals [See Also Female Criminals, Male Criminals, Mentally Ill Offenders] 19276, 19318, 20334, 20333
Criminology 19333
Crippled [See Physically Handicapped]
Crises 19796
Crisis (Reactions to) [See Stress Reactions]
Crisis Intervention [See Also Suicide Prevention] 19796, 20683
Crisis Intervention Services 20218
Critical Flicker Fusion Threshold 18173
Critical Period 18581
Critical Scores [See Cutting Scores]
Cross Cultural Differences 17515, 17600, 17623, 17629, 17671, 17683, 17922, 18509, 18551, 18626, 18696, 18751, 18753, 18765, 18865, 18867, 18932, 18949, 19004, 19189, 19302, 19303, 19372, 19527, 19748, 19881, 20120, 20227, 20358, 20488, 20500, 20504, 20511, 20524, 20555, 20570, 20583, 20677, 20720, 20777, 20866, 20921, 20967
Cross Disciplinary Research [See Interdisciplinary Research]
Crowding 18677, 20912
CRT [See Video Display Units]
Cruelty 20844
Crying 18963
Cuban Americans [See Hispanics]
Cues 17692, 17717, 17775, 17793, 17856, 17890, 17903, 17910, 17944, 18542, 19911, 20887

Cultural Assimilation 18758, 19809, 20677
Cultural Deprivation 18645
Cultural Differences [See Cross Cultural Differences]
Cultural Psychiatry [See Transcultural Psychiatry]
Cultural Test Bias 17655
Culturally Disadvantaged [See Cultural Deprivation]
Culture (Anthropological) [See Also Society] 18759
Culture Change [See Also Cultural Assimilation] 18764
Culture Shock 20677
Curriculum [See Also Drug Education, Health Education, Language Arts Education, Mathematics Education, Psychology Education, Reading Education, Science Education, Sex Education, Social Studies Education, Spelling, Vocational Education] 20354, 20366, 20462, 20577, 20681, 20867
Curriculum Development 20586
Cursive Writing 20475
Cutaneous Sense [See Also Tactual Perception, Vibrotactile Thresholds] 17726, 17738, 18232, 18257
Cutting Scores 20619
Cybernetics 17776, 17854, 19967
Cycloheximide 18395
Cynicism 20947
Cystic Fibrosis 19541
Czechoslovakia 20721

Daily Biological Rhythms (Animal) [See Animal Circadian Rhythms]
Dance 20979
Dance Therapy 20102
Dangerousness 20371
Dark Adaptation 17979
Data Collection 18720, 19773, 20423, 20705, 20760, 20948
Dating (Social) [See Social Dating]
Daughters 18771, 18773, 19283, 20043
Day Camps (Recreation) [See Summer Camps (Recreation)]
Day Care (Child) [See Child Day Care]
Day Care (Treatment) [See Partial Hospitalization]
Day Care Centers 20117, 20140, 20150
Day Hospital [See Partial Hospitalization]
Daydreaming 17859
Deaf 19534, 19593, 19626, 19627, 19633, 20037, 20623
Death and Dying 18765, 18766, 18777, 18784, 18802, 19031, 19038, 19149, 19173, 19221, 19277, 19353, 19583, 19584, 19812, 20159, 20340
Death Anxiety 18977
Death Attitudes 18954
Death Rate [See Mortality Rate]
Death Rites 18705
Deception [See Also Cheating, Faking] 18057
Decision Making [See Also Choice Behavior, Group Decision Making, Management Decision Making] 17688, 17709, 17830, 17831, 17832, 17833, 17838, 17849, 17864, 17873, 17874, 17887, 17913, 18140, 18640, 18822, 18828, 18835, 18971, 19443, 20162, 20205, 20350, 20417, 20458, 20502, 20733, 20747, 20778, 20793, 20807, 20923, 20941, 20960
Decoding [See Human Information Storage]
Decortication (Brain) 18261
Defense Mechanisms [See Also Identification (Defense Mechanism), Projection (Defense Mechanism), Repression (Defense Mechanism)] 19529, 19740, 19859

Figure 3.1(a)
Using psychological abstracts. (a) An investigator interested in the scientific literature on creativity should look under that heading in the index of *Psychological Abstracts*. Several 5- or 6-digit numbers are printed there, and the front of the volume lists the numbers sequentially. The first listing under creativity in the index is circled.

These citations are reprinted with permission of the American Psychological Association, publisher of *Psychological Abstracts* and the PsycINFO Database (Copyright © 1967–1989 by the American Psychological Association) and may not be reproduced without its prior permission.

PSYCHOLOGICAL ABSTRACTS

GENERAL PSYCHOLOGY

17501. **Rao, K. Ramakishna.** What is Indian psychology? *Journal of Indian Psychology,* 1988(Jan), Vol 7(1), 37–57. —Argues that there is a need for psychology in India to grow out of and be in tune with Indian cultural milieu, native concepts, and Indian intellectual inheritance and philosophical ethos and discusses why this is important to the progress of psychology in India.

17502. **Sujatha, A. K.** (Osmania U, Hyderabad, India) **A classification of problems in the sciences.** *Journal of Indian Psychology,* 1986(Jul), Vol 5(2), 1–20. —Classifies all problems that scientists seek to solve into 12 categories. In the process of making this classification, an analysis was made of the process by which a problem originates for an individual. A definition is given for the term "problem." Lines of enquiry are proposed for enhancing understanding of the nature of creativity in the sciences.

Parapsychology

17503. **Anderson, Rodger.** **Channeling.** *Parapsychology Review,* 1988(Sep–Oct), Vol 19(5), 6–9. —Discusses the phenomenon of channeling, formerly called automatism, which is behavior, usually spoken or written, for which the S denies voluntary control. Arguments for skepticism by psychic researchers are put forth.

17504. **Sadhakar, U. Vindhya & Rao, P. Krishna.** (Andhra U, Waltair, India) **Belief and personality factors of participants: A study in an ESP/ganzfeld setting.** *Journal of Indian Psychology,* 1986(Jul), Vol 5(2), 21–45. —Used the ganzfeld technique for ESP elicitation to examine the role of belief in ESP. personality factors, and the interaction between the 2 in 50 university students (aged 19–30 yrs) who acted as experimenters or Ss. The ganzfeld technique facilitated the manifestation of ESP. The personality factor A of the Sixteen Personality Factor Questionnaire (16PF) (aloof vs good-natured) of experimenters was significantly and positively related to the ESP scores they elicited from their Ss. The length of Ss' mentation reports and their ESP scores as elicited by the experimenters were found to be related positively and significantly. Ss who reported more spontaneous, dreamlike mental activity during the ganzfeld obtained the highest ESP scores. Ss' expectations of success at the beginning and end of the ganzfeld session correlated negatively with their ESP performance.

History & Philosophies & Theories

17505. **Burkett, Timothy.** **The hermeneutics of Hans-Georg Gadamer and Paul Ricoeur.** *Saybrook Review,* 1988(Spr), Vol 7(1), 67–97. —Compares the anti-methodological stance toward the activity of interpretation set forth by H.-G. Gadamer (1975, 1976) with alternative positions described in the works of J. Habermas (1972) and P. Ricoeur (published works 1966 to 1981). Topics discussed include foundations laid by M. Heidegger (1962) for Gadamer's work; the hermeneutics of Gadamer; Habermas's critique of Gadamer, a foundation for Ricoeur's hermeneutics; and the hermeneutics of Ricoeur.

17506. **Cunningham, Jacqueline L.** (U Texas, Austin) **A comparison of Psychology: An Elementary Textbook by Hermann Ebbinghaus with modern introductory textbooks.** Ebbinghaus Memorial Symposium of the Southeastern Psychological Association (1985, Atlanta, Georgia). *Revista de Historia de la Psicologia,* 1986(Jan–Mar), Vol 7(1), 59–70. —Compared H. Ebbinghaus's (1908) textbook, in translation by M. Meyer called *Psychology: An Elementary Textbook,* with 7 recent introductory textbooks and the *Thesaurus of Psychological Index Terms* (American Psychological Association, 1982). The comparison showed considerable overlap between the topical content of the Ebbinghaus text and recent texts. However, the categories comprising 65% of the Ebbinghaus text comprised only 45% of modern texts. (Spanish abstract)

17507. **Das, R. C.** (David Hare Training Coll, Calcutta, India) **Science and subjectivity.** *Samiksa,* 1987, Vol 41(4), 95–101. —Argues that there is an attitude of caste-mindedness within science that characterizes some of the sciences as subjective (more personal and speculative) and others as objective (free from personal bias, more reliable and accurate). It is suggested that the social status of a science is determined by the degree of objectivity it can command. It is further argued that the differentiation of sciences on the basis of objectivity is untenable in that subjectivity is inherent in the human perception of scientists. The perceived world has an objective basis in reality but is subjective in the sense that it is transformed by the human system.

17508. **Dazzi, Nino.** **The problem of concept formation in Vygotsky.** *Revista de Historia de la Psicologia,* 1986(Jan–Mar), Vol 7(1), 3–12. —Offers a critical analysis of the interaction between Western and Soviet psychology as reflected in the work of L. S. Vygotsky (1966, 1974, 1976). Examined in particular is the influence of N. Ach (1921), mainly through the work of L. Sakharov (1930), and H. Werner (1926 [1970]). It is asserted that Vygotsky ought not to be included in the tradition of Würzburg School but should be viewed as an applied-oriented person also inclined to great theoretical syntheses. (Spanish abstract)

17509. **Early, Charles E.** (Mobile Coll, AL) **Hermann Ebbinghaus and American psychology: A historiographic study.** Ebbinghaus Memorial Symposium of the Southeastern Psychological Association (1985, Atlanta, Georgia). *Revista de Historia de la Psicologia,* 1986(Jan–Mar), Vol 7(1), 81–90. —Contends that English-language history of psychology textbooks present a selective view of H. Ebbinghaus (1885 [1964], 1908 [1973]). Even a composite from multiple sources leaves gaps in knowledge of his life (e.g., no mention of his wife or family). Most of the materials examined used relatively few sources, and many relied heavily on E. G. Boring (1950). Additional research using primary source materials such as archival data is needed and is now being conducted. An outline of important dates in Ebbinghaus's life is provided. (Spanish abstract)

17510. **Frank, Alvin.** **Pervasive assumptions, the points of view, and models of metapsychology: Review and revision.** *International Journal of Psycho-Analysis,* 1988, Vol 69(4), 483–494. —Identifies shortcomings of D. Rapaport and M. Gill's (1959) view of metapsychology. The author presents his definition of metapsychology as the collection of higher level theories underlying and explaining the less abstract and experientially based theories of psychoanalysis. These theories are considered and formulated in interrelated contexts of force, excitation, time, organization, and adaptation. These principles are sometimes conceived of in the form of models. Certain psychoanalytic metatheoretical assumptions are so pervasive as to be treated within the viewpoints rather than specifically cited. They include (1) provision of the basis for a psychological theory of mind in the biopsychological human entity, (2) psychologically determined mental phenomena, and (3) a lawful sequential logic. (French, German & Spanish abstracts)

17511. **Holland, Dorothy C. & Valsiner, Jaan.** (U North Carolina, Chapel Hill) **Cognition, symbols, and Vygotsky's devel-**

Figure 3.1(b)
Turning to the front of the volume, the reader would find the listing (also circled). The listing includes an abstract and the information necessary to locate the original article.

Studying the Problem Area in Depth

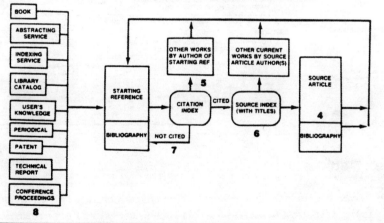

Figure 3.2(a)
An outline of the steps in a literature search with *Social Science Citation Index*.

Reprinted with permission from the Institute for Scientific Information®. Copyright 1989 Social Science Citation Index®. Philadelphia, PA.

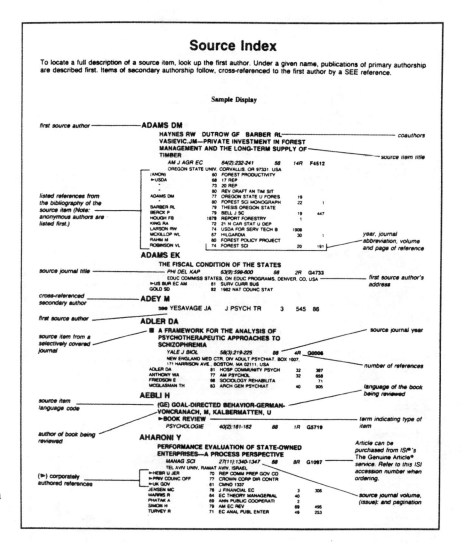

Figure 3.2(b)
A sample display showing how to locate a full description of a source item.

Index Medicus

Index Medicus is an index to the world's biomedical literature (which includes many topics of interest to social scientists such as drugs and drug abuse, sleep and dreams, biofeedback, hypnosis, and psychotherapy). Separate volumes have alphabetical listings for both subject and author. Each entry lists the article title, author, and name, volume, page, and date of the journal. (Summaries are not included.) In addition, the front section of each issue lists recent review articles; if you find a comprehensive review article, your library search will be greatly simplified.

Reviews of Research

In many fields, eminent scientists write reviews of recent research for publication in annual collections. There are annual reviews of psychology, sociology, biology, and so forth. They are listed in library card catalogues under *Annual Review of*. . . . The reviews cover primarily empirical rather than theoretical material, and they include extensive reference lists. *The Psychological Bulletin* publishes evaluative reviews and interpretative articles on methodological and research problems.

Current Contents

Current Contents is a weekly publication that reproduces the tables of contents of many journals. One part covers the life sciences and one part the social sciences. Many researchers skim through *Current Contents* and then either write away for interesting-sounding articles (authors' addresses are included) or find the articles in libraries.

Other Strategies

When you have collected a number of references, you may notice that many come from the same journal. Then, check all recent issues of that journal. Many journals have letters sections in which correspondents comment on articles recently published in that journal. The comments usually take the form of criticisms, and the original authors are given an opportunity to reply. Readers of such correspondence may learn a great deal about both the substantive topic and methodological issues.

For researchers interested in advancing a field, books are rarely as valuable as recent journal articles. Still, they may at times be helpful. *Contemporary Psychology* is a monthly journal of book reviews, and the *Mental Health Book Review Index* tells where reviews of psychology books can be found. References to reviews of books on psychological measurement and testing are found in the *Mental Measurements Yearbook*.

We're in the computer age. Most university libraries provide inexpensive computer searches for both popular and scientific articles. MaryLu Rosenthal wrote an important paper, reproduced as Appendix A, which can be used collaboratively by social scientists and librarians to help them with such searches. If you have a personal computer and a modem, you can sign up for an inexpensive computer search service from your home.

How Should You Read?

Although both amount and depth of reading are largely matters of taste, virtually all active researchers devote much of their professional lives to reading the literature. So they learn techniques for systematizing and simplifying. You should do the same. Learn to skim read. And read a general source such as an introductory book or good encyclopedia article before plunging into original research articles. That way, you'll be introduced to technical jargon as painlessly as possible, and you'll have a perspective from which to read the more detailed material. If you're fortunate and find a recent review article, start there; it may summarize and critically evaluate other articles you'll be reading.

But always read the originals, as secondary sources may be inaccurate. For example, Watson and Rayner's (1920) conditioning of the infant Albert B. is one of the most widely cited experiments in psychology textbooks. Harris (1979), after surveying introductory texts in general, developmental, and abnormal psychology, and advanced texts in behavioral therapy, reported that most referred to the Watson and Rayner study and most contained inaccuracies. Some authors made minor errors involving Albert's name, age, and whether he was initially conditioned to fear a rat or rabbit. Many made errors of a more serious nature, including erroneous lists of stimuli to which Albert's fear had generalized and false reports (sometimes in detail) that Watson eventually removed Albert's fears.

Most scientific articles include abstracts. Read them first, as they provide an overview and may indicate whether the article is suitable for your needs.

Make a card index of important articles. Put the location information at the top of each 5×8 card, and summarize the article on the rest of the card. Some people use different colored cards to classify according to topic, but one of the values of index cards is that they can be readily rearranged to accommodate different ways of classifying.

Journal articles vary in quality. Throughout this book, and especially in the next chapter, suggestions will be given for critical reading. For now, keep in mind that inferences from published reports are subject to the following limitations:

1. The data may be inaccurate because of various kinds of errors.
2. The data may have occurred by chance rather than because of the specific experimental treatment.
3. The data may not be accurate because the researcher was dishonest.
4. The data may be accurate but apply to only a restricted set of subjects or conditions.
5. The data may be accurate but the conclusions drawn from them may be inappropriate.
6. The data may be accurate and the conclusions justified, but the research may still be trivial.

Summary

Reading the scientific literature helps researchers refine their problems, reduces their chances of repeating work that has already been done, and enables them to keep abreast of advances in technology and methodology.

Readers should be guided by a research hypothesis and should not restrict their reading to works within a narrow specialty. They should familiarize themselves with several reference materials:

Psychological Abstracts: provides summaries and location information for finding articles on a particular topic or by a particular author

Citation Indexes: give titles and location information of articles that refer to a selected article

Index Medicus: lists titles and location information of articles on a particular topic or by a particular author

Reviews of Research: have comprehensive reviews of recent research in specific subfields

Computer searches are widely available and allow researchers rapid access to publications within a field.

Key Terms

Citation Index Lists titles and location information for all current articles that cite a specific previously published article.

Current Contents Lists the tables of contents of journals in a broad range of fields.

Index Medicus Lists titles and location information for the world's biomedical literature.

Psychological Abstracts A collection of summaries of research articles, with an extensive index to the summaries. Each summary tells where to find the original article.

CHAPTER 4

Evaluating What You Read: Single Studies*

By the time you finish reading this chapter, you should be able to answer the following questions:

Why would it be unacceptable to assign volunteers to one group and nonvolunteers to another?

What is the purpose of control groups?

How should experimental and control groups be chosen?

Why is it sometimes necessary to have more than one control group?

Should the experimenter know the groups to which subjects have been assigned?

Under what circumstances should researchers generalize beyond their immediate data?

What is meant by statistical significance and how does it differ from scientific significance?

Why do most scientific journals rarely publish articles in which statistical significance has not been demonstrated?

If an experiment designed to test a theory yields statistically significant results, why does that not necessarily provide evidence in favor of the theory?

---

*Methods for analyzing groups of studies are presented in chapter 24.

There is a paradox: Scientists are widely portrayed as impartial seekers of truth and the scientific method as the most powerful tool ever invented for establishing it. Yet research findings often fail to create consensus about what the truth is. Despite thousands of studies published in scientific journals of high repute, experts still disagree about the effectiveness of psychotherapy, the dangers of marijuana, the value of LSD in enhancing creativity, the reasons for differences between boys and girls in mathematical and verbal abilities, the use of special diets in the treatment of hyperactive children, the mechanism by which moral judgment develops, the existence of ESP, and a great many others. Why should this be?

Scientists may arrive at an erroneous conclusion for any of six broad categories of reasons. These are mentioned below and expanded upon later in the chapter. They are presented from the perspective of readers of research reports but apply as well to designers and conductors of research.

- A flaw in research design may lead him to attribute results to the actions of one variable when they are actually due to those of another. When this happens, a study is said to lack *internal validity*.
- He may define his variables in a way that is either misleading or inconsistent with their use by other scientists. This is a problem of *construct validity*.
- She may without warrant generalize beyond the circumstances of the study. Unwarranted generalization to other subjects, places, times, or conditions of testing is a problem of *external validity*.
- The results may not be reliable, but due only to chance. This is a problem of *statistical validity*.
- She may make an inferential error in going from data to conclusions.
- He may cheat.

The following hypothetical experiment is offered to help you learn more about the six categories.

A researcher hypothesizes that hunger is caused by activity of the neurotransmitter norepinephrine. He is unable to test norepinephrine injections directly, because the drug is inactivated before it gets into the brain. Injected amphetamine does get into the brain and promotes the release of norepinephrine, so he uses amphetamine as a substitute. He trains six pairs of rats to press a bar to get food. When deprived, they bar press more. After they've been trained, he reaches into their cages just prior to feeding time and removes one member of each pair. Each of the six is then tested for bar pressing and returned to its cage. Then the second animal of each pair is removed, injected with amphetamine, and tested for bar pressing. The results are evaluated statistically, and it is found that the amphetamine-injected rats press significantly more. The experimenter concludes that amphetamine increases hunger by increasing the activity of norepinephrine.

An experiment designed along such lines might actually turn out as described. Yet, amphetamine is often sold as a diet pill. What flaws in the experiment might account for the apparent discrepancy?

Problems of Internal Validity

Hint One. The subjects were not assigned to groups properly.

Explanation. The purpose of an experiment is to find out how the independent variable, in this case amphetamine, affects the dependent variable. But if the subjects assigned to one treatment differ initially from those assigned to another, the results might reflect the subject differences rather than any effect of the independent variable. For example, suppose the untreated subjects were 8-year-old children and the ones injected with amphetamine were sumo wrestlers?

Experimenters can never be sure that subjects assigned to one treatment are exactly the same as those assigned to another. In fact, it would be a remarkable coincidence if they were. But if subjects are assigned properly, the average differences between them will usually be small—and will grow smaller as more subjects are added. Statistical analyses take into account the likelihood of some pretreatment differences.

Proper assignment means random assignment. The rats in this experiment were not assigned randomly to receive either amphetamine or nothing. Instead, the first animal pulled out of a cage always received nothing. Might the first animal have differed in some way from the second? Possibly. It might have been less active, so more easily caught; it might have been friendlier, sicker, more aggressive, or had some other trait we're unaware of. We need not try to guess, since a simple procedure eliminates all concerns along those lines. Flip a coin or use some other randomization procedure to decide which rats get amphetamine, which get nothing.

PRINCIPLE ONE: Assign Subjects Randomly to Groups.

Hint Two. The groups were treated differently in ways other than exposure to the independent variable.

Explanation. The independent variable was amount of amphetamine received. But animals receiving amphetamine were picked up by the experimenter, had a needle inserted in them, and were tested for bar pressing later in the day. All these factors might make a difference. Injection procedures, even if no drug is squeezed through the needle, affect behavior; there has been much written on the effects of biorhythms on behavior.

The researcher should have injected one group with amphetamine and the other group with an inert substance, a placebo. The order of testing of animals should have been randomized.

Variables such as time of testing that are not deliberately manipulated by an experimenter but systematically bias results are called confounds. Control groups are used to rule out possible confounds, both known and unknown. Without a control group, a researcher cannot know if changes in behavior are caused by the independent variable or something else. But if two randomly assigned groups are treated identically except that only one of them receives the IV, subsequent differences between them are best accounted for by that variable. Following the logic further, it is sometimes necessary to use more than one control group. For example, a physiological psychologist might insert an electrode into the brain of a rat and then pass electric current through the electrode to make a lesion. Suppose that, following such an operation, the rat's behavior changes. Before concluding that destruction of the specific brain area caused the change, two control groups would be necessary. First, some rats would be sham-operated—they would be anesthetized, have their skulls shaved, have small holes drilled into the skulls, and have electrodes inserted into the same brain area. But they would receive no electric current. Second, other rats would have electrodes inserted into different brain areas and current passed through; otherwise, the possibility could not be ruled out that passage of current anywhere in the brain causes the behavior change.

PRINCIPLE TWO: Use Control Groups to Eliminate Possible Alternative Explanations of Data. Treat Experimental and Control Groups Exactly Alike Except for Application of the Independent Variable. If the Independent Variable has Several Components, More Than One Control Group may be Needed.

Hint Three. The researcher's expectations and biases may have affected the results.

Explanation. No mention was made of special precautions taken by the experimenter to ensure that his biases did not influence the results. As Rosenthal (1977) has amply documented, such precautions are necessary. When experimenters know the groups to which subjects have been assigned, errors are most frequently in the direction favoring the research hypothesis; and, whether deliberate or not, errors do occur. To prevent them from systematically influencing results, neither the experimenter nor the subjects should know who gets what treatment. An assistant can randomly assign subjects to groups, and the code should remain unbroken until the experiment is completed.

PRINCIPLE THREE: Neither the Experimenter nor the Subjects Should Know Which Subjects are in Which Groups.

Problems of Construct Validity

Hint Four. Bar pressing may not be a good measure of hunger.

Explanation. Researchers cannot always directly measure what they are interested in, so they resort to indirect measurements. The procedure is often justifiable, as when astronomers estimate the temperature of a star from the wave length of its emitted light. Psychologists too must often resort to indirect measurements, as when they make predictions about a person's voting behavior from his answers to a questionnaire. But indirect measurement was inappropriate in the present context. The experimenter was concerned with hunger but measured bar pressing. And amphetamine, which increases overall activity, might increase bar pressing for food even though hunger is reduced and the animals don't eat the pellets.

PRINCIPLE FOUR: Make Sure You Measure What You Say You are Measuring.

Problems of External Validity

If none of the above errors had occurred, the researcher might reasonably have concluded that the administered dose of amphetamine had reduced hunger for the particular food in the particular strain of rats at the specific time and place of testing. Obviously, good researchers go beyond such trivial conclusions. They make generalizations, that is, test external validity, without which science would be pointless. But some generalizations are reasonable and others aren't, and the only way to find out the correct ones is to test them. Commonsense should guide. Thus, for example, people's nutritional requirements vary with their gender, age, race, and activity level; still, we're sufficiently alike that nutritionists make dietary recommendations for all people based on experiments with volunteers. The same experiments would yield more cautious conclusions if the subjects were rats—and more cautious still if they were dung flies.

In the amphetamine study, there would be two major concerns about generalizations:

1. Independent variables often have different actions at different dose levels: A low dose of barbiturate may cause agitation, whereas high doses put subjects to sleep; a little bit of anxiety may improve performance, a lot may be paralyzing. Since only a single dose of amphetamine was used, conclusions about its general effects would be premature.

    Changes in an independent variable other than dose may make generalizations precarious. Weller and Livingston (1988) asked college students to read three vignettes, each describing a murder or rape, then to answer questions about their reactions to the crimes. The vignettes were identical for all students, but the questionnnaires were printed on either pink, blue, or white paper. Several differences were found in anger aroused by the crimes and judgments about probable guilt and appropriate punishments.

2. Rats are not people. (My friend I. Z. says—I think jokingly—that studies with humans are alright, because we can often learn a great deal about rats that way.) Findings with rats may not even generalize to mice, nor of mice to other mice. Thompson and Olian (1961) injected pregnant females of three different mouse strains with a drug; then they compared the offspring of each strain with offspring of the same strain that had undrugged mothers. Offspring activity was elevated above control levels in one strain, depressed in the second, and unchanged in the third. So, if the amphetamine researcher had concluded that norepinephrine increases hunger in people, she might have overgeneralized.

External validity concerns extend beyond that of generalizing from particular samples of subjects to wider populations. Cone (1977), elaborating on ideas put forth by Cronbach, Gleser, Nanda, and Rajaratnam (1972), suggested six additional universes of generalization: scorer, item, time, setting, method, and dimension.

Scorer

To what extent are obtained data due to the use of a specific scorer? For some methods of data collection, such as observational studies, the extent to which different scorers agree is generally computed. For other methods, such as experiments and questionnaires, interscorer agreement is generally presumed high and not necessary to verify. But the presumption is sometimes incorrect. If scorers disagree, then at least one of them needs retraining or the events being scored need redefinition.

In addition to how they score, investigators differ in age, race, gender, voice, manner of dress, mannerisms, enthusiasm for research hypotheses, and dozens of other ways that may affect outcomes. See page 241 for some examples.

Items

Although many tests are made up of items designed to measure knowledge of a single narrow topic, both zero and 100 percent scores are rare. That is, success varies on the individual items. So generalizations to tests with different sets of items must be made cautiously.

If the items of a test measure the same thing, that is, are internally consistent, generalizations to another set of internally consistent items over the same material are reasonable. The split-half reliability coefficient (see p. 181), an index of internal consistency, is computed routinely by people who evaluate standardized tests but not often enough in other fields; Cone suggested that it be used whenever people lump several behaviors together to get a single score.

Time

To what extent are data collected at one time representative of those that might have been collected at other times? One method for measuring this is test-retest reliability (see p. 183). Cone suggested other methods as well.

Setting

To what extent are data obtained in one situation representative of those obtainable in others? The question may have practical significance. For example, many reports of improvements following therapeutic interventions are based solely on gains within laboratory settings. To test the generality of findings across settings, Lick and Unger (1977) urged researchers to routinely use more than one of the following four methods of collecting data: unobtrusive observation (subjects don't know they are being observed), obtrusive observation, informants (asking friends and relatives about the subject's behavior in critical situations), and self-monitoring (subjects are asked to keep records of their behaviors). Each approach has both advantages and disadvantages. The important point is that the test setting matters.

Volunteers enter laboratory settings with expectations about the nature of the research. These expectations, called demand characteristics, may profoundly influence behavior and restrict generalizations to the particular laboratory setting. In one study, Orne (1962) tested the hypothesis that demand characteristics contribute greatly to the unpleasant effects of sensory deprivation. (Subjects for early sensory deprivation studies were treated very cautiously: They were carefully screened for both mental and physical disorders, asked to sign frightening release forms, and shown a panic button that they could press if the strain became too great.) Orne required his experimental subjects to sign release forms, and he showed them how to use a panic button. He told other subjects that they were controls in a sensory deprivation experiment and did not show them a panic button, but otherwise treated them identically.

The experimental task did not involve sensory deprivation; the subjects sat in a well-lit room. Yet the experimentals, much more than the controls, behaved like subjects in a sensory deprivation experiment. (As further proof that demand characteristics play a major role in sensory deprivation work, consider that Suedfeld (1981) has used deprivation to induce relaxation and calm.)

Method

To what extent are data obtained by one method representative of those obtainable by others? On the basis of a questionnaire, Lamb (1978) identified students in a public speaking class as highly anxious about giving speeches or not very anxious. Then the students were required to give brief speeches, and the following additional measures of anxiety were taken independently of each other: evaluation by graduate students of the anxiety of each speaker, continuous telemetric measures of their heart rates, and their scores on a questionnaire designed to measure brief anxiety states. Lamb reported high reliabilities for each measure. Yet no consistent relationships were found between scores on one measure and scores on the others.

As discussed on page 92, the mere act of measuring may change the behavior of the observed. Different measures may produce different kinds of reactivity and thus restrict generality.

Dimensions

To what extent are data on a particular behavioral dimension representative of those obtainable from other behavioral dimensions? That is, to what extent are two traits, for example heterosexual assertiveness and interpersonal anxiety, related?

Statistical Validity

If the researcher had reported that amphetamine does not affect bar pressing, the study would have lacked statistical validity. That is, because there were only six rats per group, the study would not have provided a fair test of the relationship between amphetamine and bar pressing. Since the results were positive, criticism on those grounds would be unwarranted. To show why this is so, I must give a lengthy digression on the concept of statistical significance.

Suppose you believe that a coin with a slight nick in it is biased toward coming up heads. You flip it and get heads. That hardly constitutes proof, so you flip again. Another head. Of course, you still aren't satisfied, so you flip eight more times and record the following:

				Trial					
1	2	3	4	5	6	7	8	9	10
H	H	H	H	H	T	T	H	T	H

Is the coin biased? How confident would you be that, if you flipped it another 10 times, heads would again outnumber tails? This type of concern confronts scientists whenever they analyze data. They must weigh the likelihood that their results are due to chance rather than the actions of manipulated variables. They use statistics, such as the binomial formula described in Box 4.1 and applied to the present data, to find probabilities of various occurrences. The probability that a perfectly fair coin will show at least 7 heads on 10 flips is about .16; this means that 7 or more heads would be expected 16% of the time when a fair coin is tossed 10 times. For most scientific purposes, .16 is too large. The convention among scientists is this: only if the probability is .05 or less that chance factors have caused the result, will the result be assumed true. (In some scientific journals, the convention requires probabilities no greater than .01, and other conventions are occasionally used.) The result is then said to be statistically significant.

Several points should be noted:

1. The strategy of testing for statistical significance is under severe attack. Eminent methodologist Paul Meehl (1978) wrote that there are five noble traditions in clinical psychology (descriptive clinical psychiatry, psychometric assessment, behavior genetics, behavior modification, and psychodynamics), and all five were originally developed with negligible reliance on statistical significance testing.
2. Although scientists must have some convention for deciding whether to accept results, the .05 level is only a convention.
3. Chance factors can never be eliminated. The probability of being dealt a royal flush in poker is far less than .05, but even honest card players occasionally show royal flushes.
4. Results may be statistically significant but unimportant. Statistical significance measures the likelihood that the results were due to chance—and that is all it does. If you were dedicated enough to flip

BOX 4.1

Calculating Binomial Probabilities

When an event can occur in one of two forms (heads/tails, success/failure, normal/defective), the binomial formula can be used to calculate the probability of a specified number of occurrences of each form. Let the probability of success on a single trial be p; then the probability of failure is $1 - p$. If the probabilities remain constant over a series of n trials, the probability of x successes is given by

$$\frac{n!}{x!(n-x)!} p^x (1-p)^{n-x}$$

Thus, the probability of 3 heads in 4 tosses of a fair coin is equal to

$$\frac{4!}{3!1!} (.5)^3 (.5)^1 = .25$$

and the probability of 2 or more heads in 4 tosses is

$$\frac{4!}{2!2!} (.5)^2 (.5)^2 + \frac{4!}{3!1!} (.5)^3 (.5)^1 + \frac{4!}{4!0!} (.5)^4 (.5)^0$$

$$= .375 + .250 + .0625 = .6875$$

The probability of rolling 6s twice in 5 tosses of a fair die is

$$\frac{5!}{2!3!} (.167)^2 (.833)^3 = 10 \times .028 \times .581 = .163$$

As long as n is small, the binomial formula is simple to apply.

---

the coin one million times, statistical significance would be reached if you observed 501,000 heads. Having proven that the coin is biased, you wouldn't want to bet your house on the next flip.

5. To get some feeling for the likelihood of reaching the .05 level, consider only the first five coin flips. They were all heads. The binomial formula shows that the probability of getting five heads in a row is .031. Since this is less than .05, the results were statistically significant after five trials.

If a coin is biased, the likelihood of reaching statistical significance increases with the number of tosses. With five coin flips, heads must turn up 100 percent of the time for statistical significance; with one million flips, 50.1 percent heads suffices. In one sense, a given probability level means the same whether based on five or five million flips; it indicates the likelihood that the effect (bias in the case of a coin, difference between groups for an experiment) is real rather than due to chance. But in a more important sense, the number of trials is crucial. The fewer the number, the more powerful the effect must be to reach statistical significance. By the same token, a nonsignificant probability level does not necessarily mean absence of an effect; it may mean only that there were too few trials.

Example:

Suppose a coin is so badly damaged that it comes up heads 90 percent of the time; if a researcher tests the coin for bias by flipping it seven times, what is the probability of getting statistically significant results?

From the binomial formula, with an unbiased coin ($p = .5$), the probability of getting six or more heads in seven flips is $[(7!/6!1!) + (7!/7!0!)](1/2)^7 = .06$. That just fails to reach statistical significance; for the .05 probability level, a perfect seven heads out of seven would be required. But even if the probability of getting a head on a single toss is .9, as in this example, the probability of getting seven out of seven is only $.9^7 = .478$. So, despite such a powerful bias, with only seven flips the probability of reaching statistical significance would be slightly less than one-half.

A related issue concerns the disposition of negative results (no statistically significant differences found). Most scientific journals rarely accept submissions in which a researcher reports that a treatment has had no effect. But negative results may be useful; they may call into question previous positive results or offer disconfirmation of a theory. One reason for the policy of the journals is that negative results, much more than positive ones, may be caused by ineptitude. One type of ineptitude is use of too few subjects. Others include careless administration of the independent variable or measurement of the dependent variable, selection of an insensitive dependent variable, or use of a setting with excessive background noise. B. F. Skinner once taught pigeons to play versions of ping-pong and bowling. His work is not called into question because I'm unable to do the same.

Positive results may lack statistical validity for any of at least three reasons:

1. Unless data are of the right form, statistical tests will not yield accurate probability levels. The requirements differ for each test.
2. A given level of significance is accurate only if set before the researcher looks at any data. Studies in which some findings are reported at the .05 level and others at the .01 probably violate this principle.
3. "Fishing" for significant results is unacceptable. To return to the well-worn coin, a researcher who flipped it five times and got all heads could legitimately conclude that the coin was biased. (As noted above, the probability of getting five of five heads is .03.) But if the researcher flipped 50 different coins five times each and one came up heads five times, that would not indicate a biased coin. In fact, the probability is almost .80 that at least one unbiased coin in 50 will turn up heads five times out of five. By the same reasoning, a researcher who does 50 tests of significance at the .03 level has a .80 chance of finding at least one spuriously significant result—and even higher than that if he tests at the conventional .05 level.

Although it is good policy to test more than one dependent variable, researchers must use the appropriate statistics for analyzing multiple tests or their conclusions will be invalid. (One acceptable strategy is to do several significance tests but suspend judgment until interesting findings have been replicated.)

The Relationship Between Data and Theory

Hint Five. What effects does amphetamine have other than increasing norepinephrine level?

Explanation. As indicated above, brain levels of norepinephrine are increased more by injection of amphetamine than of norepinephrine. So, the decision to use amphetamine was reasonable. But amphetamine (and all drugs and all independent variables) has many effects; it changes levels of other neurotransmitters; increases heart rate, body temperature, and blood pressure; elevates mood; dilates the pupils of the eyes; changes sensitivity to pain; and so forth. Before she could conclude that amphetamine changed behavior because of its actions on norepinephrine, the experimenter would have to do many more studies.

PRINCIPLE FIVE: Independent Variables have many Effects. Always Consider the possibility that an Effect not Under Consideration has Caused a Change in the Dependent Variable.

Hint Six. Just because data are positive does not mean that a theory is proven.

Explanation. No simple relationship exists between data and theory. In this case, the theory led to the prediction that amphetamine would increase hunger. But even had it done so, the theory would not have been proven. An infinite number of incorrect theories lead to the same prediction. (All white powders increase hunger. All drugs beginning with the letter "a" increase hunger.) So, no matter how statistically significant data are, that doesn't mean that the experimenter's conclusions from those data are correct or even reasonable. Lykken (1968) gave an interesting example, summarized in Box 4.2.

PRINCIPLE SIX: The Relationship Between Data and Theory is Complex.

Cheating, Distortions, and Related Problems

Another, unpalatable, point to consider in reading the research literature is that scientists have strong motivations to be right—so much so that they sometimes lie. Broad and Wade (1982) documented many cases of scientific fraud. They argued persuasively that the reward system for scientists and the editorial practices of journals assure the inevitability of future frauds. In a recent 15-year period, 16 major scientific research frauds were exposed and many more probably went undetected (*The Economist*, 1987, February 28). Rosenthal (1977) documented many cases of error that may have been unintentional but that in the vast majority of instances favored the researchers' hypotheses.

> **BOX 4.2**
>
> ### Frogs
>
> Sapolsky (1964) developed a theory that some psychiatric patients believe in a "cloacal theory of birth" that involves the notions of oral impregnation and anal parturition; they should therefore, according to him, become compulsive eaters if they wish to get pregnant, and anorexic if they wish not to. They should also be more likely than other patients to interpret ambiguous inkblots (the Rorschach test) as cloacal animals like frogs. So, Sapolsky predicted that a higher proportion of Rorschach frog responders than nonresponders would have eating disorders. He tested 31 frog responders and 31 controls, and found a significant difference between them: 19 in the first group, only 5 in the second, had eating disorders. Lykken (1968) did not accept that the experiment had justified the conclusion, so he asked 20 of his colleagues, most of them clinicians, to estimate the extent to which they believed Sapolsky's theory prior to reading the experiment. Then they were given a summary of the findings and asked to reassess their beliefs. Lykken reported that they didn't believe it beforehand and remained unchanged in disbelief afterwards.
>
> Note that Sapolsky's data may well be accurate; Lykken did not criticize the study as lacking in internal validity. The problem was in going from data to theory. Lykken suggested more plausible explanations for the data, such as that frog responding and eating disorders are both symptoms of immaturity, or that squeamish people might tend to both see frogs and have eating problems.

Among 24 scientists who published extensively, there was an almost perfect correlation between advocacy of conservative philosophies and research that emphasized hereditary influences on behavior; and liberal or radical philosophies and an emphasis on environmental factors. This led Pastore (1949) to comment: "This inner relationship suggests that it would be as reasonable to classify the nature-nurture controversy as sociological in nature as to classify it as scientific in nature." Whether deliberate fraud is involved or not, the results make clear that scientists do not simply report objective reality.

Williams (1977) pointed out other reasons for readers to proceed with caution:

- In order to receive continuing grant money, researchers must be productive. They may be tempted to overgeneralize and emphasize doubtful results if nothing better materializes.
- Both grant committees and journal editors tend to play safe; they accept work more readily if it is consistent rather than inconsistent with previous research. Strong arguments can be marshalled in favor of conservatism in science, but there are two negative consequences: (a) scientific innovators and breakthroughs do not gain ready acceptance; and (b) erroneous information, once it has been published, is overthrown only with great difficulty.
- Researchers are selective in what they submit to journals; they are influenced by what the journals have previously published. This promotes conservatism even more.

> **BOX 4.3**
>
> ## Checklist for Experiments
>
> I. Internal Validity: Might differences between groups be accounted for by something other than the different values of the independent variable?
>   *A. Were subjects randomly assigned to groups so there was no systematic bias in favor of one group over the other?
>   B. Were control groups used, and were all subjects treated exactly alike except for administration of the independent variable?
>   **C. Were measures taken to prevent the intrusion of experimenter bias?
> II. Construct Validity: Did the experimenter give his independent and dependent variables reasonable interpretations? The answer may depend on the specific conditions of testing. For example, the number of times a rat bar presses for food is generally a satisfactory indicator of the rat's hunger, but not if the rat has received a high dose of amphetamine; the answer to a question, "What is your annual salary?" may be taken at face value in many situations, but not when it is asked by an agent of the Internal Revenue Service to a suspected tax evader.
> III. External Validity: What types of generalizations can be made?
>   A. Subjects. From what population did they come? Do they have characteristics not shared by most people? For example, are they all of one sex, all volunteers, all inmates of a mental hospital, all Democrats, all within a narrow age range? If the subjects are nonhuman, is there justification for generalizing to humans?
>   B. Experimental conditions. Did any of the variables that were *not* of interest reach unusual values that might have affected the results?
>   C. Did the experimenter use enough values of the independent variable to get a meaningful relationship between the IVs and the DVs?
> IV. Statistical Validity: Were the data analyzed properly?
>   A. Was the statistic used appropriate to the data?
>   B. Were the results statistically significant at an acceptable level?
>   C. What did the experimenter do about missing values and about subjects who dropped out before the experiment was completed?

- Few active researchers publish all their work. By picking some studies rather than others, they have considerable opportunity to distort.
- Textbook writers, especially of introductory books, often take simplistic positions. They do not discuss findings that clash with the dominant view.

Two additional concerns were pointed out in an editorial in *The Economist* (1987, February 28).

- Because the scientific literature has grown so rapidly in recent years, it has become difficult for journal editors to find qualified people to referee manuscripts. The referees are unpaid. In one study of a set of published papers, an average of 12 errors per paper was found; these

V. Going from Data to Conclusions:
   A. If the study was designed to test a theory, does the theory clearly predict one experimental outcome over another?
   B. Can all the steps and all the assumptions from theory to prediction be clearly stated?
   C. Can all the steps and assumptions from data to conclusions be clearly stated?
VI. Of what value is the experiment? (No attempt was made to order the list below in terms of importance.)
   A. Does it provide an answer to a practical problem?
   B. Does it have theoretical significance?
   C. Is it heuristic, that is, does it suggest further experiments? (All good research should be heuristic. Einstein wrote: "As a circle of light increases, so does the circumference of darkness around it." But some research and some theories are much more heuristic than others.)
   D. Does it demonstrate a previously unnoticed behavioral phenomenon?
   E. Does it explore the conditions under which a phenomenon occurs?
   F. Does it represent a methodological or technical advance?

*It does little good to assign subjects randomly to groups if some of them fail to complete the experiment: no matter what a researcher may say, subjects are highly unlikely to drop-out randomly—unless, perhaps, some statistically-minded vandal flips coins before deciding which rats to let out of cages. The problem of nonrandom attrition bedevils attempts to compare different forms of therapy. Suppose, for example, that 70 of 100 patients who undergo therapy X are pronounced cured at the end of treatment; and that 100 patients undergo therapy Y but only 50 of them stay beyond a week; of those 50, 40 are cured. Which therapy is better? Therapist X has cured 70 percent of his patients. Therapist Y has cured, according to X's reasoning, 40 of 100 or 40 percent. According to Y's reasoning, however, the 50 drop-outs should not be counted as they did not give therapy a chance.

**Virtually all methodology books assert the importance of double blind experiments, in which neither the researcher nor the subjects know which values of the independent variable are assigned to each subject; yet in a random sample of American Psychological Association journals (the most recent 1987 issue that was available in my school library of the *Journals of Comparative and Physiological Psychology Consulting and Clinical Psychology, Experimental Psychology: Human Perception and Performance, Experimental Psychology: Animal Behavioral Processes, Professional Psychology: Research and Practice, and Counseling Psychology*), I found that double blind procedures were described in only 4 of 52 cases. Double blinds are expensive and time-consuming, and when phenomena are robust, perhaps they are unnecessary. At least experimenters, if not textbook writers, think so.

....

ranged from glaring mistakes that called the conclusion into question to minor errors such as discrepancies between text and graphs.
- In a 1985 study of six leading journals, two radiologists and a librarian found that authors were misquoted in 15 percent of all references. In another 1985 study, of biological journals, Sabine found that 2 percent of articles were later amended by corrections, some so major that they contradicted the originals.

Box 4.3 is a checklist for evaluating experiments—your own and those of others. As indicated throughout the book, experiments are not the only way to do research. But sensitivity to items on the checklist will help you evaluate other research strategies.

Questions for Consideration

Four studies are described below, each limited in a major way. What would you do next if you came across such a study in your field? My answers are given at the end of the chapter.

1. An important conclusion is drawn from a study that does not adequately support the conclusion: In 1981, the results of a large-scale epidemiological study (MacMahon, Yen, Trichopoulos, Warren, & Nardi, 1981) were made public; it showed a direct relationship between the amount of coffee people drank and their risk of getting pancreatic cancer. The conclusion, given much coverage in the press, was that heavy coffee drinking causes pancreatic cancer. This conclusion was inappropriate, since correlational studies never prove causal relationships (see p. 103).

2. An interesting phenomenon appears in a tiny proportion of subjects: Weiss et al. (1980) tested the idea that ingestion of food dyes causes problem behaviors in normal children. They gave children special drinks every day for 77 days. On eight of the days the children received drinks containing food dyes, on the other days they received placebo drinks. On the eight food-dye days, one child had a small increase in two problem behaviors; and the problem behaviors of another increased dramatically. The rest of the 22 children were unaffected.

3. Subjects cannot be randomly assigned to groups, and experimental and control groups cannot be treated exactly the same except for the independent variable: In efforts to find out if the death penalty deters violent crime, researchers have compared the incidence of violent crime in countries with and without the death penalty; and changes in the incidence of violent crime in countries that changed their policy with respect to the death penalty. The procedures violate several of the principles listed at the beginning of this chapter.

4. The idea on which the study is based, and which it is supposed to support, is incorrect: Schuster and Schuster (1969) derived from evolutionary theory the prediction that the parent under less stress at the time of conception would be more likely to have offspring of its sex, and they tested their idea with both rats and humans. They stressed either male or female rats by keeping them immobilized in bandages for 24 hours prior to mating; and they asked people to fill out questionnaires about how stressed they were at the time of conception. Both types of data supported their idea: when the father was more stressed at the time of conception, the majority of offspring were female; when the mother was more stressed, male.

Yet the underlying idea is almost surely wrong. In fact, from an evolutionary perspective, a parent is probably more successful when it has offspring of the opposite sex. Success is measured in terms of the number of genes contributed to the next generation. Female mammals contribute an X chromosome to both their daughters and sons; males contribute an X to their daughters and a Y to their sons. But the X chromosome contains many more genes than does the Y, so daughters inherit relatively more genes from their fathers than sons do.

Summary

Scientific studies should have four important types of validity: internal, construct, external, and statistical.

To maximize internal validity, assign subjects randomly to experimental and control groups; treat both groups exactly the same except for administration of the independent variable; and take steps to prevent experimenter bias from influencing the results.

To maximize construct validity, carefully choose the methods by which you measure your variables.

To maximize external validity, be cautious about the generalizations you make. Test your generalizations.

To maximize statistical validity, make your tests powerful enough to detect relationships and use the right statistical tests. Don't fish for significant results.

Key Terms

Confound   A variable not deliberately manipulated by a researcher that systematically biases the research.

Construct Validity   Evidence that the measure of a construct used in a study (a) distinguishes between groups believed to have different levels of the construct, (b) yields similar scores to accepted measures of the construct, and (c) does not measure other constructs.

Control Group   Subjects treated exactly the same as the experimental group except that they don't receive the experimental treatment.

Demand Characteristics   Expectations of subjects about the nature of the research; they influence how subjects behave and thus restrict external validity.

Experimenter Bias   Experimenters who know the groups to which subjects have been assigned often behave differently toward them in ways that increase the likelihood that the research hypothesis will be supported.

Experimenter Effects Includes experimenter bias, but also any characteristics of experimenters that influence how subjects behave.

External Validity The extent to which the results of a study can be generalized to other subjects, scorers, measures, times, settings, methods, and behavior dimensions.

Internal Validity The extent to which conclusions from a study about the relationships between variables are justified.

Random Assignment Each subject in an experiment should be assigned to experimental or control groups by means of some chance process. Alternatively, each subject can be exposed to each different level of the independent variable, with the order of exposure randomized.

Statistical Validity The extent to which conclusions about the degree of relationship between two variables are justified.

Answers

My answers are certainly disputable. My hope is to initiate discussions.

1. Researchers should strive to conduct flawless studies that allow only a single, unambiguous interpretation—but they should recognize that as an impossible ideal. Even the most carefully controlled experiments may be bedevilled by chance factors or by extraneous variables reaching unusual values. So, though scientists must distinguish between well- and poorly-designed studies, they should reserve some doubt about the former and recognize that the latter may provide some useful information.

    The coffee/pancreatic cancer study was correlational, and such studies are inadequate for proving causation; some third factor, such as stress, might have caused both heavy coffee drinking and pancreatic cancer. Nevertheless, when two variables are correlated, and in absence of evidence to the contrary, the relationship may be causal. Scientists and laypeople alike sought additional evidence concerning the link between caffeine and pancreatic cancer, so the original study was heuristic. (The most recent studies included a failure to replicate (Nomura, Stemmermann, & Heilbran, 1981) and a suggestion of a possible third factor (Kinler, Goldblatt, Fox, & Yudkin, 1984): Cancer of the pancreas is usually associated with high levels of blood sugar and often with overt diabetes; these are likely to increase thirst and intake of all beverages.)

    Some studies, so poorly done that alternative explanations are as plausible as the one given by the researcher, are best ignored. Others, if the results are interesting enough, should not be.

2. Science deals with universals, not with individual events. Therefore, unless results are reproducible, they are not part of science. A major reason for the reluctance of most psychologists to accept ESP research is that the findings are not reproducible. But the

requirement that results be reproducible from one laboratory to another is different from the requirement that results found with one subject should also hold for others. In the study with food dyes, only one child had severe behavioral problems on food-dye days; but the problems occurred reliably in that one child, so the results are meaningful. In fact, the study leads to interesting questions revolving around the unique characteristics of that child.
3. Donald Campbell has been a leading figure in the move to get psychologists to realize that they must often try to solve important problems without the luxury of controlled experiments. They must do their best. See chapter 18 for a discussion of quasi-experiments.
4. Data that seem to support a theory should not be discarded just because the theory is incorrect. The data and theory must be judged independently. Wynne-Edwards 1962 book, in which he theorized that many animals perform altruistic acts for the good of their species, was a progenitor of the important field of sociobiology; yet most modern biologists believe that Wynne-Edwards was wrong. His data were not discarded but reinterpreted to fit with current beliefs.

CHAPTER 5

Asking Valuable Questions

By the time you finish reading this chapter, you should be able to answer the following questions:

What are the features of good scientific questions?

"What are the effects of stress on personality?" is not a good scientific question. Why not?

What are the values of basic research?

Which types of questions lead to deep analysis of phenomena?

What is meant by a ruthless test of a theory?

Imagine walking along a deserted country road and coming upon a stranded UFO. Being a person of great courage, you boldly enter and see several little green people, distraught because their craft is disabled and they are faced with the prospect of being stranded forever on earth, spending their declining centuries watching reruns of "Cheers." You diagnose their problem as a clogged fuel pump and, after a few minutes of light banter, say goodbye. They show their gratitude by giving you a small creature whose fur, rubbed across a bald head, grows luxurious hair. (They also leave a great recipe for strawberry blintzes.) You promise to look them up if you're ever out Alpha Centaurus way, and now you face the prospect of testing the fur ball before putting it on the market and making your millions. What would you do?

You might get two groups of bald-headed men and rub alien fur on the heads of one group, cat fur on the heads of the other. In barest essentials, that's how experiments are designed. Of course, you'd have to be sensitive to many factors: How many men should you use? Which men would get alien fur, which cat fur? How often and with how much pressure should the fur be rubbed? How would you measure hair growth? How much difference between the two groups would convince you that the fur works? How would you test for possible side effects? How would you decide if the procedure is likely to work for all men or for just those tested?

When the research question is clearly delimited—Does alien fur help bald men grow hair?—people who learn the principles of experimentation described in chapter 14 should be able to design a method for answering it. Such research is often vital to the scientific enterprise. Moreover, understanding of the principles helps people become intelligent consumers of science, capable of making sense of scientific journals and the science sections of newspapers. But if they aspire to become innovative researchers of the first magnitude, they must develop another skill—the ability to ask good questions. Asking meaningful and answerable questions is often the most important phase of a research project. Albert Einstein said, "The formulation of a problem is far more often essential than its solution, which may be merely a matter of mathematical or experimental skill. To raise new questions, new possibilities, to regard old problems from a new angle, requires creative imagination and marks real advance in science." And Nobel Prize winner Harold Urey, when asked how his parents had contributed to his scientific abilities, answered that his father had never asked him "What did you learn in school today?" but "Did you ask a good question today?"

Features of Good Questions

Questions can be practical or impractical, answerable or impossible to answer. Hans Selye contrasted good questions—"What would happen if I did this? How much of a substance must be administered to produce a specific effect? How are the nerves distributed?"—with unanswerable statements such as "I wish I knew more about the adrenal glands" and "I wish I could find a cure for cancer." The good questions are limited in scope and suggest methods for testing them. Peter Medawar made the related point that nobody applauds scientists who nobly attempt to solve problems that are too difficult. He said scientists should practice the "art of the soluble," seeking nontrivial problems that they have the competence and resources to solve.

Kurman (1977) noted that ". . . any research question limits and defines the range of possible answers. If one asks a 'why' question one is likely to come up with explanatory and not descriptive material as the major result. Questions contain directions for their answers . . ." On the same theme, Nobel laureate Werner Heisenberg wrote, "What we observe is not nature itself but nature subject to our questions."

Good scientific questions share three characteristics: they can be answered by empirical means, they indicate the nature of acceptable answers, and they are worth answering.

Good Scientific Questions Must be Answerable Empirically

The following are not scientific questions:

- Is it ethical to experiment with animals? (This question is important and merits discussion. But questions of value cannot be answered empirically, so they are not scientific.)
- Is there an afterlife? (At the present time, this question is also unanswerable empirically. If procedures are developed for testing for the existence of an afterlife, the question would become scientific.)
- What is science? (Definitions are agreed upon by convention, not by empiricism.)

A prerequisite to finding empirical answers to questions is that the procedures for producing and measuring key variables be made explicit. The importance of these procedures, called operational definitions, is discussed in the next chapter.

Good Scientific Questions Indicate the Nature of Acceptable Answers

A question such as "What are the effects of stress on personality?" is too broad and abstract to be useful. An answer satisfactory to one scientist may be irrelevant for another. Compare: "How do a series of loud noises affect exploratory activity in hamsters?" with "What are the mechanisms of voodoo death?"

Good Questions are Worth Answering

Selye (1964) told of a man who, having developed a technique for accurately measuring the iron content of rat feces, encouraged his colleagues to do research that would require use of the technique. Although the man was uniquely qualified to answer questions about rat fecal iron content, the value of the information was not immediately apparent.

Scientific questions are sometimes judged by the extent to which they are likely to generate answers to pressing, practical problems. On this criterion, applied science fares much better than basic research. But, as Selye noted, applied projects stick close to the commonplace, to what is already known; hence, they rarely lead to new heights of discovery. On the other hand, when questions are asked about the fundamental nature of the universe—as exemplified in the works of such scientists as Newton, Pasteur, Mendel, and Einstein—the answers may play a major role in advancing technology. G.H. Hardy was a pure mathematician and proud of it. He asserted that the beauty of mathematical topics was directly related to their uselessness, and that applied mathematics was repulsively ugly and intolerably dull. During Hardy's lifetime, his work was quite useless. Yet since then it has had many practical applications.

Lederman (1984) claimed that bright young people are more likely to be attracted by the opportunity to do basic science than by the lure of making practical contributions. Basic science raises the spirit of the scientific community, sets standards throughout science, and provides a shared body of knowledge that simplifies interactions between workers in different disciplines. He noted that instruments and techniques developed to solve questions in basic science often lead to the creation of goods and services. Much of the theoretical work can be used in other fields; for example, solutions to certain kinds of equations that occur in elementary particle interactions have been applied to the study of polymers. But the major benefit of asking fundamental questions, in his view, is cultural. Scientists seek a coherent account of the world and our place in it, and as Box 1.2 shows, the results pervade society.

Types of Questions

Questions That Lead Directly to Research Projects

Most studies reported in social science journals provide answers to one or more of the following types of questions:

Descriptive Questions Descriptive questions lead to answers about the frequency, duration, and intensity of the behaviors of interest or about the frequency, duration, intensity, and range of behaviors performed by the subjects of interest. Like newspaper reporters, researchers interested in descriptive questions ask who?, what?, where?, and when? They sometimes ask about behavior sequences, that is, how various behaviors go together. An investigator studying the sexual behavior of rats might ask, "Which behaviors reliably precede, which reliably follow, copulation in rats?"

Mahrer (1988) argued that descriptive questions currently play a meager role and should play a prominent one in psychotherapy research. His suggested strategy can be applied regardless of field of interest. First, an investigator should pick a type of event that interests him, such as a patient's erupting into hearty laughter during a therapy session. Then, he should obtain records of such events, for example, by studying audiotapes and videotapes of therapy sessions. Next, the investigator should describe each instance of the event in detail, paying careful attention to the context in which it occurred. He should form provisional categories of contexts and refine the categories as new instances are collected. He should then organize and display the data and look for anything exceptional, surprising, challenging, or disconcerting. In this way, the descriptive questions may lead to interesting discoveries.

Correlational Questions   A correlation indicates the degree of relationship between variables. Correlational studies never conclusively prove that one variable causes the other, so many psychologists prefer other research methods. This is unfortunate, since correlations may be valuable idea generators.

If variables X and Y are correlated, then in most cases X causes Y or Y causes X or some other factor(s) causes both. If any of the three are true, a causal link exists between X and Y. A researcher interested in relating two seemingly separate areas might analyze several measures for reliable (repeatable) correlations. Then she would seek testable explanations for the correlations. This scatter-gun approach may seem at odds with what was written in chapter 1 about data collecting, but it is suggested as only a first step, as a method of exploration. As mathematician George Polya (1954) wrote, playing with data can be both enjoyable and a great spur to meaningful research questions.

Scientists who study animal behavior often ask correlational questions. They compare individuals or groups and gain insights that can't be reached by focussing on single subjects. For example, most mammals are polygynous (a single male mates with more than one female, and many males don't mate). Most bird species are monogamous, but some are polygynous and a few polyandrous (a single female mates with more than one male). Some fish and insects are monogamous, others are not. Scientists like Gordon Orians (1969), wondering why, looked at correlations between mating system and features of the environment. They now believe that monogamy occurs if any of the following conditions are met:

1. Receptive females are scarce and widely distributed.
2. Successful reproduction requires the cooperation of both parents. (In mammals, where females nurse their young, males would be a handicap—they would compete for food and attract predators. This is especially true for vegetarians; unlike carnivores, they cannot share food efficiently with non-nursing offspring. Almost all the monogamous mammals are carnivores.)
3. A valuable resource, such as a nesting site or rich food source, is scarce and must be defended.

Whether the speculations are accurate awaits further testing. Whether they have any relevance to human marital systems is widely debated. But one thing is clear: they would never have been developed without comparative, correlational studies.

Another example of the potential of correlational studies for uncovering interesting relationships comes from the work of Norman Geschwind (1984). He administered questionnaires to nearly 3,000 people and found correlations between learning disabilities, left-handedness, being a male, and certain diseases of the immune system; the left-handed, learning-disabled men were also likely to have exceptional athletic, artistic, or mathematical talents. Geschwind developed a theory, involving the role of testosterone on the developing nervous system, to account for his findings.

Questions with the Behavior of Interest as Dependent Variable The dependent variable is a behavior measured to see if it changes as another variable—the independent variable—is changed. For example, many scientists have studied biological rhythms. With a particular biorhythm as dependent variable, the basic form of the question has been, "Does X modify or disrupt the rhythm?" In various studies, X has been drugs, lighting conditions, diet, stress, group living, solitary living, and removal of various endocrine glands and brain structures.

Many studies are published even though they do little to advance science; a high proportion of such studies involve testing the effects of whatever is available on a popular dependent variable. But creativity of a high order is possible, as illustrated in the work of Zucker and his students (1976) described later in the chapter.

Questions with the Behavior of Interest as Independent Variable Any variable can be dependent or independent. It depends on the research question. Questions with biological rhythms as the independent variable are of the general form, "How does time of testing affect X?" In various studies, X has been physical strength, endurance, resistance to disease and other forms of stress, mental performance, pulse rate, sensory acuity, job skills, susceptibility to drugs, mood, and episodes of abnormal behavior. The comments made about the previous category apply here too.

If an independent variable does affect a dependent variable, an important follow-up question is "Which component(s) of the IV are responsible?" A drug may increase the activity of rats not because of its pharmacological properties, but because the injection procedure induces stress. A psychotherapist may help a client work through a difficult problem not because of his many years of training, but because he listens sympathetically. These plausible alternatives can be tested by fractionating the IV into its component parts and analyzing them separately (administer the drug by putting it in the rat's food; inject the rat but give it saline solution instead of drug; place clients with untrained, sympathetic listeners).

Questions About How to Reach a Desired End-Point Two neglected research questions, according to Greenwald, Pratkanis, Leippe, and Baumgardner (1986), are "What are the conditions under which a particular known finding can and cannot be obtained?" and "What are the conditions under which previously unobtainable results can be produced?" This involves careful specification of the dependent variable (the end-point) and creative or systematic manipulating of the independent variable. For three extensive criticisms and a spirited defense of the Greenwald et al (1988) proposal, see *Psychological Review,* 1988, *95,* 559–579.

Questions that Organize Research Programs

The several classes of questions described above share an important feature—they can be answered by a single, well-designed study. Researchers at a second level proceed differently. They too ask descriptive and correlational questions, and questions with the behavior of interest as either IV or DV. But not in isolation. Instead, they plan comprehensive research programs for penetrating deeper aspects of phenomena. Perhaps the skills for establishing such programs can also be taught. As of 1968, two-thirds of the American Nobel Prize winners in science had been trained in the laboratories of other Nobel laureates (Merton, 1968). This shows, according to Leopold (1978), that creative science can be learned through training from others who are highly skilled. (There is a plausible alternative explanation; can you think of it? See p. 75, answer 4.) If Leopold is right, part of the skill is probably that of asking deep questions.

What are the Boundaries of a Phenomenon? Research that initially appears highly-specialized may have widespread applicability. Questions about boundaries test the possibility. For example, John Falk (1969) fed food-deprived rats at approximately 1-minute intervals for a few hours per day, and found that they drank enormous quantities of water—during 3-hour test sessions, they drank 10 times their normal 24-hour totals. From this starting point, Falk asked several questions about the boundary conditions:

1. How crucial is the 1-minute interval? (Very. If the animals are fed every 20 seconds or every 2 minutes, they don't drink excessively.)
2. Does excessive drinking occur in animals other than rats? (Yes—with mice, pigeons, monkeys, apes, and humans.)
3. Will animals drink excessively if they are deprived of something besides food, and that something is given at 1-minute intervals? (Yes, deprivation of a variety of commodities, including sawdust and running wheels, is effective.)
4. If water is unavailable, will an animal show a different excessive behavior? (Yes—wheel running, pecking, and attacks against a second animal. And humans show various excessive behaviors if given intermittent opportunities to solve problems or access to games.)

Because of the considerable generality of the phenomenon, Falk believes it may be relevant to an understanding of drug addiction, overeating, excessive gambling, and many social rituals (Falk, 1984, 1986).

How Does a Mechanism Work? Falk's work illustrates another type of creative question. He considered several hypotheses to account for the cause of the excessive drinking, such as that the dry pellets used as standard food make the rats thirsty (so he tested them with liquid food and found little change in their drinking); and that initially infrequent drinking, since it is always followed within 1 minute by food, promotes a superstitious association between drinking and being fed (so he delayed feeding them each time they drank, but they continued to drink). Falk eventually developed a promising theory: Animals often resolve conflicts most effectively if they delay resolution as long as possible. A hungry rat fed infrequently is in conflict between staying in the situation and trying to escape, and any behavior performed to excess serves to keep it there.

As a second example, consider the work of Zucker and his students (1976). They sought the physiological mechanism that maintains biological rhythms; they might have systematically stimulated or lesioned selected areas of the brains of laboratory animals, but even simple animals like rats have millions of neurons and other cells in their brains. Random testing might have taken centuries. Instead, they limited the possibilities by asking what features an animal must have if it is to maintain a rhythm.

All known mammalian rhythms are synchronized by the light-dark cycle, so the animal must be able to detect light. But Zucker's rats maintained rhythms even after he eliminated all known visual pathways and they could no longer make simple visual discriminations. So Zucker considered alternative pathways to the eye. In certain fish, light penetrates directly through the skull; so light was shined on the skulls of rats—without effect. Search of the anatomical literature revealed the existence of a pathway, function unknown, between the retina and a part of the brain called the suprachiasmatic nucleus of the hypothalamus. Zucker and students confidently lesioned the suprachiasmatic nuclei of several rats—and the rhythms of wheel running and drinking were disrupted.

The research of Falk and Zucker exemplifies the attitude of most good scientists—they are not content with merely demonstrating interesting phenomena, but they organize comprehensive programs to get at the underlying mechanisms. Their attitude is that good research almost invariably, and perhaps as part of its definition, stimulates further research.

How Does a Behavior Help the Behaver Survive? Nobel Prize winner Niko Tinbergen (1963) stressed the importance of asking how the frequency, intensity, and exact form of a behavior helps an individual in its natural environment. This requires knowledge of the naturally-occurring behavior, which underscores the importance of descriptive studies. Tinbergen's classic studies on the survival value to black-headed gulls of removing eggshells from their nests are described on page 6.

How Can a Theory be Tested? Scientific theories often represent major advances and offer new ways of looking at the world. But they must be tested. Philosopher of science Karl Popper encouraged scientists to theorize boldly, but he also insisted that they ruthlessly test their theories and discard those that fail. (By "ruthless tests" he meant those likely to be failed if the theory is false; ruthless tests are widely used in the physical but not in the social sciences.) Paul Meehl (1970) has argued convincingly that the nonruthlessness of social scientists works to the serious detriment of their fields. The matter is discussed further on page 330.

Motley (1985) devised a clever procedure for testing Freud's belief that anxiety, especially sexual anxiety, causes slips of the tongue. To create simple anxiety, he told volunteer male subjects that they would receive electric shocks; to create sexual anxiety, he used a provocatively dressed female experimenter. To induce slips of the tongue, he showed subjects pairs of words on a screen; a buzzer signalled the subjects to read the next pair out loud. One sequence of word pairs was:

toy/dog
flat/tire
could/knock
cold/nuns
nosey/cooks

with the last pair to be read aloud. Because they had been set up by the prior two pairs, about one-third of the subjects said "cozy nooks" instead of "nosey cooks." When sexual anxiety was created, the subjects made slips such as "fast passion" for "past fashion" and "happy sex" for "sappy hex." As Freud would have predicted, the men who scored highest on a pre-test of sexual anxiety had the most slips of the tongue.

Motley's work, although interesting, does not ruthlessly test Freudian theory. (To learn why, see the discussion of significance testing, p. 330.) And, because useful theories have many implications and therefore many ways of being tested, his experiment is not conclusive. No single test could be. In fact, Fisher and Greenberg (1977) reported that several thousand studies have been conducted to test Freudian theory, and the results have been mixed.

Be creative by developing theories and by deducing interesting theoretical implications and subjecting them to ruthless, unambiguous, and economical tests. An efficient procedure, one that virtually ensures the usefulness of a study, is to test plausible competing theories against each other. Platt (1964), building on an idea of Chamberlin (1904), advised researchers to construct competing explanations for a phenomenon, design and carry out studies so that each possible outcome eliminates at least one explanation, and then refine the remaining possibilities and start again.

Can Diverse Areas be Unified? Great scientists try to bring unity out of diversity. Bronowski (1956) said: "The progress of science is the discovery at each step of a new order which gives unity to what had long seemed unlike.

Faraday did this when he closed the link between electricity and magnetism. Clerk Maxwell did it when he linked both with light. Einstein linked time with space, mass with energy, and the path of light past the sun with the flight of a bullet; and spent his dying years in trying to add to these likenesses another, which would find a single imaginative order between the equations of Clerk Maxwell and his own geometry of gravitation."

Down's syndrome and Alzheimer's disease may be linked. Down's syndrome is a hereditary disorder characterized by moderate to severe mental retardation, a short flattened skull, and slanting eyes. Alzheimer's disease, which causes severe and permanent memory loss, confusion, and depression, probably accounts for more than half of all diagnosed senility. Epstein (1986) noted, among other things, that almost all people with Down's syndrome who live beyond the age of 35 develop pathological changes in their brains similar to those in victims of Alzheimer's disease. Epstein expressed the hope that useful insights will emerge from research suggested by the possible connection.

Can a Paradox or Conflicting Findings be Explained? Great scientists have been interested in unusual or puzzling events. "Treasure your exceptions," said early geneticist William Bateson. Suppose a theory predicted that people would respond in a certain way to an experimental situation, and 99 of 100 did so. Most researchers would be delighted at the strong confirmation, and they'd probably go on to other aspects of the theory. The exceptional researcher, however, might ask questions about the exceptional subject. How did that 1 out of 100 differ from the others?

John Garcia, who helped induce a minor revolution in the field of learning, was inspired by the following paradox:

1. Learning, according to psychology texts of 20 years ago, occurs only if rewards or punishments follow responses by less than a second.
2. The texts also said that many pairings of response and reinforcement are required for learning.
3. Rats are difficult to poison.

The paradox is this: Rats are hard to poison because they sample cautiously when they discover a novel food. If they get sick after their single exposure, even though sickness typically doesn't occur until hours later, they learn not to eat that food again.

Garcia, Hankins, and Rusinak (1974) devised experiments to make sense of the paradox. They exposed rats, while the rats were eating, to a dose of radiation that made the rats nauseated hours later. After a single trial, the rats refused to eat again in the same room. Yet when the rats were allowed to run between areas that were and were not exposed to radiation, they entered the exposed area repeatedly and accumulated high, often fatal, doses.

In another study, thirsty rats were given sweetened water, and each time they drank they heard a clicking sound. Then some were punished with radiation and some with electric shocks. Rats given radiation drank much less of the sweetened water when they were retested, but they did not acquire an aversion for the clicks. Rats that received electric shock retreated from the clicking sound, yet continued to drink sweetened water.

Garcia et al. reasoned that the behavioral strategies necessary to cope with the external environment (represented by shock) are different from those needed to discriminate between nutritious and harmful foods. Clicks and shocks, both being part of the external environment, are readily connected; clicks and the gastrointestinal upset caused by radiation are not. Similarly, distinctive tastes are easily associated with feelings of nausea but not with electric shock.

Their findings were not immediately accepted. In fact, they quoted one skeptic as saying, "These results are as likely as finding birdshit under a cuckoo clock." But they have been replicated many times and forced psychologists to revise some strongly held beliefs. Psychologists now recognize that there are constraints on learning. Animals learn some tasks rapidly and others, from a human standpoint just as easy, with much more difficulty.

What if the Conventional Wisdom is Wrong? Campbell, Daft, and Hulin (1982) suggested that researchers test the accepted beliefs in their fields. They should list as many conventional wisdoms as they can, and then state the converse of each. Acting as though the converse is true may lead to some interesting research questions. These should be explored even if they seem unreasonable. As an example, Campbell et al. listed 12 conventional wisdoms concerning organizational behavior. One of them was, "Job stress is bad. It leads to dysfunctional behaviors." They suggested research designed to analyze the idea that job stress is good, because it makes jobs more stimulating and interesting.

What is the Cause of a Condition? Identifying the cause(s) of an effect may help scientists produce more or less of it. But linkages between causes and effects are often difficult to establish. Causes may be immediate or remote. The immediate cause of syphilis is a spirochete, *Treponema pallidum,* transmitted by sexual intercourse. But the spirochete can be transmitted only by people who harbor it, so the sexual partner is also a cause, on a different level. And the partner must have been previously exposed to a carrier, a more remote cause. The analysis can continue back to Adam and Eve, each blaming the other. Causes can always be pushed back a step, and this at times is a productive research strategy. Having found that X is the cause of Y, seek the cause of X.

Causes may be necessary but not sufficient. An adequate supply of oxygen is needed for good health, but oxygen by itself is insufficient. Causes may be sufficient but not necessary. Being hit on the head by a giant tortoise may be fatal, but death occurs even without tortoises.

Table 5.1. *Causes of Eating*

Why We Start Eating

Biological Factors

Activity within the lateral hypothalamus
Low levels of glycerol in the bloodstream
High levels of fatty acids in the blood
Stomach contractions (of relatively minor importance)
Setpoint (to some extent inherited) for the amount of fats stored in the body; when below the setpoint, we feel hungry

Psychological Factors

Specific hungers for foods containing needed substances (for example, rats that have their adrenal glands removed can no longer retain sodium; within minutes of waking from the operation and before they have become sodium deprived, they drink more concentrated salt solutions than they did before the adrenalectomy)
Stress or depression
Social stimulation
 family and cultural eating rituals, symbolically significant food; many animals eat more when in a group than when alone
Temporal and situational factors: We eat at certain times of day and in certain places and situations

Factors of Mixed Origin

Sensory Cues: How food looks, smells, and tastes
Anticipatory activities: We eat to prevent depletion

Why We Stop Eating

Biological Factors

Activity within the ventromedial nucleus of the hypothalamus
High levels of blood glycerol
Low levels of blood fatty acids
Pressure receptors within the stomach are stimulated when it is full
Bitter foods induce a rejection reflex
Level of body fats exceeds the setpoint

Psychological Factors

Fear
Conditioned food aversions
Cultural pressures toward slimness, dieting
Mental disorders such as anorexia

A cause may contribute only a small part to the effect. Children exposed to violent television programs may increase their levels of aggression. But television violence is not the only cause of aggression and is not necessary for it.

There are no spirochetes in psychology, no single, unequivocal causes. Social scientists must generally make do with causal analyses of the television violence/aggression type. Even in relatively advanced sciences like physiology, multiple causes are common. See Table 5.1, adapted from Zimbardo (1985), for the results of more than 80 years of research on why people start and stop eating.

Summary

Researchers should devote a great deal of their attention to the kinds of questions they ask. Good scientific questions can be answered empirically, indicate the nature of acceptable answers, and are worth answering.

Descriptive questions lead to answers about the frequency, duration, and intensity of the behaviors of interest; or about the frequency, duration, and range of behaviors performed by the subjects of interest.

Correlational questions ask about the degree of relationship between variables.

Much scientific research is aimed at answering questions about the effects of one variable on another. When an experiment is designed to answer a question of the type, "What is the effect of X on Y?" the X is called an independent variable, the Y a dependent variable.

Some questions are aimed at revealing deeper aspects of phenomena. These include the following: questions about the boundaries of a phenomenon; how a mechanism works; how a behavior helps the behaver to survive; that test a theory; that create bridges between areas that had seemed different; that are aimed at resolving a conflict; that challenge conventional wisdom; and about the causes of an event.

Key Terms

Dependent Variable   A variable whose values are measured at different levels of the independent variable.

Independent Variable   The variable manipulated by an experimenter.

1. Can you arrange the letters in 'questions' to make a three-word phrase?

2. Medical researchers seek cures for diseases and psychologists seek ways to promote psychological health. The medical researchers have in many cases been successful. Can you think of any analogies between the two systems that might prove useful to psychologists?

3. I stated previously that any variable can be used as either an independent or dependent variable. Can you think of a study in which alcohol consumption is used as an IV? As a DV?

Answers

1. quest is on
2. Researchers who wish to modify a condition should be able to recognize it. Medical researchers have no trouble recognizing polio and malaria, but psychologists have difficulty with creativity and happiness. Careful definition of terms is a first step in finding solutions.

Finding the cause of a disease is often a major step toward finding a cure. Medical researchers have developed extensive checklists of possible causes. Iivanainen (1974) considered the factors below in trying to determine causes of mental retardation. The same or a modified series of factors might be studied as possible contributors to mental health and mental illness.

- When did the individual become retarded? Was it prenatal, perinatal, or postnatal? Did it involve multiple or unknown timepoints?
- Was it caused by infection, intoxication, trauma, or a physical agent? Disorder of metabolism, growth, or nutrition? New growths? Unknown prenatal influence, unknown or uncertain cause with obvious structural reactions, uncertain or presumed psychological cause with the functional reaction alone obvious?
- How do retardates and other hospitalized people compare with respect to the following: social class, family history, previous abortions and stillbirths, maternal age, abnormalities during pregnancy, primiparous mothers, delivery without skilled assistance, obstetrical procedures, birthweight, maturity of the subject at birth, multiple pregnancy, condition of the subject immediately after birth, abnormalities during the late neonatal period, early manifestations of mental retardation, epileptic seizures, head weight and circumference, handedness, major malformations, cutaneous findings, cerebral palsy, electroencephalogram results, hearing ability, and chromosome karyotyping.

Note that Iivanainen's study was correlational, so even if he had found a relationship between, for instance, birthweight and retardation, he would not have been able to say with certainty that low birthweight causes retardation. But the finding would have generated useful ideas for additional research.

3. The possibilities are limitless. Here are two:

   IV: What are the effects of consumption of a six-pack of beer within a one-hour time period on time to run the 100 meter dash?

   DV: What are the effects of running a 100 meter dash on consumption of beer during the 30 minutes immediately following?

4. People are not assigned randomly to work in laboratories. Nobel laureates are probably able to select the brightest students for their research assistants. Then, even if the training is quite ordinary, the students are likely to have productive careers.

CHAPTER 6

Selecting and Measuring Variables

By the time you finish reading this chapter, you should be able to answer the following questions:

What is an operational definition, and why are operational definitions an essential part of scientific work?

What are the different scales of measurement?

Does measurement always require numbers?

What rules should be followed when setting up a system of classification?

What are the criteria for a good method of measurement?

How is reliability computed?

How is validity computed?

How can the reliability, sensitivity, and validity of tests be improved?

How are indexes and scales constructed?

Why should researchers usually resist constructing their own scales?

Can a test be valid if it is constructed without any regard for theory?

"If you cannot measure, your knowledge is meagre and unsatisfactory."

Inscription on Social Science Research Building at University of Chicago

The ultimate aim of all research questions is to discover relationships between variables, and this requires that the variables be measured. Physics is a powerful science in large part because its basic terms, such as length, mass, and electric charge, are clearly defined and quantifiable. Psychological terms have no such clarity, and investigators often disagree in their assessments of qualities such as self-esteem, creativity, aggressiveness, and paranoid tendencies.

Upon beginning a research project, consult the literature to see if relevant terms have already been defined. In addition, ask how the terms are measured in ordinary life. Why, for example, are some people considered highly aggressive and others not aggressive at all? Try to imagine specific instances in which they would be likely to behave differently and see if these can be put in the form of a measure.

Or turn to a dictionary. Suppose a researcher wants to find out whether stress affects the consumption of alcohol. Consider "stress": The American Heritage Dictionary defines it as a mentally or emotionally disruptive or disquieting influence. Armed with that definition, she might conduct an experiment and observe that stressed college students drink more alcohol than do unstressed ones. But her colleague, using the same definition, might find that stressed students drink less. On the assumption that both experiments are conducted flawlessly, how might the conflict be explained?

Even though they used the same dictionary, the difficulty may lie in their definitions. Dictionaries are legitimate starting points for definitions, but researchers must then find ways to measure the concepts so defined. In fact, and regardless of what the dictionary says, the method by which a scientific concept is produced or measured becomes its definition. When different methods of measurement are applied to the same concept, conflict often results. See Box 6.1 for two examples.

I dislike loud music so I might create stress by arranging for subjects to sit through a Grateful Dead concert. And you, more conventionally, might give electric shocks. (Or we might agree on how to induce stress but measure drinking differently.) The result might be two distinct descriptions of the effects of "stress" on "drinking." We might then increase our own stress levels by debating the merits of our definitions, but such debates rarely advance science. On the other hand, careful description of measurement procedures often does. If electric shock and Grateful Dead concerts have different behavioral consequences, then they don't both produce stress; or they produce different types or levels of stress; or the behaviors are influenced by factors besides level of stress. Analysis of the possibilities may lead to valuable insights.

Stress can be defined as the reaction to 60 minutes in the front row at a Grateful Dead concert or to 10 one-half-second shocks of 45 volts delivered to the right forearm. Some researchers, extending the popular use of the term, list positive events like winning a lottery or getting a promotion as stressors. Some future Chicken Little may even define stress as the reactions of subjects to an announcement that the sky is falling. The important thing is that the definitions be clear. Readers of a scientific study should never be in doubt about how the concepts were measured. As Francis Bacon wrote, "Truth is more

BOX 6.1

The Dependent Variable in Psychotherapy; Development of Morality

In 1952, H. J. Eysenck published an apparently devastating critique of psychotherapy. He surveyed the literature and reported that about two-thirds of neurotics improve substantially within two years, and this is as true for those who never enter therapy as for those who do. His conclusion, based on data from the German Psychoanalytic Association in which patients were classified according to the extent to which they had improved under therapy, was that psychotherapy is useless. The number of patients in each category, out of 721 analyses, were as follows:

241 prematurely terminate

117 still in progress

47 uncured

116 improved

89 very much improved

111 cured

In 1971, Bergin published a reanalysis of Eysenck's data; Bergin challenged Eysenck on many grounds, the most important having to do with the definition of "improved." He quoted the original definitions of the psychoanalysts (from Fenichel, 1930):

"We were most particular in what was to be understood as 'cured.' Included were only such cases where success meant not merely the disappearance of symptoms but also the manifestation of analytically acceptable personality changes and, wherever possible, confirmative follow-up. This strictness of definition demands that most of the 'very much improved' cases must, for all practical purposes, be closely coordinated with the 'cured' ones. 'Improved' cases are those who upon termination still lack in one aspect or another, including such cases as those which had to settle, for external reasons, with an only partially successful outcome. . . ."

Eysenck included only the "cured" and "very much improved" in his calculations of percent improved. Bergin wrote: "A decision like this crucially affects the ultimate results and it illustrates well how subjective bias is the only way of resolving ambiguities in the raw material." To Bergin's comment, I wish to add that controversy over Eysenck's critique persists today, partly because readers of secondary sources are unaware of his definition of the dependent variable.

Both Piaget (1965) and Kohlberg (1976) studied the development of moral reasoning in children by presenting them with moral dilemmas. For example, the children were asked if a man should steal a drug to save his wife's life. Their answers led to the conclusion that moral reasoning develops in stages, and older children have a higher level of morality than younger ones. But when children were given opportunities to cheat on achievement tests, older ones did so the most (Hartshorne and May, 1928). Different method of measurement, different conclusion.

---

likely to come from error than from confusion." So, scientists should always describe the operations for measuring their variables. These measures are called operational definitions.

Operational definitions, detailed and specific, are essential for scientific communication. But science also requires abstractions, or else it would consist of nothing but descriptions of specific events—the effects of 45 volts of shock, of a Grateful Dead concert, of hearing that the sky is falling—and no connection between them. So scientists create high-order abstractions, called constructs, to represent phenomena and be used in theories; and they move frequently between levels of abstraction. Operational definitions tell how to measure constructs like "stress," "aggressiveness," and "authoritarianism" for use in the formulation and testing of laws and theories; and as the theories are tested and modified, the operational definitions are often revised.

One cautionary note: The meaning of a concept should dictate the nature of the operational definition, not the other way around. But unfortunately, in psychology the tail sometimes wags the dog. For example, because standardized IQ tests are easy to administer and score, some psychologists proclaim that "Intelligence is what intelligence tests measure." For many years, they have used the two or three digit numbers that summarize test performance as an index of a person's intellectual capacity. Skills more difficult to measure, such as ability to interact well with others and to develop insights into one's own behavior, have been neglected. Gardner (1983) contended that standardized tests measure linguistic and logical mathematical intelligence, which are only two of seven facets we know about (the others being spatial, musical, bodily-kinesthetic, interpersonal, and intrapersonal intelligence).

Scales of Measurement

Below are eight operationally defined variables. They can be grouped into four pairs based on how they are measured. Can you find the correct groupings?

1. Students are sorted into four groups: freshmen, sophomores, juniors, and seniors.
2. Babies are ranked from least to most active.
3. People's scores on the Scholastic Aptitude Test are recorded.
4. People's cholesterol levels are obtained.
5. Students are classified as either under or overachievers.
6. The order in which people finish a marathon race is recorded.
7. The number of times a patient's eye twitches during an interview is recorded.
8. Students' scores on a test of introversion are obtained.

The pairings should be as follows: 1 and 5; 2 and 6; 3 and 8; and 4 and 7.

1 and 5: These measurements assign subjects to categories (class standing, under and overachiever) that serve as labels and nothing more. The categories are like numbers on the jerseys of football players, with no necessary implication that one is better than another. This type of measuring scale is called nominal.

2 and 6: Subjects are rank-ordered on a single variable. The degree of difference between ranks is not considered. The scale is called ordinal.

3 and 8: This type of measurement is called interval. It tells not only the rank-order of subjects, but also by how much they differ. The intervals between scores represent equal quantities of the variable measured, so that a score of 10 is midway between a score of 5 and a score of 15. That's not necessarily true of ordinal scales; the 10th place finisher in a race may be closer to 1st than to 11th. (Educational and psychological tests such as the SAT are not really interval scales but can be used as such without creating serious problems.)

4 and 7: Most 3-year-olds would probably get no items correct on the Wechsler Adult Intelligence Scale, but that wouldn't mean they have zero intelligence—or how would we classify 1-year-olds and goldfish and daisies. There is no established zero point for intelligence. Zero points do exist for many attributes, primarily physical ones like height, weight, speed, and temperature. If a scale has interval properties plus a zero point, it is called a ratio scale.

Being able to recognize the different scales helps scientists understand and analyze their data. They generally prefer measuring at the interval or ratio level but cannot always do so. And numbers are sometimes irrelevant to the research purpose. For example, pollsters are usually uninterested in how much passion people feel for their political convictions but only whether they intend to vote Democratic or Republican. So, there is value in learning how to measure at the nominal level, that is, how to classify.

Principles of Classifying

Classifying is arranging things into categories. A classification scheme should be exhaustive, that is, every item should fit into a category. Classifying television programs into comedies, dramas, musical specials, movies, and children's shows would be inadequate in this respect, because of the omission of programs such as documentaries and sporting events. The categories should be mutually exclusive, that is, every item should fit into only one category. The television classification fails here, too; some programs are both comedies and children's shows, both movies and dramas, and so forth. Categories should be created according to whatever principle best suits the research question, and that single principle should be applied consistently. A different principle may be used at each division into subcategories.

Lazarsfeld and Barton (1951) suggested making classifications with several broad categories, each broken down into detailed subclasses. In that way, fine distinctions can be preserved and data analyzed at whatever level is most appropriate. Their advice for developing categories was to collect observations, organize them into a preliminary scheme, apply the scheme systematically to the data, refine the scheme, reapply it to the data, and so on.

Once categories have been formed, items can often be assigned to them with little difficulty. Such classifications as the rank of soldiers, gender of respondents, types of crimes reported in newspapers, and types of public transportation are objective and straightforward. But many concepts are difficult to apply. Lazarsfeld and Barton asked when ". . . is a news story sensational, a child maladjusted, a leader authoritarian? Is the army characterized by a caste system? Has union leadership become bureaucratized? Are the rank and file apathetic? Is the atmosphere of a group permissive, cooperative, competitive?" As a way of dealing with this problem and minimizing subjectivity, they suggested that indicators be specified from which judgments can be made. The judgments can be broken down into segments, with separate indicators for each, and then combined into a single score.

To illustrate the procedure, Lazarsfeld and Barton indicated how a researcher might measure children's "adjustment." First, he would specify broad areas of data for raters to attend to, perhaps including appearance, response to interviews, attitude toward others, and attitude toward self. Among researchers with a body of common training, such as child psychologists, the specification would help focus observations and increase their reliability. The researcher might give more detailed indicators for each area, which would increase reliability even more. For appearance, these might include excessively untidy hair and clothing, chewed fingernails, rigid facial expression, abnormally rigid or slack posture, and so forth. Then would come the difficult task of combining indicators. "Are badly chewed fingernails and untied shoelaces equally bad? Should each count one point on a maladjustment score? How much weight should be assigned to an extremely submissive attitude toward one's parents? What if the latter is combined with an extremely aggressive attitude toward one's schoolmates?"

Properties of Good Measures

Whatever the scale of measurement, variables should be operationally defined. They should have certain other qualities as well:

- A good measure is affected only minimally by fluctuations in anything other than the characteristic being measured. It is **reliable**.
- A good measure is highly sensitive to changes in the characteristic being measured. It is **sensitive**.
- A good measure measures the right thing. It is **valid**.

Described in the following pages are several hypothetical methods for measuring thirst. Each is an operational definition, though only a textbook writer would ever consider any of them; and each has a serious flaw, noted in parentheses. The discussions following expand on the meanings of reliability, sensitivity, and validity.

Measures Must Be Reliable

1. Subjects are given a wide-mouth bowl full of beer and told to pour the beer into a shot glass and drink from the glass. The bowl is weighed before and after a 10 minute drinking period, and the difference is recorded as amount drunk. (More beer would end up on the floor than in anybody's mouth, and the spillage would be largely random. So people with the same degree of thirst might earn very different scores. The measure would therefore be unreliable—unstable when repeated.)

Reliability is a composite term, and in addition to stability may refer to agreement between scorers or consistency among component items of a single test. A good measure is reliable in all respects. To find a test's stability over time, administer it twice to a group of individuals. See if their scores change, and also see how much their relative rankings change; that is, compute a correlation between test and retest (see Box 13.1). Be aware, however, that although the test may be stable, the subjects may not be. People who score high on thirst because they quickly drink an enormous quantity of liquid may score low if tested five minutes later.

To measure agreement between scorers, have several individuals score the test and compute the extent to which they agree (see pp. 147–148). To assess the internal consistency of multi-item measures, see Box 13.2.

Unreliable measures may lead to unjustified conclusions. For example, Hartshorne and May (1928), working in diverse settings such as schools, homes, sports events, and parties, unobtrusively presented children with opportunities to act dishonestly; and they found low correlations between dishonest acts in each setting. Their classic study has been used as proof against the view that people have enduring personality traits. Instead, so the argument goes, behavioral consistencies occur only because people are generally observed in similar situations; when tested for the same trait in different situations, the consistency disappears. But Block (1977) reported that Hartshorne and May's measures of honesty were unreliable. An unreliable measure has a low correlation with itself, and it cannot have a higher correlation with another measure. (Note: The critique of their study does not mean that trait theorists won the day; many recent writers have emphasized that situational factors and traits interact with each other.)

Identifying the cause of a problem is often the most significant step toward its solution. The many causes of unreliability can be grouped as follows:

- The measurement situation. Measurement of the same property may yield dissimilar results if made under different conditions. Conditions can differ in infinite variety, so the possibilities can't be enumerated. Besides, variations that affect one type of measurement may be irrelevant to another. Here is a list of some factors that have affected responsiveness to drugs (Leavitt, 1982): the route of administration;

Computing Changes in Reliability as the Number of Items in a Test is Increased

Suppose that a test has a reliability, r; and that comparable items are added to increase its length by a factor, n; then the reliability of the new test equals

$$\frac{nr}{1 + (n - 1)r}.$$

So, if a twenty item test with a reliability of .50 is lengthened to 100 items of comparable quality, the new reliability equals

$$\frac{5 \times .50}{1 + 4(.5)} = .83$$

---

quantity of food in the stomach; intestinal contents; presence of other drugs; time of day; acidity of urine; temperature of the room; temperature of the drug; temperature of the subject; amount of light; noise level; what the subjects were doing at the time; in rats, whether their cages had been cleaned daily or once a week; in mice, whether their bedding was of ground corncobs or red cedar chips; in people, whether the drug administrator was enthusiastic or not, and who administered the drugs. If not held constant, these factors reduce reliability. And, response to drugs is not a uniquely sensitive behavioral measure.

An additional example may help emphasize the importance of keeping conditions of measurement uniform. An assistant of Ivan Pavlov trained mice to run to a feeding place when he rang a bell. Then he trained their offspring, then their offspring, continuing for several generations. He was testing the belief that acquired characteristics are inherited, and his results seemed to offer strong support—the number of trials before they reached the criterion for learning averaged 300 for the first generation, then 100, 30, 10, and 5 for generations 2 through 5. But the assistant's interpretation that each generation had gotten better was in error; as Pavlov later showed, the mice hadn't changed, but the assistant had become a more efficient trainer (Razran, 1958; Zirkle, 1958).

- The task. Ambiguous test questions yield lower reliabilities than do lucid ones. More generally, ambiguous tasks are less reliable than ones that are clearly-defined. Tests comprised of many items are more reliable than shorter tests. See Box 6.2 to learn how to compute changes in reliability as the number of items in a test increases.

- The data recorder. People often record data directly as generated; but compared to various electronic devices, we're not very good at it. We have fallible memories and slow reflexes, and we make inconsistent judgments. So, if your research budget allows, buy timers, counters, and related devices, and automate your recording as much as possible. If people must be used, help them by making scoring categories unequivocal. Train them and make sure that interobserver agreement is high (see pp. 147–148 for calculations). Check periodically to ensure that their scoring criteria don't shift over time.

Measures Must Be Sensitive

2a. The subjects, all heavy drinkers, have three shot glasses of light beer placed in front of them on a hot day. The researcher calls those subjects thirstiest who empty the most glasses. (All the subjects would probably drink all three glasses. The measure wouldn't permit a sufficient range of responses to allow for discriminations between them.)

2b. The subjects have 10 glasses placed in front of them, and again the measure is amount drunk. The drink is urine-flavored vodka. (The problem is the same as in the previous example, but from the other side. Nobody would take more than a sip, so the measure would not discriminate and would therefore be useless.)

2c. The subjects are asked to choose between a bottle of beer and a five dollar bill, and their choices are duly recorded. (With only two alternatives, no gradations in thirst can be detected. It is generally useful to be able to detect subtle as well as gross differences. But don't go beyond the level of precision needed to answer your research question. Don't invest in a scale sensitive to milligrams if you intend to weigh elephants.)

Using Indexes and Multiple-Item Scales Scores based on many items are more reliable and sensitive than those based on a few, because they allow a wider range of scores. Indexes and multiple-item scales are formal methods for combining items. An index is made by combining two or more scores, sometimes weighting them, to yield a composite. For example, the Dow-Jones Industrial Index is obtained by summing the share prices of 30 selected stocks. The selection of items for inclusion in an index is subjective, and items are not separately analyzed for effectiveness. By contrast, scales are constructed according to specific procedures, and unsatisfactory items are winnowed out.

Bonjean, Hill, and McLemore (1967) asserted that social scientists are too quick to construct their own scales, while making insufficient use of published ones. They examined all articles and research notes in four major sociological journals over a 12-year period: there were 3,609 attempts to measure various phenomena by the use of scales and indexes, and 2,080 different measures. Only 589 were used more than once. There are at least 15 scales that measure hostility in dream reports, and they correlate poorly with each other (Winget & Kramer, 1979) (which is further evidence of the importance of operational definitions). The policy of creating scales for each new study increases the difficulty of measuring validity and of comparing results between studies. Therefore, before trying to construct a scale, search the literature to see if an appropriate one already exists. The books listed below evaluate scales:

- Buros, O. (Ed.). *The Mental Measurements Yearbook*. Highland Park, N.J.: Gryphon.
- Buros, O. (Ed.). *Tests in Print*. Highland Park, NJ: Gryphon.
- Comrey, A. et al. (1973). *A Sourcebook for Mental Health Measures*. Los Angeles: Human Interaction Research Inst.
- Goldman, B. & Busch, J. (1974, 1978, 1982). *Directory of Unpublished Experimental Mental Measures*. New York: Human Sciences Press. (Vol. 1–3).
- Johnson, O. & Bommarito, J. (1971). *Tests and Measurements in Child Development*. San Francisco: Jossey Bass.
- Lyerly, S. (1973) *Handbook of Psychiatric Rating Scales*. Rockville, Maryland: NIMH.
- Miller, D. (1982) *Handbook of Research Design and Social Measurement*. New York: Longman.
- Robinson, J. & Shaver, P. (1973). *Measures of Social Psychological Attitudes*. Ann Arbor, Michigan: Institute for Social Research, University of Michigan.
- Rosen, P. (Ed.). (1973). *Measures of Self-Concept*. Princeton, N.J.: Testing Service.
- Simon, A. & Boyer, E. (1974). *Mirrors for Behavior: An Anthology of Classroom Observation Instruments*. Research for Better Schools, Center for the Study of Teaching.

People who must construct lots of scales should familiarize themselves with the various types. Dawis (1987) described them in an excellent review article that has many references. Beginning researchers need not be so eclectic and can focus on a single type. The most widely used scale for measuring attitudes is called a Likert Scale. The steps in its construction are as follows:

1. Write lots of potential items. These should be statements that seem to express either strongly favorable or strongly unfavorable attitudes, and they should cover the entire domain of the attitude. For each item, have five possible answers (such as strongly agree, agree, undecided, disagree, strongly disagree), so the item can receive a score between 1 and 5. Some statements should be worded such that "strongly agree" indicates a favorable attitude, some such that "strongly disagree" is favorable. But the most favorable responses should always receive a score of 5, the least favorable a 1.
2. Administer the items to at least 100 subjects chosen randomly from the population for whom the questionnaire is being designed.

3. For each subject, sum the scores on each item to get a total score.
4. Analyze the ability of each item to discriminate between high and low scorers. One way is to make a two-column table for each item. In the first column, list the total score of each respondent; in the second, list the respondent's score on the item. Compute the correlation between them. See Box 13.2 for a worked example. Discard items that have low correlations with the total score.

Although multiple-item measures are generally more reliable and sensitive than single items, the items must be combined properly into a single global assessment. Meehl (1954) and Faust (1984) showed that statistical formulae are superior to clinical judgments for combining. For example, three highly trained pathologists predicted the survival times of patients diagnosed as having Hodgkin's disease. The three identified nine characteristics of cells that had diagnostic significance, but they could not integrate the information from all nine to get an aggregate score, so they did not predict accurately. Computers can integrate information from large numbers of variables. When a computer was fed data from the first 100 cases about the nine characteristics and subsequent survival times, it assigned weights to each characteristic; formulae were developed and applied independently to the next 93 cases, and predictive accuracy was high. Meehl and Faust gave many more examples of the superiority of statistical over clinical judgments.

Measures Must Be Valid

3. Subjects pour from a large, heavy keg into a glass. The measure is the weight difference in the keg from beginning to end of the drinking period. (The researcher might find that heavy subjects "drink" less than petite ones. The petite subjects would not handle the keg easily, so would spill more beer. The measure would not be valid, since it would be tied more closely to strength than thirst. Valid measures measure what they are supposed to. Note: Unreliable measures cannot be valid, but invalid measures can be reliable. Stone's measure of sexual activity in rats, described in Box 6.3, was reliable but not valid.)
4. The researcher does away with alcohol entirely and simply asks subjects to estimate how much they would drink. (Indirect measurements, such as answers to questions about thirst, should not be rejected out-of-hand. In fact, all measurement is indirect. A quality as thoroughly understood as temperature is measured by such indirect means as the expansion of mercury in hollow glass tubes, or changes in the color or electrical resistance of materials. Questionnaires, widely used by social scientists, are often highly informative even though they do not directly measure what people believe or how they act. But as can be seen from Box 6.4, answers to questionnaire items often depart greatly from the truth. Unless

BOX 6.3

Rat Sex

Many years ago, Calvin Stone (1939) wanted to find out the effects of castration (removal of the testes) on the sexual behavior of male rats. He measured the number of times per test session that males mounted females, both prior to and after castration. To his surprise, there was little difference. Can you think of any reason?

When an active male and a receptive female rat are placed together in an observation box, they copulate in a stereotyped way. The male mounts from the rear, and if he has an erection and achieves vaginal penetration, he thrusts. This is called an intromission. Mounts without penetration are simply called mounts. There are typically 5 to 15 mounts and 5 to 10 intromissions before the male ejaculates. Then both animals lie down, light up cigarettes, turn on the TV, and relax for 4 or 5 minutes. (They do lie down and rest for a few minutes.) Then they start over again and repeat for as many as 10 ejaculations, with the pause following ejaculations getting progressively longer.

Stone was a pioneering researcher and had to make do with the information available at that time. He did not distinguish between mounts, intromissions, and ejaculations. He called all of them copulations. The castrated rats were unable to achieve intromissions or ejaculations, but they continued to mount; since they didn't ejaculate, there were no long post-ejaculatory pauses, and their total number of copulatory responses was equivalent to that of intact animals. Stone's misleading results were, as is so often the case, due to poor measurement of variables.

---

the purpose of a study is to compare responses to questionnaires with more direct measurements, the latter are preferred if available and inexpensive.)

The question "Is measure X valid?" can't be answered in the abstract. Three general reasons for measuring are to classify, to predict, and to incorporate a construct into a scientific network; and separate types of validity (called concurrent, predictive, and construct validity, respectively) apply to each. The first two are considered together under the heading criterion-related validity.

Criterion-Related Validity   Physicians order laboratory tests to help them diagnose (classify) their patients, for only then can the patients be given the best treatments. The accuracy of a test in classifying is the test's concurrent validity. Other measures predict, as when personnel managers use tests to screen candidates for various jobs, college admissions officers use high school grade point averages to predict success in college, and football scouts weigh prospects, time them in the 40-yard dash, and give them agility and strength tests before deciding whether to draft them. Accuracy in these cases is called predictive validity. To evaluate either concurrent or predictive validity, test results must be correlated with some criterion, hence the more general name, criterion-related validity.

BOX 4

Surveys

Surveys are widely used in the social sciences for the collection of data, and they provide lots of useful information; but researchers should be aware that answers are influenced by specific wording of questions and by their context, and that respondents, deliberately or otherwise, often stray from the truth. The subject is discussed in detail in chapter 11. Below are a few examples for now:

- Sixty-four percent of people whose scores on a fear questionnaire put them in the high fear group toward snakes, were willing to actively touch a snake; 16 percent of the low fear group were not (Bernstein, 1973).
- In 1941, Rugg asked similar groups of people either (a) "Do you think the United States should forbid public speeches against democracy?" or (b) "Do you think the United States should allow public speeches against democracy?" Thirty-nine percent answered "Don't forbid," 21 percent answered "Allow."
- In a 1937 study by the American Marketing Association, women answered more favorably toward advertising when questions about advertising followed questions about dresses than when the order of the identically worded questions was reversed.
- Both cigarette smokers and heavy drinkers tend to underreport their consumption (Pernanen, 1974).
- Information obtained from 900 people by experienced interviewers was later checked against the records of various agencies (Parry & Crossley, 1950). Forty percent of the respondents falsely said they had contributed to the United Fund; 25 percent that they had registered and voted; 10 percent that they had a driver's license; and 17 percent gave inaccurate information on their age.

....

Tests with high criterion-related validity may be useful even if the connection between the test and the criterion is obscure. That's the rationale behind drug screening with animals. Pharmacologists have learned that drugs that increase the activity of rats are likely to suppress appetite in people; drugs that retard activity are often effective in treating psychotic patients; and drugs that increase the body temperature of rabbits often cause people to hallucinate. Even if nothing more were known about the tests, they would be valuable in drug screening.

Suppose 10,000 people were asked to name their favorite color, and 50 of them said lavender. Suppose further that, within a year, each of the 50 and few of the others had suffered a fatal heart attack. Would the test be a good one for screening people with heart problems? In fact, psychologist Starke Hathaway and neurologist John McKinley used a similar strategy in developing the most widely used psychological test in the world, the Minnesota Multiphasic Personality Inventory (MMPI). They asked both normal people and psychiatric patients to respond to a large number of statements with a "yes," "no," or "neither." Answers to more than 500 statements differentiated the normals from people with various disorders, and these were compiled into

PSYCHOLOGICAL MEASUREMENT OF THE FUTURE

a final version. Many of the statements have no obvious relevance to psychopathology, but the MMPI nevertheless has high criterion-related validity (actually, validities, since a separate validity coefficient must be calculated for each diagnostic category). That accounts for its popularity.

A researcher looking for an interesting scientific problem might try to figure out why a particular measure has high criterion-related validity. What, for example, is the reason for the relationship between increased activity in rats and appetite suppression in people?

The criterion is often straightforward: grade point average in freshman year of college, making the football team, having a heart attack, reporting hallucinations. But measures can't be validated so easily in most areas of psychology. Criteria are disputed and unreliable for evaluating creativity, neurotic tendencies, motivation, and so forth; and a measure's coefficient of reliability sets an upper limit on its correlation with anything else. That's why personnel tests are often of little value; the criteria for job success (supervisor's rating, attendance, grades in training courses, accident frequency, quality of output) are disputable, often subjective, and thus, unreliable.

Construct Validity Despite the lack of a reliable, agreed-upon criterion for, say, creativity, thousands of research publications have been devoted to the topic. Books and courses promote methods for enhancing creativity, tests are available for measuring it, and several analyses have been made of the lives of creative people (whose identification, of course, presupposes a method of measurement). A new test, to be accepted, would have to fit into the established network. It would have to correlate highly with currently available tests and be sensitive to manipulations believed to affect creativity. It would have to distinguish between individuals and groups believed to differ in creativity; published poets would be expected to score higher on average than students in lower level English classes. To the extent the test behaved according to expectations, it would have construct validity.

Considerations in Selecting an Independent Variable In addition to picking reliable, sensitive, and valid IVs, researchers should test them at several values. The effect of electric shock on drinking depends partly on the intensity of the shock; a mild shock may increase drinking, a strong one decrease it. Similarly, low doses of alcohol may stimulate behavior, high doses cause sleep. A rat deprived of food for twenty-four hours will be quite active; a rat deprived for twenty-four days will not be. Review the literature and, perhaps, do a pilot study to get an idea of the appropriate range of IV levels.

Try to verify that your IVs have worked as intended. Sometimes they don't. Sudden loud noises don't frighten deaf people, and a researcher who shouts that the sky is falling is unlikely to create stress (unless you're locked in a room with him and think he may be armed). Consider doing the following as checks on the effectiveness of your IVs:

- At the conclusion of a study, ask your human subjects for feedback.
- If your IV is designed to produce a particular state, look for other indicators of that state. Suppose, for example, you use loud noises to induce stress in rats, perhaps to see how stress affects drinking. Even though **not interested in** other behaviors, you might monitor heart rate, activity, urination, defecation, and vocalizations to see if they show their normal patterns to stressful stimuli.

Considerations in Selecting a Dependent Variable Developers of standardized tests routinely modify them until high standards of reliability and validity are achieved. Social scientists should do the same, yet they often use procedures that minimize the reliability and thus the validity of their empirical work. Bahrick (1977) illustrated the problem through an experiment in which people were asked to try to learn a long list of words during three exposures to it. Then they were shown a new list of 120 words. Some subjects were told that it contained 60 words from the original list and to try to identify them, while others were not told how many remained from the original list, but to try to identify as many as possible. Only some of the subjects were told to guess if in doubt about answers. When corrections were made for guessing, the groups were equivalent in correct answers. But variability was much lower in subjects who were told how many responses to make. These subjects would have been more sensitive to the effects of an IV.

A frequent measure in learning experiments is time to reach a given criterion level, and the most widely used criterion is a perfect score. Bahrick showed that this is a poor choice, as it causes more subject variability than criteria such as 8 out of 10 correct responses.

Use Multiple Measures   Webb, Campbell, Schwartz, and Sechrest (1966) wrote that all psychological measurements tap multiple processes and are theoretically complex. So, all operational definitions provide only one out of a potentially infinite number of ways to measure a construct. Even in physics, Webb et al. noted, "no meter ever perfectly measures a single theoretical parameter." The score on any measure embodies assumptions. In their more recent book (1981), they gave the example of a boy who scores 125 on an intelligence test. An obvious conclusion is that the boy is bright. But there are other possibilities (called rival hypotheses by Webb, Campbell, Schwartz, Sechrest, and Grove): (a) the boy had been coached; (b) the examiner made a mistake in arithmetic; (c) the examiner liked the boy so was extremely generous in his scoring; (d) the boy had taken the test several times previously; and (e) the examiner was the boy's mother.

Rival hypotheses can be entertained about any measurement. As their plausibility increases, confidence in the measure diminishes. So scientists should measure with the specific aim of reducing both the number and the plausibility of rival hypotheses. One way to do this is to use multiple measures.

In contrast to historians, who analyze unique events, scientists seek universal laws. If a scientist finds a relationship between attending a Grateful Dead concert and beer drinking the next night, she should seek other evidence of the proposition that stress affects drinking. This requires multiple measures. Sidman (1960) and Lykken (1968) made a similar point in arguing that researchers should always replicate their results, but not exactly; phenomena, unless robust enough to appear under somewhat altered conditions of testing and measuring, are poorly understood and probably trivial.

Neal Miller (1956) was an early advocate of multiple measures. He cited work with his students on a relatively simple system—thirst in rats. Thirst was induced by administering a saline solution via stomach tube. Fifteen minutes later, given the opportunity to bar press for water, the rats pressed no more than did uninjected controls. But they drank more and tolerated stronger dilutions of the water with quinine. Three hours later, the difference between the groups in drinking pure water reached its maximum; but differences in bar pressing and drinking quinine-diluted water reached their maxima at least three hours after that. So, the influence of saline on thirst depended considerably on how thirst was measured.

There are three types of dependent variables in psychological research: verbal, behavioral other than verbal, and physiological. They do not always correspond. A person may say he's unafraid of snakes but refuse to step into a room with one, or vice versa; and his heart may or may not start racing at the sight of a snake. So, researchers should try to measure their variables in more than one way. The extra work is worth it, even if measures conflict; instead of being discouraged, an enterprising researcher might in such cases launch an exciting new project to discover the reasons.

Use Nonreactive Measures   Physicists since Heisenberg have accepted that the act of measuring often changes the object measured. That is, the process is often reactive. For example, the rays of light required for observing the location of an electron change its momentum. Social scientists confront a similar problem on a molar level, but we have an advantage over physicists, as people are more easily tricked than electrons. Social scientists can use nonreactive measurements, and Webb et al. (1981) urged that we do so whenever possible. Rather than obtrusively assess the effects of an air disaster on next-day travellers by means of interview or questionnaire, a scientist might unobtrusively and nonreactively analyze flight cancellations and new ticket sales, changes in trip insurance policies, and business at the airport bar. (The name *Unobtrusive Measures,* was changed to *Nonreactive Measures* for the second edition. The change reflects the authors' realization that it is not the noticeability of the observational procedures that matters, but how the behavior of the observed is affected by them.)

The following examples, all referenced in *Nonreactive Measures,* are instructive for at least three reasons: (a) they indicate feasible alternatives to surveys and experiments (b) they indicate feasible methods by which surveyors and experimenters can obtain multiple measures, and (c) they illustrate the ingenuity with which good scientists approach their problems.

Physical Traces: Erosion and Accretion   Erosion measures are those for which the degree of selective wear on some material yields the measure.

Examples:

1. Floor tiles are often coated to resist wear. Uncoated tiles have been used in museums to measure the relative popularities of the exhibits; the more eroded the floor, the more popular the exhibit.
2. Mosteller measured the wear and tear on different sections of an encyclopedia by noting dirty edges of pages, frequency of dirt smudges, finger markings, and underlinings. This provided him with a measure of the frequency with which the different sections had been read.

Accretion measures are those that provide evidence in the form of some deposit of materials.

Examples:

1. Police techniques include many accretion measures, such as soil analyses from shoes and clothing. The murder of Napoleon was uncovered on the basis of arsenic traces in remains of his hair.
2. Liquor consumption has been estimated by counting the number of empty liquor bottles in trash carted away from homes.

Archives: The Running Record   Various ongoing, continuing records of a society (votes, city budgets, births, marriages, deaths, weather reports, newspaper and magazine articles) can be exploited by social scientists.

Archives: The Episodic and Private Record   Scientists can seek out materials that occur periodically and are not part of the public record: insurance sales, suicide notes, diaries, the nurse's record on a bedside clipboard.

Simple Observation   Clothing, jewelry, shoe style, tattoos, tribal markings, calluses, and scars are examples of exterior physical signs that have been used as nonreactive clues to people's behavior and status. Here is another example:

The hypothesis to be tested was that the length of hair of psychologists is a good predictor of their interests. At a professional convention, investigators classified the meetings by probable appeal to "tough-minded" or "tender-minded" psychologists and recorded the hair lengths of the attendees. The results showed clearly that tough-minded psychologists have shorter hair.

The significance of expressive movements such as frowning, baring of teeth, and erection of hair was discussed by Darwin in 1872. Expressive movements convey a great deal of information.

Investigators attentive to subtle cues have made plausible inferences about status, mood, friendliness, and other behavioral dispositions from observing the physical positions of individuals.

Examples:

1. Where seating is voluntary (in classrooms, lunch counters, and public transportation), the degree to which people of different types (ages, races, men and women) mix has been used as an indicator of the significance of those variables for acquaintances and friendships.
2. Snippets of conversation have been used as evidence that men and women think differently. Moore walked the streets of New York every night for several weeks and jotted down every bit of audible conversation. He found that 8 percent of conversations between two men were about women, whereas 44 percent of conversations between two women were about men.
3. People's interest in objects can be gauged by the amount of time they spend working on, watching, or otherwise attending to the objects. And differences in people's time perception, as estimated from the accuracy of their watches, has been related to the importance they attach to achievement (more high- than low-need achievers have watches that are fast).

Measures can be nonreactive even though they involve hardware or are contrived; many episodes of the television show "Candid Camera," were prepared by filming people secretly while they responded to the strange situations created by the show's writers.

Example:

A woman stopped men on the street and asked if they would help her carry a suitcase on the ground next to her. She said she was tired after having carried it a long way. It was filled with metal. Filmed abroad, the study provides comparisons of men of different cultures. Frenchmen shrugged, Englishmen kept at it.

Summary

At the beginning of any research project, scientists should carefully consider how to operationally define the variables of interest. Measures should be picked for reliability (stability and consistency), sensitivity, and validity (make sure they measure what they are supposed to measure). Scientists should use multiple measures and test several levels of their independent variables.

Nominal measurement means assigning variables to categories, that is, classifying them. Classifiers should follow three rules: (a) every item should fit into a category (b) every item should fit into only one category and (c) a single principle should be used as the basis for each division.

Multiple-item scales are widely used in the social sciences. Before constructing a scale, read the reference works listed on page 85 to see if any published ones are suitable for answering the research question. If not, go through the steps listed on pages 85–86.

Key Terms

Construct An abstract term, such as motivation or intelligence, that is created because it helps tie together several pieces of empirical data.

Construct Validity Evidence that a measure (a) distinguishes between groups expected by theory to have different levels of a construct (b) yields similar scores to accepted measures of the construct and (c) does not measure other constructs.

Criterion-Related Validity Evidence that scores on a measure correlate with scores on a criterion.

Index A composite measure of two or more items.

Likert Scale A scale made by subjecting items to analysis and excluding those that don't correlate highly with the total scale. An individual's score on a Likert scale is obtained by adding the individual item scores.

Nominal Measurement Measurement in which objects are assigned to categories. Quantity is not involved, that is, there is no implication that one category is more or less than another.

Operational Definition The definition of a variable by the procedures used to produce or measure it.

Reliability The resistance of a measure to changes caused by changed conditions of testing; the consistency and stability of the measure.

Sensitivity The extent to which a measure detects changes in the variable being measured.

Validity The extent to which a measure does what it is supposed to do. See above under criterion-related and construct validity.

PART III

Developing a Strategy for Answering the Questions

CHAPTER 7

Developing a Strategy for Testing the Speculations

By the time you finish reading this chapter, you should be able to answer the following questions:

What are the advantages to social scientists in knowing several ways to conduct research?

What data collection methods are available to social scientists?

What are the five major research strategies available to social scientists?

What are the reasons for choosing one strategy over another?

Which research strategy is best for answering questions about incidence and prevalence of a condition in a population?

Why are one-shot case studies unsatisfactory for establishing causes?

What are retrospective and prospective studies?

What are the defining characteristics of experiments?

How can tobacco company spokespeople argue that cigarette smoking has not been shown to cause cancer?

"Life is the art of drawing sufficient conclusions from insufficient premises."

Samuel Butler, English author (1835–1902)

"Every method is imperfect."

Charles-Jean-Henri Nicolle,
physician (1866–1936)

In their attempts to solve mysteries, the great detectives of fiction and real life have a major advantage over most social scientists. The detectives seek out whatever clues are appropriate under the circumstances of the case, whereas the social scientists restrict themselves to one or at most a few methods. Sherlock Holmes checked for fingerprints and footprints, cigarette and pipe ashes, and bits of cloth torn from the coat of the murderer that remained embedded in the victim's nails. He interviewed friends, relatives, and rivals of both victim and suspects. His genius lay less in his ability to apply previously established methods as in broad-mindedness in their choice and invention of new ones as necessary. His modern counterparts rely heavily on recent developments in voice, blood, and hair analysis. Both modern and old have sought explanations to invalidate seemingly airtight alibis and have proposed mechanisms, modus operandi, for crimes.

But many social scientists gather evidence from a very narrow base. They rely exclusively on a single type of apparatus such as a Skinner box or computer. They always observe, or always interview, or always experiment. Their subjects, whether rats, baboons, or volunteers from introductory psychology courses, are their sole subjects. Their crucial concepts are tied to a single type of measurement, be it electric shock as a stress inducer or standardized IQ tests as a measure of intelligence. Their research is always designed to be analyzed by the same statistics, whether that be a t test, F test, or product-moment correlation. They labor under a handicap, much as if Holmes had restricted himself to fingerprint analyses, and they are less likely than other scientists to find satisfactory answers to their questions.

Psychologists and other social scientists should look to archeologists for inspiration. Archeologists rarely ask questions that can be answered by experiments, so they have developed several creative strategies for using evidence. They infer the functions of artifacts from size, shape, wear patterns, and contexts in which discovered. They analyze fossil pollen grains found at sites for clues about the prevailing types of vegetation when the sites were inhabited, which in turn indicates what the climate was like. Plant and animal

A SCIENTIST SHOULD BE LIKE A DETECTIVE.

remains provide clues about diet and territorial range of primitive groups. And the density of artifacts, amount of floor space within living enclosures, and number of burials at a site are used to estimate populations. Although the conclusions from such kinds of evidence are far less certain than those from experiments, they have enabled archeologists to give plausible accounts of primitive societies. Without them, our understanding would be diminished.

The methods available to social scientists for collecting data can be categorized as indicated below and are discussed further in later chapters.

- Observation: Behavior is studied in the subject's natural environment (naturalistic observation) or in special environments; the researcher may or may not manipulate variables.
- Participant observation: The researcher interacts with and collects data unobtrusively in the environment of the individual or group being studied.
- Questionnaire: A printed form containing a set of questions is distributed to subjects.
- Interview: Either face-to-face or by telephone, a researcher asks subjects a series of questions.
- Objective tests: Subjects are evaluated on the basis of their answers to clear, unambiguous, objectively scoreable test items.
- Projective and other indirect tests: Subjects are evaluated on the basis of their responses to ambiguous stimuli.
- Secondary analysis: Existing data are reanalyzed for a different reason than the one for which they were originally gathered.
- Physical evidence: From physical traces, such as the contents in a person's garbage, an investigator draws inferences about the person's behavior.

The collection of data is done in a context, as part of an overall research strategy. Five major strategies are available to social scientists, and each has several variants.

- Survey: A sample of individuals is studied and generalizations made from them to a larger population.
- Case Study: Individual subjects are investigated in detail.
- Descriptive Study: The goal is to describe behavior as it occurs naturally.
- Correlational Study: Scores are obtained from a group of subjects (people, animals, vegetables, minerals, institutions) on two different measures; the researcher determines if scores on the measures are related.
- Experiment: A researcher administers a treatment to one group, a different treatment to one or more other groups, and measures to see if the groups have different scores on a dependent variable. Alternatively, subjects are exposed in random order to each IV.

Certain data collection methods are associated with specific research strategies, as questionnaires and interviews are generally associated with surveys. But that leads to an unnecessary restriction of options. In fact, data, by almost any method collected, can be analyzed with any of the research strategies. Suppose, for example, that with no concern about ethics, the following experiment is conducted: Several people are randomly picked and sent a letter saying they have won a small prize, while several others, also randomly picked, are sent a different letter. If a researcher manages to befriend the people and record their reactions without their knowledge, he is a participant observer. If he chooses instead to watch them unobtrusively with binoculars and a listening device, he is doing naturalistic observation. Similarly, he can ask them for interviews, to fill out questionnaires, to submit to psychological tests, or he can study their garbage.

Surveys need not be restricted to people who dutifully fill out questionnaires. A sample of anything—observations, projective tests, secondary analyses, physical evidence—can if properly chosen provide information about the population from which it came. Similarly, correlations can be made between any two sets of data, no matter how collected, on the same subjects. And good case studies invariably involve several types of data collection.

Little is to be gained by evaluating the research strategies hierarchically. They differ in costs, are not always equally feasible, and answer different types of questions. Most importantly, they are complementary rather than competitive. Different strategies that point to the same conclusion strengthen belief in that conclusion; and if they lead to conflicting results, a researcher may gain valuable insights trying to figure out why. The proper direction is from question to research strategy rather than the other way around.

If the research question involves quantitative descriptions of the characteristics of a population, such as the mean height of all females or the proportion likely to vote for a certain candidate, surveys are best. They are the most effective way of measuring attitudes on a large scale. Surveys require random sampling (see chapter 11), which is not always possible. Case studies can also be used to provide information about populations; the problem of generalization from individuals to populations is discussed on pages 220–221.

Questions about relationships between variables can be answered by correlational designs (including correlations computed from survey data), case studies, or experiments. If a researcher cannot manipulate the variables, the only possible approach is correlational. That would be the method for answering questions such as "What is the relationship between birth length and weight?" For many questions about causal relationships (for example, "What is the effect of a woman's anxiety during pregnancy on the birth weight of her child?"), the researcher would also be limited to correlational data. Then conclusions would be offered tentatively, as correlational studies cannot conclusively establish causal relationships. Certain types of case studies provide convincing evidence of cause and experiments are designed to do so.

Questions of the type, "What is the effect of x on y?" are about causes, so the comments of the previous paragraph apply.

Debate over the relationship between cigarette smoking and lung cancer in people illustrates the issues involved in choosing research strategies and selecting evidence. Despite thousands of studies, many prominent scientists and statisticians remain unconvinced that smoking causes lung cancer. (However, when the studies are evaluated collectively, they make an overwhelming case against smoking and they show the value of multiple methods.)

1. *Statements a and b below are not equivalent. On the assumption that both are true, which presents a better case for getting smokers to quit? Which of the methods described below lead justifiably to the more powerful statement?*

a. Smoking causes lung cancer.

b. People who smoke are more likely than nonsmokers to get lung cancer.

2. *If you were an attorney defending the tobacco industry against the charge that cigarette smoking causes lung cancer, how would you attack the various types of evidence given below?*

Surveys

Surveys help answer questions about incidence and prevalence of a phenomenon, as when people are asked how many cigarettes per day they smoke; their age of onset of smoking; how deeply they inhale; whether they use filters; and their favorite brand. The conclusion from a United States survey (Pinney, 1979) was that 54 million Americans smoke. Such data can be used to establish relationships (although not causal ones) between variables. In 1951, the British Medical Association sent a questionnaire to all British physicians about their smoking habits; more than 40,000 replies were received (Doll & Hill, 1954) and correlated with subsequent mortality data.

Questionnaire data must be interpreted cautiously. Respondents may misunderstand questions, or they may have inaccurate memories. Warner (1977) found that widespread publicity on the dangers of cigarette smoking affected self-reports of smoking more than it did actual smoking behavior. Many people don't return questionnaires—in the study of British physicians, more than 30 percent did not respond—and respondents and nonrespondents may differ greatly.

Case Studies

The case study approach has a long tradition in clinical medicine and psychology. When an individual with an unusual set of symptoms presents himself to a clinical investigator, the investigator may seek out special circumstances that might have caused them. A report in 1761 by Dr. John Hill that two snuff users had swellings of the nose, which he believed cancerous, is often cited as the first evidence of a smoking/cancer link.

Case studies stimulate questions of the "how" and "why" type. "How and why did the swellings of the nose occur?" As exemplified by the experiments of Falk and Zucker described in the preceding chapter, "how" and "why" questions often lead to productive research programs. But, too often, case studies are of the one-shot variety: an investigator observes that many patients with a particular set of symptoms have been exposed to certain stimuli and concludes, on that basis alone, that the stimuli caused the symptoms. But one-shot case studies are inadequate for showing that one event causes another. Suppose, for example, that a pathologist learns that 1,000 people have been given a procedure called LR, and 995 of them die within a week. Should she publish a case study identifying LR as one of the deadliest things in existence? No—it's just my abbreviation for last rites.

Not only are one-shot case studies incapable of showing a causal relationship (like statement *a*, previously), they cannot even show an association (like statement *b*). The reason is that they don't use enough information. There are, in fact, four classes of subjects:

1. people who smoke and develop cancer
2. people who smoke and do not develop cancer
3. nonsmokers who develop cancer
4. nonsmokers who do not develop cancer

Hill observed two smokers with cancer. Suppose the numbers in groups 2, 3, and 4 were 1 million, 1 million, and 0, respectively? Or 0, 0, and 1 million? The conclusions would change. And even 1 million smokers with cancer, without data from the other groups, would not justify a conclusion that smoking and cancer are linked.

Although one-shot case studies do not establish causal relationships, they may, as they did with John Hill, provide clues. And intensive study of individual cases, in which investigators use varied data collection methods over an extended time period, may yield causal conclusions as persuasive as those reached by Holmes.

Correlational Studies

Data gathered from a variety of sources were evaluated for a possible correlation between smoking and cancer. The survey data showed large differences. Lung cancer patients were more likely than control subjects to report having been smokers (Doll, 1955; U.S. Public Health Service, 1964); and the intensity of smoking, as measured in several different ways, varied directly with likelihood of getting cancer (Kunze & Vutue, 1980; Rimington, 1981). Postmortem analyses indicated a greater incidence of premalignant tumors of the bronchial epithelium in smokers (Auerbach, Stout, Hammond, & Garfinkel, 1962).

Surveys that ask about past events are called retrospective, whereas prospective studies look to the future: people are asked to indicate their current habits; then, at some later time, their conditions are studied. An attempt is made to correlate some current habit with a future condition. The British Medical Association study was prospective. Of the 40,000 physicians who responded in 1951, more than 11,000 had died by 1973, and mortality rates from cancer were considerably higher among the smokers (U.S. Department of Health and Human Services, 1982). In other prospective studies, smoking cessation was found to reduce the risk of developing cancer (Doll & Hill, 1956).

Several countries publish twin registries, and researchers have used them as a data source for both retrospective and prospective correlational studies. Among monozygotic (identical) twins discordant for smoking (only one twin smoked), the smokers were more likely to develop lung cancer (Cederlof, Friberg, & Lundman, 1977).

Secondary analysis is the use of data previously collected for different purposes. Using records of cigarette sales as an index of amount smoked, researchers have established a strong correlation between a country's per capita consumption and its lung cancer deaths (U.S. Department of Health and Human Services, 1982).

Correlational studies are designed to answer questions about relationships—Do x and y occur together? Does x reliably precede or follow y?—but they are incapable of demonstrating causality (x causes y). And retrospective studies do not even show relationships unambiguously. The statement that a certain percentage of heavy smokers get cancer tells little unless the percentage of similarly afflicted nonsmokers is established. In other words, a control group is needed. But there is no acceptable way to pick one, and whatever method is used will lead to nondefinitive conclusions. For example, suppose that hospitalized smokers are examined for possible cancer. Should hospitalized nonsmokers be picked as controls? Or hospital personnel? Or people who respond to a newspaper advertisement for paid volunteers? The choice would influence the results.

The use of hospitalized nonsmokers might seem satisfactory at first glance: for each smoker, a nonsmoking patient could be sought of similar age, sex, race, socio-economic status, and any other variables deemed important. But Paul Meehl (1970) argued persuasively that the procedure is unacceptable. The reason is that smokers and nonsmokers (and any two groups formed in a nonrandom way) *must* differ even before any of them smoke (or do whatever constitutes the basis for assignment to groups). Then, matching makes at least one of the groups unrepresentative. Suppose, for example, that smokers on average were taller, smarter, and wealthier than nonsmokers; it might be possible to find sufficiently tall, smart, and wealthy nonsmokers, but they would not be typical.

Control group problems also plague prospective studies—and all other research designs (that is, all nonexperimental designs) in which subjects are not randomly assigned to groups. For example, a researcher who advertises for volunteers in a health magazine might not fill his quota of smokers; he might then place a second advertisement in another magazine. The two sets of recruits might differ in many ways other than smoking behavior, such as mean age, race, general lifestyle, and so forth. Then, even if smoking were an irrelevant habit, the groups might differ in incidence of lung cancer.

If all subjects come from the same source, as in the study of physicians, the control group problem is lessened although never eliminated. Another concern, indicated above, is that many people do not return questionnaires, those who fail to do so in one group may differ from both the participants and the nonparticipants in the other. For example, physician smokers with health problems may have preferred to remain anonymous.

Secondary Analysis

Public documents are an excellent source of data and probably underused by social scientists. However, appropriate secondary sources may not be available for answering specific questions. In addition, secondary analysis is subject to serious bias—researchers may be highly selective in their choice of documents.

Experiments

Experiments have two defining characteristics: (a) subjects are assigned randomly to treatment or control groups or exposed to each IV in random order and (b) the experimenter manipulates the independent variable. For obvious reasons, both legal and moral, no experimenters have randomly assigned human subjects to either a smoking or nonsmoking group and recorded deaths from lung cancer. But many experiments have been conducted with animals. Rhesus monkeys, dogs, and baboons have been trained to smoke, with inconclusive results to date; whereas rats and Syrian hamsters exposed to air diluted with smoke were more likely than control animals to develop tumors (Dalbey, Nettesheim, Griesemer, Caton, & Guerin, 1980; Dontenwill & Wiebecke, 1966).

A researcher who wishes to analyze the causes of a phenomenon should, if possible, conduct experiments. Because subjects are assigned randomly, the experimental and control groups can be presumed equivalent initially, within tolerable error. Because both groups are treated exactly the same except for application of the independent variable, any post-treatment differences are attributed to the IV. The IV is said to be the cause of the differences. But, here too, there are difficulties with interpretation.

1. Experimenters try to hold constant all variables but the IV. But outside the lab, variables interact in complex ways.
2. What is true for rats and Syrian hamsters forced to inhale cigarette smoke may have little relevance for human voluntary smokers.

Why Choose One Method Over Another?

Given that all methods can under certain circumstances be helpful in answering scientific questions, at least three reasons govern people's choices. First is temperament. Some scientists like running laboratory experiments, some prefer poring over documents in the National Archives, and some like observing wildebeest on the Serengeti Plain.

The second reason is that different resources are required for each method. Secondary analysis may involve no expenses other than transportation to a documents library, whereas a well-equipped biomedical laboratory may contain hundreds of thousands of dollars worth of equipment. A small naturalistic observation study on behavior of students in the college library can be conducted by a single person; but many interviewers and a great deal of money are required for large-scale national surveys.

The third reason for choosing one method rather than another is that, for a given question, some methods promise more definitive answers. But as Webb et al. (1966) emphasized, all methods have assumptions built in and all have weaknesses. Only experiments can conclusively establish causes, but experiments in which laboratory animals are forced to smoke may be irrelevant to the smoking of humans. Our metabolic pathways differ; the effects of voluntary and involuntary smoking may differ; and the effects on animals forced by confinement into inactivity may differ from those on active subjects. (The belief of most scientists that the animal experiments provide powerful evidence, the denial by some that the experiments are relevant to humans, shows that subjective factors are part of scientific decision-making.) Experiments are better than questionnaires for causal analyses. But a researcher who has found that Syrian hamsters that smoke get cancer, may learn more from questioning cancer patients about their smoking habits than from a replication with Mongolian gerbils.

Additional Comments

H. J. Eysenck (1980) vigorously dissented from the view that smoking causes cancer. His major argument was that there have been no experiments with human subjects. His book raises certain other methodological issues that shall be considered briefly:

1. The parallel increases in cigarette smoking and lung cancer during the past 70 years have been used as evidence for a link between them. Eysenck cited Burch's contention (1976) that lung cancer has not increased, but there have been tremendous improvements in its diagnosis. Whether correct or not, this position shows the importance of proper measurement of the dependent variable.
2. Eysenck argued that no conclusions should be drawn from the study on British doctors, "because doctors are not a typical sample of the general population, differing from the rest of the population

in education, socio-economic status, professional knowledge, and in many other ways." But, although scientists should generalize cautiously, their generalizations are what make science worthwhile. It would certainly have been inappropriate to generalize from the incomes of doctors, or their political beliefs, to those of the general population; but as to changes in their lungs due to cigarette smoke, doctors are probably similar to the rest of us.

3. Eysenck cited findings by Tokuhata (1972) that non-smoking relatives of smokers were more likely than other non-smokers to die of lung cancer. This evidence suggests a genetic predisposition to lung cancer, and it provides an additional demonstration of the danger of arguing from correlation to causation.

4. The smoking controversy illustrates the tendency of many scientists to cite and interpret the literature selectively. Tokuhatu's work, cited in the previous paragraph, was also mentioned in the Surgeon General's Report (1982) but with a different conclusion. First, Cederlof et al. (1977) were quoted: "The well-documented evidence of a causal association between smoking and lung cancer found in other subjects has been further supported." Then, referring to Tokuhatu's work: "Similar conclusions were reached in a retrospective study of families of lung cancer patients." (Eysenck also quoted Cederlof et al.: "Even if quantitative assessments presently cannot be made of the relative roles of smoking, other risk factors and genetic disposition, it seems that epidemiological investigations that have not controlled for factors referred to above may have overestimated the role of smoking as the causative factor in coronary heart disease and also, considering for example the distribution of suicides among smokers and non-smokers, that such studies have more than duly made smoking responsible as the causative factor for all excess deaths seen in smokers.") The books by Eysenck and the Surgeon General, though they were published within two years of each other and dealt with the same topic, differed substantially in the references cited and the conclusions drawn. Taken together, they show the danger of reliance on secondary sources for information.

Summary

Scientists should be eclectic in their approaches to research. Rarely can a question be answered definitively by a single approach, and all methods can provide some useful information. The different methods of data collection and strategies for research were listed and defined.

Selected research on the effects of cigarette smoking on lung cancer were discussed. They illustrate the strengths and weaknesses of the various methods and show the value of using a combination of methods.

Key Terms

Case Study   Individual subjects are investigated in detail.

Correlational Study   Scores are obtained from a group of subjects on two different measures; the researcher determines if scores on the measures are related.

Descriptive Study   Individuals or groups are observed so that their naturally occurring behavior patterns can be recorded.

Experiment   A researcher administers a treatment to one group, a different treatment to one or more other groups, and measures to see if the groups have different scores on a dependent variable.

Interview   Either face-to-face or by telephone, a researcher asks subjects a series of questions.

Naturalistic Observation   Behavior is studied in the subject's natural environment, with no manipulation on the part of the researcher.

Nonnaturalistic Observation   The researcher observes after introducing some form of manipulation.

Participant Observation   The researcher interacts with and collects data unobtrusively in the environment of the individual or group being studied.

Questionnaire   A printed form containing a set of questions is distributed to subjects.

Retrospective Study   A study in which people are asked about their pasts.

Secondary Analysis   Existing data are reanalyzed for a different reason than the one for which they were originally gathered.

Survey   A sample of individuals is studied and generalizations made from them to a larger population.

Answers

1. Statement *a* is stronger than statement *b*; *a* implies that smokers are more likely than nonsmokers to get cancer, and that the reason they will get cancer is because they smoke. So, if *a* is true, *b* must be true; but the reverse relationship need not hold. For example, the statement that people with gray hair are more likely to get cancer does not imply that gray hair is the cause. The obvious explanation is that older people are more likely than younger ones to be afflicted, and gray hairs are a sign of age. For the same reason, widows and widowers and people receiving social security benefits are in the "more likely" category. So are close relatives of cancer victims (to some extent, cancer has a genetic basis), and people who own both a television set and a car (because cancer is more prevalent in industrialized countries).

All of the evidence for a smoking/cancer link, with the exception of the experiments, is of the statement *b* variety. Each type of nonexperiment demonstrates that smokers of a given age and sex die more frequently than nonsmokers of lung cancer; but in no case do they show (because nonexperiments *cannot* show) that smoking is the cause.

2. Attorneys for the tobacco companies have emphasized the correlational nature of the case against cigarettes. They sought (and found) evidence that smokers differ statistically from nonsmokers, before they start smoking, in such traits as blood type, personality characteristics and morphology. Even before any studies had been done, attorneys could have persuasively argued for a pre-smoking difference between the groups. To deny such a difference is to claim that smoking behavior is uncaused. And the difference(s), whether or not known, might cause the higher incidence of lung cancer in smokers.

CHAPTER 8

Developing an Ethical Strategy

By the time you finish reading this chapter, you should be able to discuss the following issues:

What types of conflict exist between the interests of researchers and the rights of individuals?

What harms might be inflicted on society-at-large because of deceptive research practices?

Are deceptive research practices ever justified?

Do researchers have the right to harm subjects to advance science?

Should scientists ever suppress research findings?

What are the ethical responsibilities of scientists toward animals?

Terrible things have been done to people in the name of scientific research. The most infamous examples occurred during World War II, when Nazi scientists conducted experiments such as measuring survival times of concentration camp prisoners forced to ingest poisons or sit in ice water. But research conducted in the United States by reputable scientists both before and after that period does not seem strikingly different. For example, starting in the 1930s and continuing for four decades, 399 men with syphilis were led to believe they were being given excellent medical treatment when in fact they were deliberately not treated; penicillin, effective against syphilis, was withheld so scientists could study the course of the disease (Jones, 1981). Additional examples include the following:

- In the 1960s, live cancer cells were injected into elderly patients without their knowledge (Lear, 1966).
- During heart surgery, 11 youngsters aged three-and-a-half months to 18 years had their thymus glands removed only because the surgeon was interested in how the thymus gland affects the immune response (Beecher, 1966).
- Radioactive plutonium was injected directly into the bloodstream of eighteen men, women, and children without their knowledge (Diener & Crandall, 1978).
- In a collaboration between the United States Army and the CIA, the drug LSD was secretly administered to several subjects; one of the victims became severely disturbed and eventually committed suicide (Diener & Crandall, 1978).
- To test the interchangeability of antipsychotic drugs, 40 formerly hospitalized schizophrenic patients were abruptly switched from their doctors' choice medication to equivalent doses of another drug. Twenty control patients were continued with the original choice. The patients had responded well enough to the originals so they could be discharged and live in foster homes. During the study, which continued for two years, many more of the switched rather than the control patients deteriorated and had to be rehospitalized (Gardos, 1974).

Social scientists, unlike biomedical researchers, rarely expose subjects to life-threatening consequences; still, the potential for harm may be considerable. Kelman (1967) gave several examples:

- Small town grocers were told that a large supermarket was about to be built in their neighborhood. The purpose was to study reactions to stress.
- In another naturalistic study of stress, Army recruits were led to believe their plane was about to crash.
- Male undergraduates were falsely informed that they had been homosexually aroused by photographs of men.

NO I CAN'T GIVE YOU EXTRA CREDIT...THAT WOULD BE UNETHICAL!

- Male alcoholics were given a drug that temporarily stops respiration. All said the experience was horrible and they thought they were dying.

Watson and Rayner's (1920) classic study on Little Albert was published 70 years ago and is still reported in introductory books. To study the development of phobias, the authors repeatedly terrified a little boy in the presence of a white rat.

During the 1960s and 1970s, a series of powerful articles by psychologists and physicians exposed and denounced such research (Baumrind, 1964; Baumrind, 1971; Beecher, 1966; Kelman, 1967). Several authors noted that much of social science research produces harms that, though more subtle, are quite real. Since then, bioethics has become a subject for public forums and college courses, and authors have added ethics chapters to revised editions of methodology books. Professional and governmental organizations have published ethical regulations and guidelines such as those of Box 8.1—the guidelines of the American Psychological Association (APA) on research with human subjects. Most universities have human subjects committees to ensure that the guidelines are met, and grant agencies require letters from them.

Key Ethical Issues Facing Scientists Who Use Human Subjects

In this section I discuss four types of conflict between the interests of researchers and the rights of individuals. Readers must resolve the issues according to their own criteria for evaluating morality. Resolution may be easy for those whose lives are governed by a broad ethical principle such as "Do unto others as you would have others do unto you," "The end justifies the means," "Knowledge is the greatest good," or "Seek the greatest good for the greatest number." Alternatively, consideration of the separate issues may help people formulate a broader principle (Beauchamp & Faden, 1982; and Diener & Crandall, 1978).

Box 8.1

Summary of American Psychological Association Guidelines for the Ethical Treatment of Human Subjects, 1982*

The decision to undertake research rests upon a considered judgment by the individual psychologist about how best to contribute to psychological science and human welfare. Having made the decision to conduct research, the psychologist considers alternative directions in which research energies and resources might be invested. On the basis of this consideration, the psychologist carries out the investigation with respect and concern for the dignity and welfare of the people who participate and with cognizance of federal and state regulations and professional standards governing the conduct of research with human participants.

A. In planning a study, the investigator has the responsibility to make a careful evaluation of its ethical acceptability. To the extent that the weighing of scientific and human values suggests a compromise of any principle, the investigator incurs a correspondingly serious obligation to seek ethical advice and to observe stringent safeguards to protect the rights of human participants.

B. Considering whether a participant in a planned study will be a "subject at risk" or a "subject at minimal risk" according to recognized standards, is of primary ethical concern to the investigator.

C. The investigator always retains the responsibility for ensuring ethical practice in research. The investigator is also responsible for the ethical treatment of research participants by collaborators, assistants, students, and employees, all of whom, however, incur similar obligations.

D. Except in minimal-risk research, the investigator establishes a clear and fair agreement with research participants, prior to their participation, that clarifies the obligations and responsibilities of each. The investigator has the obligation to honor all promises and commitments included in that agreement. The investigator informs the participants of all aspects of the research that might reasonably be expected to influence willingness to participate and explains all other aspects of the research about which the participants inquire. Failure to make full disclosure prior to obtaining informed consent requires additional safeguards to protect the welfare and dignity of the research participants. Research with children or with participants who have impairments that would limit understanding and/or communication requires special safeguarding procedures.

E. Methodological requirements of a study may make the use of concealment or deception necessary. Before conducting such a study, the investigator has a special responsibility to (1) determine whether the use of such techniques is justified by the study's prospective scientific, educational, or applied value; (2) determine whether alternative procedures are available that

Is It Ever Appropriate to Harm Anybody for the Advancement of Science?

Many scientists believe they have a moral obligation to pursue knowledge and advance their fields. For example, Denzin (1968) wrote that they should try to minimize harms inflicted upon subjects but should not discontinue research because it might cause harm. He did not specify the amount of permissible harm, which presumably could include any of those described earlier. Risk-benefit analyses have been used to justify potential negative consequences of do not use concealment or deception; and (3) ensure that the participants are provided with sufficient explanation as soon as possible.

F. The investigator respects the individual's freedom to decline to participate in or to withdraw from the research at any time.** The obligation to protect this freedom requires careful thought and consideration when the investigator is in a position of authority or influence over the participant. Such positions of authority include, but are not limited to, situations in which research participation is required as part of employment or in which the participant is a student, client, or employee of the investigator.

G. The investigator protects the participant from physical and mental discomfort, harm, and danger that may arise from research procedures. If risks of such consequences exist, the investigator informs the participant of that fact. Research procedures likely to cause serious or lasting harm to a participant are not used unless the failure to use these procedures might expose the participant to risk of greater harm or unless the research has great potential benefit and fully informed and voluntary consent is obtained from each participant. The participant should be informed of procedures for contacting the investigator within a reasonable time period following participation should stress, potential harm, or related questions or concerns arise.

H. After the data are collected, the investigator provides the participant with information about the nature of the study and attempts to remove any misconceptions that may have arisen. Where scientific or humane values justify delaying or withholding this information, the investigator incurs a special responsibility to monitor the research and to ensure that there are no damaging consequences for the participant.

I. Where research procedures result in undesirable consequences for the individual participant, the investigator has the responsibility to detect and remove or correct these consequences, including long-term effects.

J. Information obtained about a research participant during the course of an investigation is confidential unless otherwise agreed upon in advance. When the possibility exists that others may obtain access to such information, this possibility, together with the plans for protecting confidentiality, is explained to the participant as part of the procedure for obtaining informed consent.

*The above are guidelines, not laws. Researchers do deviate from them.
**If some subjects withdraw from a study, the ones who remain are no longer representative. So, internal validity is compromised. In addition, telling subjects they can withdraw may affect their behaviors; thus, external validity is also affected.

research, but typically neither risks nor benefits can be estimated with any degree of accuracy; and the benefits are usually diffuse, whereas the risks are borne by specific people.

Research-created harms are not restricted to individual subjects. Warwick (1982) identified several more encompassing ones, both real and potential. First, frequent deception research may cause widespread loss of trust in once-respected institutions. In the words of Kelman (1982), ". . . the mere violation of the rules against lying or breaking promises entails a social cost in that it weakens these rules and hence reduces the predictability and trust necessary for the fulfillment of human potentialities within a society." A second harm is that people may become increasingly reluctant to help others. In one field experiment (Piliavin & Piliavan, 1972, see also p. 226), a man deliberately fell down in a subway train and released a small trickle of "blood" from an eyedropper. He repeated the performance about fifty times to test people's reactions to medical emergencies. These and many variants, repeated often enough, may explain incidents such as occurred at the University of Washington following a campus shooting. Students on their way to class did not stop to help the victim or pursue the assailant. When campus reporters asked why, students said they thought it was just a psychology experiment (Gay, 1973). A third harm of some social science research is that it moves us closer toward a society in which people must be constantly vigilant against the possibility that others are covertly watching or listening.

Subgroups within society may be harmed by research. Data collected on radical political organizations may alter public opinion about them or be used to bring legal or extralegal sanctions against their members. Cross-cultural research may undermine the political sovereignty of the countries investigated, and both researchers and research professions may be harmed in a variety of ways.

On the other hand, a complete ban on potentially dangerous research would harm all of society. Tests for possible toxicity of drugs and chemicals would be disallowed. Manned space flights would be discontinued. No accepted treatment would ever be withheld to further test its value, so useless treatments would persist indefinitely. Eisenberg (1977) reported that tonsillectomies and adenoidectomies make up 30 percent of all surgery on children: about 1 million are performed each year in the United States at a cost of about 500 million dollars, plus doctor and nurse time and hospital space. The risk of mortality is small but not zero—and as recently as the 1950s was as high as 200 to 300 deaths per year. The pain and suffering are considerable. Yet, except for a few uncommon conditions, there is no evidence that the surgery is effective. Doctors disagree so much that residents of one part of Vermont are five times as likely as residents of another part to have their tonsils removed by age 20. According to Eisenberg, tonsillectomies cause more deaths per 100,000 each year than the total from all nontherapeutic research.

In chapter 18, I quote extensively from Gilbert, McPeek, and Mosteller (1977). Here is their conclusion.

The basic question involves comparing the ethics of gathering information systematically about our large scale programs with the ethics of haphazardly implementing and changing treatments as so routinely happens in education, welfare, and other areas. Since the latter approach generates little reliable information, it is unlikely to provide lasting benefits. Although they must be closely monitored, like all investigations involving human subjects, we believe randomized controlled field trials can give society valuable information about how to improve its programs. Conducting such investigations is far preferable to the current practice of "fooling around with people," without their informed consent.

Is It Ever Proper to Deceive Anybody in the Interests of Research?

Deception is widespread in social psychology research, with estimates of its occurrence being as high as 44 percent of all studies. (Note from Box 8.1, Principle E, that investigators are expected to debrief subjects—to tell them what the research was about and describe any deceptions—as soon as possible.) Diener and Crandall (1978) summarized the case for deception as follows: "Because of practical, methodological, and moral considerations, much research would be difficult or impossible to carry out without deception. Important knowledge gained in the past would have been forfeited had the practice been totally abandoned."

Then Diener and Crandall gave *methodological* reasons for minimizing deception research. For one thing, research subjects are becoming increasingly suspicious of scientists' descriptions of the purpose of their studies; the suspicions may affect behavior and invalidate results. Secondly, deception limits the range over which variables can be administered. An experimenter interested in the effects of anger might attempt to provoke it by insulting subjects; but since strong, unprovoked insults would arouse suspicion, the experimenter would be likely to make them indirect and subtle. As a result, some subjects might not even notice and others show only slight annoyance.

Both Kelman (1967) and Bok (1978) argued against deception on ethical grounds, calling it degrading; and according to the teachings of Christianity, Judaism, Buddhism, Confucianism, and Hinduism, deception is morally wrong. We demand truth in labeling and condemn deceptions of government officials, advertising executives, and marital partners. Should we allow scientists a more relaxed standard?

Do Researchers Have the Right to Invade the Privacy of Others?

Privacy can be invaded in many ways and expose people to a variety of harms (Kelman, 1982). First, subjects may be persuaded to disclose information voluntarily. If findings are broadcast beyond the research setting, possible consequences include humiliation and legal actions. Second, subjects may be observed secretly, which denies them the opportunity to control how they present themselves to others; control of self-presentation, important in most interactions, is also reduced when subjects are pressured to reveal sensitive personal information or confronted with unexpected or disturbing events. Third, all of us have both a physical and a psychological "private space" that we try to protect; this includes our bodies and certain possessions, biographical facts, and personal thoughts. Some research designs violate this private space.

Must Informed Consent of Subjects be a Prerequisite to Research?

The first commandment in both the Decalogue of Nuremberg and the code of the World Medical Association is that patients must be told that they are not required to accept any treatment for other than therapeutic reasons; they must also be told of any risks that nontherapeutic treatment entails. But zealous researchers have adhered to the letter of the law while ignoring patients' rights.

Gray (1975) interviewed 51 women who had signed a consent form to receive a new labor-inducing drug. Having been told that the drug was new, most assumed it was superior; only 20 understood that it was experimental. Few women realized that they were not required to participate or that they might be exposed to hazards.

More recently, 200 cancer patients who had signed a consent form for chemotherapy were asked questions about it (Cassileth, Zupkis, & Sutton-Smith, 1980). Only 40 percent had read the form carefully, only 60 percent understood the purpose of chemotherapy, only 55 percent were able to list even a single major risk, and only 27 percent could name even one alternative treatment.

Fries and Loftus (1974) pointed out two serious problems with informed consent procedures as they now exist. First, the forms are devised by lawyers to protect institutions rather than inform subjects. Following an analysis of consent forms from five hospitals, Grunder (1980) concluded that "the readability of all five were approximately equivalent to that of material intended for upper division undergraduates or graduate students. Four of the five forms were written at the level of a scientific journal, and the fifth at the level of a specialized academic magazine." Second, information is sometimes harmful: it can suggest symptoms, induce anxiety, and cause panic-related accidents or even serious physiological reactions.

Methods for Reducing Ethical Problems

Before doing research with human subjects, investigators should consider the suggestions below.

1. Make sure all the research you do is necessary.
2. Explore the possibility of using naturalistic observation or archival records to collect needed information. For example, rather than studying helping behavior by having confederates feign injuries, read newspaper accounts of people's actions in crises.
3. Unless there are compelling reasons to do otherwise:
   a. Inform subjects of all procedures and expectations.
   b. Do not deceive them. If deception is essential, consider the possibility of warning subjects that they may be deceived.
   c. Do not invade their privacy.
   d. Do not expose them to any harm.

4. If extreme procedures must be used, make them as weak as possible for testing your hypothesis.

5. Ask friends, colleagues, and potential subjects to review your research proposal and comment on possible ethical problems.

6. Do pilot studies and ask participants for their detailed reactions.

7. Screen potential subjects and eliminate any who might be at special risk. (For certain kinds of studies, a high proportion of volunteers are likely to be at risk. Lasagna and von Felsinger (1954) reported that serious problems of emotional adjustment were found in more volunteers than nonvolunteers for drug studies, and Esecover, Malitz, & Wilkens (1961) found some form of psychiatric disorder in 46 percent of volunteers for studies with hallucinogenic drugs.)

8. Keep subjects' identities confidential; as soon as names can be separated from data, do so.

9. Assess any possible harm you may have caused and remedy it.

10. Have subjects role play. For example, instead of deceiving subjects by having a confederate pretend anger at them, tell the subjects the person is a confederate and ask them to respond as if the anger were real. See Mann and Janis (1968) for an effective application of the method and Cooper (1976) for critical comments.

11. Certain statistical techniques, developed primarily to increase the likelihood that respondents to questionnaires and interviews will give truthful answers, also protect their confidentiality. I'll describe two techniques; Boruch and Cecil (1982) review many others.

    1. Respondents are asked to roll a die or use a similar randomization device out of the interviewer's sight before answering each question. The number showing on the die tells the respondent whether or not to answer the question truthfully. For example, the respondent is told to answer falsely if a one shows and truthfully otherwise. The respondent does not tell the interviewer which number shows. Under these conditions, averaged over many respondents, one-sixth of the answers will be false. Suppose that 40 percent of respondents say they have committed a particular act. To find the proportion who have actually done so:

        a) From the observed proportion of Yes responses (0.40), subtract the probability of a false positive response (one-sixth = .17).

        $.40 - .17 = .23$

b) Add the probability of a false positive response to the probability of a false negative (in this example, also 0.17) and subtract the total from one.

$$1 - (.17 + .17) = .66$$

c) Divide the value obtained in one by the value obtained in two. The result is the best estimate of the number of people who have committed the act.

$$.23 / .66 = .36$$

2. Respondents are given two questions, one sensitive and the other innocuous, each answerable with a number (for example, "How many times did you use cocaine last week?" "How many times did you use the telephone last week?") The respondents are told to add the two numbers and not to show the interviewer the individual answers. A second random sample of respondents is given the same questions and asked to subtract the second number from the first. Suppose that the answers to question one (A1) + the answers to question two (A2) average five; and A1 − A2 average one. To find the best estimate for average cocaine use, do the following:

a) Write two equations.

$$A1 + A2 = 5$$
$$A1 - A2 = 1$$

b) Solve.

$$2A1 = 6, \text{ so } A1 = 3 \text{ and } A2 = 2.$$

Respondents in this example used cocaine on the average three times last week and used the telephone twice.

An Ethical Dilemma

A few years ago, I received a telephone call from a man who asked if I would serve on a three person committee to review a grant proposal for approximately $100,000. I agreed, and the proposal was sent soon afterwards. The applicants were the head of the Department of Medicine and the head of the Department of Toxicology at a large hospital. Attached to the proposal was a document signed by the members of the hospital's ethics committee, indicating their unanimous approval.

The rationale for the study was that, once a drug is approved for human use, doctors can administer whatever dose they choose. In fact, prescribed doses vary considerably. Part of the reason is that patients vary in age, weight, severity of illness, and so forth; but doctors vary as well in their beliefs about which dose works best. Dosage recommendations are given in the literature, but a doctor's personal experiences may be more influential. The proposers wanted to establish the proper baseline dose for every drug used in the hospital (which doctors could then modify according to patients' characteristics) so that eventually all doctors within the hospital would prescribe the same dose. Beyond that, they hoped that doctors and their patients everywhere would benefit from the information. They designed their study in two parts.

Part I. They would review hospital records for the previous 20 years for a select group of drugs. They would correlate administered dosage with ultimate outcome. For each drug, they would find the dose associated with the best outcome.

1. Why wouldn't Part I have been sufficient? Suppose that 70 percent of patients who had been given 10 milligrams of a drug per kilogram of body weight had recovered, but only 50 percent of patients given 100 mg/kg and 40 percent given 5 mg/kg. Why would that not be sufficient proof that 10 mg/kg is better than either of the other doses?

Part II. They would enlist the cooperation of other doctors in the hospital. The doctors would randomly assign their patients for the next several months to get one of the three doses. Patients would not be told they were participating in a study. (If they were told, some assigned to receive either 5 mg or 100 mg might decline to participate, which would undermine the purpose of random assignment.)

2. How would you have voted on the proposal? Justify your answer.

Should Ethical Concerns Ever Lead to the Suppression of Research?

In 1951, a short story by Fredric Brown appeared in *Astounding Science Fiction*. "The Weapon" opened with top scientist James Graham sitting at home thinking about his retarded son—his only son, whom he loved dearly—playing in the next room. The doorbell rang and, answering it, Graham faced a stranger who asked to speak with him for a moment. Graham graciously let him in, and they chatted. Then the stranger said, "Dr. Graham, you are the man whose scientific work is more likely than that of any other man to end the human race's chance for survival." He begged Graham to stop working on his current project.

Graham realized the man was a crackpot and regretted having let him in. Still, he answered politely. "I'm a scientist and only a scientist. Yes, it is public knowledge that I am working on a weapon, a rather ultimate one. But, for me personally, that is only a by-product of the fact that I am advancing science. I have thought it through, and I have found that that is my only concern."

If the stranger was disappointed he didn't show it, and he accepted a drink. Finally, readying to leave, he said he'd brought a small gift and handed it to Graham's son while the drinks had been poured. After saying goodbye, Graham went into his son's room and saw him playing with the gift. That caused him to sweat, but he kept his voice calm and asked his son to hand it over. The story closed with Graham's thought: "Only a madman would give a loaded revolver to an idiot."

Some future behavioral scientist may have to make a decision like the one Graham faced. She may develop a highly addictive and debilitating drug or one that produces long-lasting homicidal tendencies. She may perfect a technique that enables psychotherapists to dramatically restructure the personalities of their patients, and which can also be used to restructure anybody else's personality. He may learn how to craft irresistible persuasive messages.

Studies of racial differences on intelligence test scores exemplify actual research that may fall in this category. The research (which has been severely criticized on methodological grounds, cf. Mackenzie, 1984), has probably painfully shaken many people's self-concept and had adverse effects on race relations.

Having raised the problem, I have no satisfactory answer to the question of whether scientists should ever suppress research. Decisions to pursue specific lines of research and to publish or suppress must be left to individual scientists. But I would have been tempted to sabotage Graham's laboratory.

Ethics of Research with Animals

Each year, in the United States alone, an estimated 70 to 90 million animals are used as research subjects (Regan, 1986). Scientists engaged in such research must treat them humanely. Box 8.2 presents the APA guidelines for care and use of animals. Box 8.3 reprints both an editorial in which I tried to evaluate the important ethical issues confronted by animal researchers and a rebuttal to the editorial.

For additional criticisms and justifications of experimentation with animals, see Coile & Miller (1984), King (1984), Pratt (1980), Regan (1986), Rowan (1984), Reines (1982), Orlans (1977), and Fox (1986). The latter two authors described humane methods for housing and handling animals and offered various alternatives to popular techniques that cause needless pain.

American Psychological Association Guidelines for Care and Use of Animals

An investigator of animal behavior strives to advance understanding of basic behavioral principles and/or contribute to the improvement of human health and welfare. In seeking these ends, the investigator ensures the welfare of animals and treats them humanely. Laws and regulations notwithstanding, an animal's immediate protection depends upon the scientist's own conscience.

a. The acquisition, care, use, and disposal of all animals are in compliance with current federal, state or provincial, and local laws and regulations.

b. A psychologist trained in research methods and experienced in the care of laboratory animals closely supervises all procedures involving animals and is responsible for ensuring appropriate consideration of their comfort, health, and humane treatment.

c. Psychologists ensure that all individuals using animals under their supervision have received explicit instruction in experimental methods and in the care, maintenance, and handling of the species being used. Responsibilities and activities of individuals participating in a research project are consistent with their respective competencies.

d. Psychologists make every effort to minimize discomfort, illness, and pain of animals. A procedure subjecting animals to pain, stress, or privation is used only when an alternative procedure is unavailable and the goal is justified by its prospective scientific, educational, or applied value. Surgical procedures are performed under appropriate anesthesia; techniques to avoid infection and minimize pain are followed during and after surgery.

e. When it is appropriate that the animal's life be terminated, it is done rapidly and painlessly.

WELL, HE'S CHOSEN MONDAY NIGHT FOOTBALL TEN TIMES IN A ROW... I'LL BET YOU A CHEESE PUFF THAT HE'LL CHOOSE THE FEMALE THIS TIME!

BOX 8.3

Valid Experiments or Animal Abuse? Oakland Tribune Editorial, March 29, 1988

Animal rights activists do a great service when they call attention to abuses of animals at the University of California and other college campuses. But they would probably gain much more support, both outside and within the scientific community, if they discriminated between the different purposes for which animals are used. Below, I describe and comment on four types of purpose:

1. Animals are used to provide vivid demonstrations of phenomena. For example, in some medical schools, the intestines of dogs are sewed up so the dogs can't defecate, and they inevitably die. Students probably remember better when they watch such agonizing deaths than if they merely read from textbooks, but that is insufficient reason to kill or torture animals; and there is certainly no reason to do repeated, live demonstrations when a single case on videotape would suffice.

2. Students gain hands-on experience from performing surgical and other techniques on animals. Instructors should consider whether the tasks are essential for the demands of their courses and needs of their students. Psychology students at Cal State, Hayward are required to take a lab course in experimental psychology. Until a few years ago, many instructors—I was one of them—assigned each student a rat to work with. This was done even though few of the students were interested in careers in animal research. Once the quarter was over, there was no further use for the rats. We couldn't keep feeding and housing them, and we certainly couldn't release them, so they were killed. To prevent that from happening, many students took the rats home as pets. I now believe, partly because the animal rights people sensitized me to the problem, that such indiscriminate use of animals is wrong. Alternative methods are available for teaching principles of experimentation, and instructors should be made aware of and encouraged to use them. But activists go too far when they claim, as some have done, that alternatives to animal laboratories exist for all important teaching objectives. Given a choice, how many of us would elect to undergo surgery with a person who had never operated on any living organism?

3. Animals are subjects in experiments designed to answer practical problems. Opponents of such research claim (correctly) that the solutions are often erroneous or not forthcoming. Despite the enormous sums spent on cancer research, scientists have not found a cure; and many drugs have proven hazardous to humans after passing safety tests with animals. But people who focus on the failures must do so for rhetorical reasons rather than in the interest of truth—animal research has helped medical science. Neurosurgeon Robert White wrote that virtually no major treatment or surgical procedure in modern medicine could have been developed without animal research. Animal models have helped scientists conquer polio, scarlet fever, smallpox, diphtheria, and many others, and have helped animals as well. Family pets benefit from vaccines for rabies, distemper, anthrax, and feline leukemia. The use of animal models by the pharmaceutical industry increases the likelihood that new drugs will be safe and effective. I, for one, am glad that new medications are tested in animals before being released for human use. Although animal models are imperfect, they are essential to the work of many scientists; and to maximize our chances of finding

satisfactory solutions to our problems, we must allow scientists to choose their methods of inquiry.

It is easy to criticize animal research in the abstract, but people feel differently when the results may impinge on their lives. For example, most respondents on a British survey conducted several years ago argued strongly against the testing of cosmetics on animals. Yet in a second survey of a similar sample just a few years later, 100 percent answered "No" to the following question: "Would you use a shampoo if it had not been safety tested on animals?"

4. Animals are used in basic research. Basic research often seems bizarre to laypeople and even to scientists outside a particular field of expertise; but major scientific advances, often with far-ranging practical applications, are a frequent result. Darwin measured thousands of beaks of birds on his way to formulating the theory of evolution; Gregor Mendel painstakingly cross-pollinated pea plants to learn how characteristics are inherited; Pasteur inoculated chickens with microbe cultures and Einstein did thought experiments in which he leaped from one train travelling at the speed of light to another. By contrast, practical researchers stick close to what is known, to the commonplace, so they are less likely to make profound discoveries.

Virtually all scientists who conduct animal research require grant support. Competition is keen, and recipients must convince a jury of experts that their proposals are meritorious. It is presumptuous of nonexperts to condemn the subsequent research on scientific grounds. On the other hand, receipt of grant money does not release scientists from the responsibility of treating animals as humanely as possible.

Scientists try to create a coherent account of the world and our place in it. Their discoveries often have esthetic appeal and help mold our religions, philosophies, literature, and art. These additional benefits of basic research place an added burden on animal experimenters—they must ensure that their research is not perceived as sanctioning gratuitous cruelty.

I have argued for certain uses of animals and against others. But even if some experiments are justified, animals should never be subjected to harmful or painful conditions beyond the requirements of the experiment. The animal rights activists should remain vigilant against people who disregard this responsibility. If the activists limited themselves to uncovering such people and stopping their abuses, they would probably receive the support of most of the scientific community.

In response (printed on April 8, 1988):

Fred Leavitt asserts in his article, "Valid Experiments or Animal Abuse?" that animals should be subjected to experimentation both because such research has "practical" applications and because it has non-practical ("basic research") implications. That, however, begs the question and ignores ethical issues and worthwhile alternatives.

Leavitt effectively gives his blessing to the torture and mutilation of many rabbits so that his unnatural, chemicalized shampoo, for example, is "safety tested." Does he understand the horrors of the standard toxicity tests (the LD50 and the Draize tests) and their known, scientific unreliability? Does he even care?

Leavitt cites with approval the sewing up of dogs' intestines at some medical schools so that the dogs, unable to defecate, die in agony, thereby "enlightening" the students. He bemoans only that this practice is repeated live "when a single case on videotape would suffice." It seems to escape him entirely that such an act, done even once, is demented and perverted. Unfortunately, it reflects only too well our society's general attitude toward animals.—*Dennis Walton*

Notes: For the Draize test, immobilized but conscious rabbits have test substances (cosmetics) dropped into one eye with the other eye serving as control. This may cause blindness and pain.

LD50 stands for the dose of drug that is lethal to 50 percent of test animals. The pharmaceutical industry, which routinely determines the LD50 of new drugs, kills huge numbers of animals for meager information.

Summary

Scientific research confers many benefits including the prevention or treatment of previously debilitating and fatal diseases. But there are costs as well, so researchers who use human subjects should confront four types of ethical questions:

1. Is it ever proper to harm subjects? Several types of potential harms were noted:
    a. to individual subjects
    b. to society as a whole, because of loss of trust in once-respected institutions
    c. to society as a whole, because people become less willing to help others
    d. to various subgroups, because public opinion about them changes or legal or extralegal sanctions are brought against their members
    e. to entire countries, by undermining their political sovereignty
2. Is is ever proper to deceive subjects?
3. Do researchers ever have the right to invade the privacy of others?
4. Must researchers receive the informed consent of their subjects?

The American Psychological Association guidelines for conducting research with human subjects were presented and strategies given for minimizing the potential harms to research subjects.

The use of animals enables researchers to exert considerable control over important variables and introduces a new set of ethical concerns.

Answers

1. Remember, correlation does not imply causation. Patients might have been assigned different doses nonrandomly. Many drugs have powerful side effects, and patients too feeble to tolerate them might be given low doses (5 mg); whereas extremely sick patients might, in desperation, be given heroic doses (100 mg). The ultimate outcomes of therapy might have less to do with drug doses than with initial conditions of patients.

    Doctors as well as patients differ. Some doctors might routinely prescribe 10 mg, others 100. Suppose the scientific literature indicates that the most effective dose of a drug is probably 10 mg. Doctors who read the literature, i.e., keep up with important developments in their fields, are most likely to prescribe 10 mg. They are also most likely to be effective doctors, with high cure rates for their patients, even if the drug turns out to be useless.

2. I voted against the proposal, solely on ethical grounds. From a scientific standpoint, it had considerable merit. I was outvoted two to one. A year later, I received a copy of the results to that point and was asked to vote again on whether the study should be extended for a second year. The researchers had learned a great deal of useful information. For one of the drugs, evidence strongly confirmed previous beliefs about the optimal dosage: survival rate was much higher among patients randomly assigned to receive that dose than any other. Again I voted against the study and again I was outvoted. That ended my involvement.

   Although I would vote the same way today on any similar proposal, I wish to add two points that weaken my position considerably. First, nobody was sure what the best dose was or there would have been no need to do the study. Gilbert, Light, and Mosteller (1975) cited evidence from nonmedical fields that educated guesses about which of two treatments is superior are wrong in about 50 percent of cases.

   Second, the number of lives lost in such studies is extremely small compared with their long-range potential for saving lives. Suppose that survival rates for a particular drug, as established in Part I of the study, had been 70 percent, 50 percent, and 40 percent for 10, 100, and 5 mg doses, respectively. Suppose further that these correlational data reflected the true effectiveness of the drug and that 300 patients were subsequently given the 10 mg dose. We would expect 210 to survive. If instead the patients were randomly assigned, 100 to each dosage group, the expected number of survivors would be $(100 \times .7) + (100 \times .5) + (100 \times .4) = 160$. Fifty patients would die who could have been saved. Suppose further in this hypothetical but realistic case, that the disease for which the drug is prescribed struck 10,000 people worldwide each year, and 10 percent of doctors normally prescribed each of the suboptimal doses. Then, each year, instead of a survival rate of $10,000 \times .7 = 7,000$, the actual survival rate would be $(8,000 \times .7) + (1,000 \times .5) + (1,000 \times .4) = 6,500$.

Every year—forever, because without proper experimentation there would be no way to learn the proper dose—500 extra people would die.

CHAPTER 9

A Few Suggestions for Creative Problem Solving

By the time you finish reading this chapter, you should be able to answer the following questions:

What are some strategies for helping people achieve new perspectives on their problems?

How can people test key assumptions and perhaps unnecessary boundaries of their problems?

Should ideas for solutions be evaluated as they arise, or should evaluations be deferred until the end of the idea generation stage?

How can problem solvers use analogies?

Are any strategies available for helping select and implement ideas?

Does relaxation facilitate problem solving?

Can people improve their creative efforts by deliberately modifying their mental states?

"There was far more imagination in the head of Archimedes than in Homer."

Voltaire, French author (1694–1778)

Many recent writers have argued that scientific discovery can be classified under the general heading of problem solving. Langley, Simon, Bradshaw, and Zytkow (1987) used what is known about the heuristics (rules of thumb that guide solutions) of human problem solving to develop several computer programs. The programs successfully formulated both quantitative and qualitative laws and theories from inputted data, which suggests that scientists would profit if they learned the heuristics (Colgrove, 1968; DeBono, 1970; DeBono, 1972; Gordon, 1961; Levine, 1988; Maier, 1970; Osborne, 1963; Polya, 1957). VanGundy (1988) comprehensively reviewed and evaluated 105 different problem-solving techniques, many for use only in groups. I selected several for individual use that he specially recommended.

VanGundy classified techniques according to their applicability to four problem-solving stages: redefining and analyzing problems; generating ideas; evaluating and selecting ideas; and implementing ideas. I emphasize the first two stages because the techniques for those stages seem most useful for the problem-solving activities of scientists. (Evaluative techniques are discussed in chapter 4.)

Redefining and Analyzing Problems

The goal at this stage is to achieve a new perspective on the problem.

Boundary Examinations

Confronted with a problem, people typically assume that only certain strategies are permissible in trying to solve it and that acceptable solutions will be of a certain form. The objective of Boundary Examinations is to restructure these assumptions. The steps are as follows:

1. Write down an initial statement of the problem.
2. Examine key words and phrases for any hidden assumptions.
3. Without worrying about the validity of the assumptions, write down any important implications they suggest.
4. Write down any new problem definitions suggested by the implications.

VanGundy gave as an illustration the following problem of a pharmaceutical company for which the initial statement is "How can our company encourage customers to buy our medicine?" From the key word "company," the boundary is examined that the company must be the one to promote the product. Perhaps the government, consumer groups, or doctors could be persuaded to do so. Alternatives to "encourage," are "reward" and "punish." Instead of "buy," the problem-solver might consider "use."

Five W's and H

The Five W's and H technique provides a framework for systematically gathering relevant facts, which in turn stimulates problem redefinitions. The steps are as follows:

1. State the problem using the format, "In what ways might . . . ?"
2. Write separate lists of Who? What? Where? When? Why? and How? questions about the problem. Answer the questions while withholding all judgments.
3. Use the answers as stimuli to redefine the problem.
4. Write down redefinitions.
5. Select the redefinition that seems most useful for solving the original problem.

VanGundy gave a lengthy illustration for the problem "How can we motivate our employees to work harder?" Below are the questions he generated and the answers that led to one redefinition.

Who
    are the employees?
    doesn't work hard?
What
    is motivation?
    does "work harder" mean? (Accomplishing more with limited resources; being more efficient while maintaining quality.)
    motivates most employees?
Where
    are employees motivated to work hard?
    is employee motivation not a problem?
When
    are employees not motivated to work hard?
    do bosses try to motivate employees?
Why
    motivate employees to work harder?
How
    can employees be motivated?

The redefinition was, "In what ways can we better provide resources employees need to do their jobs?"

Considering Related Problems

Looking at related problems may give ideas for solving the current one. The related problem may be easier to solve for any of several reasons. It may already have been solved. It may have an additional element. It may be an extreme case or a simplified case. The problem solver can deliberately add new elements or analyze extreme or simplified cases. Try to solve the following from Levine (1988):

Two flagpoles are standing, each 100 feet tall. A 150-foot rope is strung from the top of one of the flagpoles to the top of the other and hangs freely between them. The lowest point of the rope is 25 feet above the ground. How far apart are the two flagpoles?

The problem seems to call for complex mathematical analysis. But it's simple if the extremes are explored. Considering one extreme, when the poles are as far apart as they can be (150 feet), does not help with the solution. Considering the other, when the poles touch, provides the answer: the rope is folded in half 75 feet on a side, and its low point is 25 feet from the ground.

Working Backwards

Visualize the solution and ask what the immediately preceding conditions must have been. Keep going to the beginning. Here is an example from Polya (1957):

How can you bring up from the river exactly six quarts of water when you have only two containers, a four-quart pail and a nine-quart pail, to measure with? Working forwards, that is, filling and emptying the two containers to see what will happen, is an inefficient way to solve the problem. It's much better to work backwards. Ask, "What must the next-to-last step be?" The way to get six quarts is to pour three quarts from the filled nine quart pail. It is possible to pour exactly three quarts only if the four quart pail already has one quart. This can be accomplished by filling the nine quart, and pouring from it twice to fill the four quart, emptying the four quart each time. Then pour the remaining quart from the nine into the four.

Generating Ideas

Science requires that methods and theories be constantly evaluated, but the discovery phase should be uncontaminated by concerns about evaluation. To facilitate creative problem solving and discovery, try brainstorming. Set aside an hour or two to write down as many ideas as you can. Don't shy away from bizarre ones, and don't censor them in any way. At some later time, go over the ideas carefully and reject the bad ones. But keep the two stages separate: production of ideas precedes and should not be hampered by evaluation.

Try this variant of brainstorming described by VanGundy: Pick somebody, real or fictional, and pretend that the person has a stake in your problem. Then generate ideas as you imagine that person would.

Seek Analogies

The value of analogies to scientists was discussed on page 29. Campbell (1974) used an analogy to explain how scientists come to hold various theories. He wrote that theories evolve just as do living organisms. In his view, the creation of ideas is largely random, and then the good ideas are retained by a selection process analogous to survival of the fittest.

I adapted the following steps for using analogies from VanGundy.

1. Identify the major principle underlying the problem. For example, if the problem is preventing school vandalism, the underlying principle is prevention. If the problem is improving a household flashlight, the principle is improvement.
2. List as many objects, people, situations, or actions that have a similar principle. For prevention, list things like birth control, spraying a car to prevent rust, etc. For improvement, VanGundy listed the following: a facelift, sessions with a psychiatrist, a heart transplant, piano lessons, attending college, a professional development conference, exercising, skin grafts, gene splicing, dog obedience classes, and learning brain surgery through a correspondence course.

    Analogies involving inanimate objects generally work best for problems involving living things and animate objects for problems with inanimate things. And playful analogies are perfectly acceptable.
3. Describe one of the analogies in detail, listing all parts, functions, or uses. While completing this step, try to forget about the problem. VanGundy, with the goal of improving flashlights, followed up on the correspondence school brain surgery analogy. One elaboration included the idea that frequent contact with instructors would be necessary.
4. Write down ideas suggested by each description, then repeat step 3 with another analogy. From the brain surgery elaboration came the idea of installing small radio transmitters and receivers for emergencies.

Cliches, Proverbs, and Maxims

1. Pick an interesting cliche, proverb, or maxim that appears unrelated to the problem. The following six are particularly good because they are familiar and likely to arouse visual imagery:

    Like father, like son.
    Don't count your chickens before they're hatched.
    You can't tell a book by its cover.
    A penny saved is a penny earned.
    You can't teach an old dog new tricks.
    Two heads are better than one.

2. Write down any implications and interpretations suggested. For example, for "Too many cooks spoil the broth," one of the implications given by VanGundy was "If physical facilities limit the number of people who can work on a task at one time, they should take turns."

3. Use the implications as stimuli for new ideas. For the problem, "How might we encourage drug-dependent employees to seek treatment?" and the "Too many cooks . . ." cliche, came the idea to have drug-dependent employees take turns role-playing a counselor.

Draw

Using large sheets of paper, represent the problem visually and draw other ideas as they come. Don't worry about the quality of either the ideas or the drawings. Just draw. Researchers generally agree that visualization, whether relaxed with eyes closed or while staring at diagrams, facilitates problem solving. In addition, making the elements of a problem accessible visually frees the mind from the distraction of trying to remember them.

Evaluating and Selecting Ideas

Highlighting

1. Review a list of ideas without concerning yourself with their workability, and mark any that look interesting.
2. Look over your marks and identify those that seem related.
3. Examine each cluster of related ideas and identify what it represents to you in terms of meanings, implications, and possible consequences.
4. Pick the one cluster that best satisfies your needs for the problem. If necessary, combine two or more clusters.

Implementing Ideas

Potential Problem Analysis

1. List each event that must occur for the project to succeed.
2. Generate a list of potential problems, that is, anything that could possibly go wrong to prevent any critical event from occurring.
3. Identify the nature of each potential problem: Ask "What?" "Where?" "When?" and "To what extent?"
4. Estimate the likelihood of occurrence of each problem and how damaging it would be. Focus on the most likely and damaging.
5. Search for possible causes of each problem, and estimate their probability of occurrence if no action is taken.
6. Develop means for preventing the causes or minimizing their effects.
7. Develop contingency plans for the most serious problems. Specify the actions to be taken if the problem occurs—but don't use contingency plans as substitutes for preventive action.

Below are some additional thoughts on problem solving in science.

Work

Novelist Hideki Yukawa said (1973):

The important thing if one is to achieve creativity, I feel, is to keep plugging away at one thing despite all the miscellaneous tasks and the superfluity of information with which everyday life seeks to claim one's attention. What is needed, in other words, is tenacity of purpose.

Many renowned scientists have been exceedingly tenacious. Copernicus was a student when he conceived the idea that the sun is the center of all planetary motions. He spent the rest of his life elaborating it into a system. Darwin had the idea of natural selection when he was 29, and he spent the remaining 44 years of his life developing it. Pasteur made several discoveries that were printed as short notes and followed by 10 or 15 years of elaboration, consolidation, and clarification. Einstein said, "When I asked myself how it happened that I in particular discovered the Relativity Theory, it seemed to lie in the following circumstance. The normal adult never bothers his head about space-time problems. Everything there is to be thought about, in his opinion, has already been done in early childhood. I, on the contrary, developed so slowly that I only began to wonder about space and time when I was already grown up. In consequence I probed deeper into the problem than an ordinary child would have done."

Try to fill in the blank in this final quote, from physicist Lord Kelvin.

1. "One word characterizes the most strenuous of the efforts for the advancement of science that I have made perseveringly during 55 years; that word is _____."

The "publish or perish" system that operates in so many academic institutions encourages rapidly done studies that have a high probability of quick payoff. Kelvin, Copernicus, Darwin, and Pasteur might have had trouble functioning under such circumstances; or they might have adapted and devoted their careers to potboilers.

Relax

Historians of science have noted that answers to problems often come when scientists are not actively thinking about them. They fill their minds with facts, work intensively for awhile, and then relax and think of other things. Platt & Baker (1931) surveyed by questionnaire 1,450 outstanding research chemists, people whose work accounted for major advances of the chemical industry during the previous 25 years. Practically all replied that their significant new ideas and insights rarely came while they were in their laboratories or poring over data. The great discoveries occurred during moments of relaxation that had been preceded by intensive involvement. They came at odd moments in strange places, as when August Kekule dreamed about the structure of the benzene ring—an event that marked the beginning of modern organic chemistry. He advised fledgling scientists to learn to dream.

As counterpoint to Kekule, and to show that science has room for people of varied temperaments, here is a story about Thomas Edison. He was seeking a solvent for hard rubber; other scientists had been unsuccessful finding one through theory and experiment. Edison, no dreamer he, immersed a piece of hard rubber in a vial of each of the many chemicals in his huge storeroom. He found his solvent.

Pretend

Colgrove (1968) devised a simple method for increasing creativity. She gave subjects the following instructions for role playing: "You have the reputation of being a very original person and of being good at coming up with answers to difficult problems. Keep this in mind while studying the problem." It worked—they developed superior solutions while role playing.

Be Happy

Isen, Daubman, & Newicki (1987) did a series of studies in which they modified students' moods and then had the students perform tasks generally regarded as requiring creativity. Mood was modified by having subjects watch comedy, neutral, or negative films, and by giving them small gifts. Creativity was measured by a psychological test (the Remote Associates Test) and a problem-solving task. The main finding was that positive manipulations improved performance.

Don't Be Smug

Maier (1970) noted that problems often resist correct solutions because an alternative but incorrect solution is readily available. The problem solver arrives at the solution and is satisfied with it, so stops looking further. If told that her solution is incorrect, she is likely to keep searching and eventually find the correct one. Similarly, scientists who stop at the first ideas for research that occur to them, may choose less important projects than if they build on or reject those first ideas. To illustrate, Maier gave this problem.

2. Trains leave Detroit for Chicago every hour, and they leave Chicago for Detroit every hour. The trip takes six hours. If you boarded a train in Detroit, how many trains coming from Chicago would you see along the way?

Summary

Scientific discovery is a type of problem solving, so several problem solving strategies are presented. Some help researchers consider problems from new perspectives. Others are designed to help with the generation, evaluation, selection, and implementation of ideas.

The generation of ideas and evaluation phases should be kept separate.

Creative problem solving is facilitated by a confident attitude and positive mood.

Answers

1. Kelvin's word was "failure."
2. You would see 12 or 13 trains. Suppose you leave from Detroit at 11:05 A.M. You would arrive in Chicago at 5:05 P.M. having seen the six or seven trains that left there between 11:05 A.M. and 5:05 P.M. You would also see the six or seven trains that left between 5:05 A.M. and 11:05 A.M.

PART IV

The Different Strategies

CHAPTER 10

Observing

By the time you finish reading this chapter, you should be able to answer the following questions:

What are the different reasons for doing observational studies?

Can naturalistic observations be used to test hypotheses?

What are the three basic forms of observational data?

How should investigators pick which behaviors to observe?

Can observational skills be taught?

Why should medical students always observe their instructors closely?

> "Most of the knowledge and much of the genius of the research worker lie behind his selection of what is worth observing. It is a crucial choice, often determining the success or failure of months of work, often differentiating the brilliant discoverer from the plodder."
>
> *Alan Gregg*
> *Director of Medical Sciences,*
> *Rockefeller Foundation*

> "You can observe a lot just by watching."
>
> *Yogi Berra, baseball player*

The ability to make careful, systematic observations has distinguished many eminent scientists. Of Charles Darwin, his son said, "He wished to learn as much as possible from an experiment so he did not confine himself to observing the single point to which the experiment was directed, and his power of seeing a number of things was wonderful." Freud took copious notes on his patients, and over a span of several years Piaget meticulously observed and recorded the behaviors of his children. Physiologist Claude Bernard wrote that "experimental reasoning always and necessarily deals with two facts at a time: observation, used as a starting point; experiment, used as conclusion or control."

Bernard observed primarily under controlled laboratory conditions, whereas Darwin observed things and events as they occurred naturally. Yet many psychologists behave as though observational studies, especially naturalistic ones, are methodologically unsound or uninformative; in any event, such studies are neglected. For example, of 1,409 child and adolescent studies published between 1890 and 1958, only 110 involved direct observation of naturally occurring things and events (Wright, 1960). That's unfortunate, as systematic observations serve several functions.

Reasons for Doing Observational Studies

To Describe Nature

Researchers in the well-developed sciences find descriptive studies invaluable: astronomers use star charts, chemists refer to handbooks on the properties of chemicals, biologists classify flora and fauna, geologists consult tables on the hardness of minerals, meteorologists describe cloud formations, and so forth. But few psychologists pay sufficient attention to behaviors as they naturally occur. Barker and Wright (1955) were among the pioneers in this area.

Several recent publications have instructed on classroom observing, and many comparative psychologists observe the naturally occurring behaviors of animals. Because variables are uncontrolled in naturalistic observations, such studies are of limited use for testing causal hypotheses. But they should not be devalued on that account, as the description of nature is a legitimate end. And, as indicated in the next section, they can provide some evidence to support or discredit hypotheses.

To Generate and Test Hypotheses

Good scientists don't record events passively; they select, organize, and try to account for their observations. Observations lead to hypotheses, which are then tested by means of additional observations and experiments.

That's what Darwin, Freud, and Piaget did and what modern epidemiologists do. Epidemiologists study incidences of diseases around the world and among people with different lifestyles. They search out connections between particular lifestyles and the likelihood of contracting a disease. Their observations, not experiments, suggested links between cigarette smoking and lung cancer, cholesterol level and heart disease, obesity and high blood pressure, and intravenous drug use and AIDS. Experiments were crucial for testing the suspected causes against possible alternatives. But the observations came first, and without them there would have been no reason for the specific experiments.

Moreover, some research questions and some hypotheses can be tested directly with observational data. For example, Block (1976) cited several observational studies in support of the hypothesis that boys are more active than girls.

To Assess Behavior

Assessment of behavior through direct observation is essential to progress in psychotherapy, since changes in behavior can't be measured or evaluated if the starting point is unknown. And since deviance is meaningful only by contrast with normality, normal behavior should be studied. Some of Tinbergen's research exemplifies this point (1972). He found that certain characteristic behaviors of autistic children also occur in normal children when they are afraid; this was the starting point for a productive research program on autism.

Some rats, mice, and fruit flies have been bred in laboratories for many generations; but for other species, including humans, the laboratory is not a natural environment. Therefore, explanations of laboratory behaviors may miss the mark unless compared with what occurs naturally. Consider this story cited by Willems (1973). An ornithologist with a zoo, wishing to add small, rare birds called bearded tits to the zoo's collection, designed an environment that closely mimicked the tits' habitat. He put a male and female inside, and they appeared happy. They mated, built a nest, laid eggs, and hatched and fed babies. But shortly after their birth, the babies were found dead on the ground. The ornithologist assumed there had been an accident or illness, as the parents continued to eat, drink, mate, and otherwise seem happy. In fact, they soon reproduced again. But again, the babies were found dead soon afterwards. The cycle repeated itself several times, until one day he observed the parents push the babies out of the nest and onto the ground, where they died. He realized that something was seriously wrong with the environment he'd created, so he went out to observe bearded tits in the wild.

After many hours of observation, he noted three clear patterns of behavior. First, parents spent much of the day finding and bringing food to their nestlings. Second, the infants cried for food incessantly throughout the daylight hours. Third, parents meticulously cleaned their nests, shoving out inanimate objects such as leaves and beetle shells.

Returning to his captive birds, he recognized the problem. The parents had been supplied so abundantly that they were able to feed their infants to satiety; so the infants fell asleep and were treated like other inanimate objects by being shoved out of the nest. The ornithologist was able to remedy the situation by making food harder to get, and the birds then produced many families.

Even within laboratory settings, observation should be an important part of assessment. Domer (1971) argued that direct observation should be routinely used when testing the effects of drugs on laboratory animals; researchers who focus exclusively on a few preselected dependent variables may miss important behavioral changes.

Because Other Methods Are Impractical

Miller (1977) noted that some research questions are impossible to answer by laboratory experiments. These include questions about the social organization of animals, the reactions of animals to naturally occurring stimuli, and many human social phenomena. Laboratory experiments to study other questions, such as the effects of loss of a parent, would obviously be unethical. For these, systematic controlled observation may be the only route to knowledge.

Becoming a Good Observer

As Sherlock Holmes fans can testify, good observing is difficult. Dr. Watson had access to the same data as Holmes but with inferior results. Still, even if few people can develop powers equal to those of Holmes, substantial improvement is possible.

The first step is to make explicit the reason for observing. Darwin wrote, "How odd it is that anyone should not see that all observation must be for or against some view if it is to be of any service." How could it be otherwise? Imagine being asked, with no further instructions, to record your observations of a group of young men and women at a party. Wouldn't the results depend on whether you are man or woman, single or married, minister, fashion designer, college recruiter, or salesperson? Wouldn't different strategies be required for the questions "Are there any eligible men present?" and "Who is the most likely prospect to buy a motorcycle?" So, before doing any actual observing, clarify the research question.

Decide on Type of Observation

There are three basic forms of observational data:

- Narratives. Narratives—running accounts of behavior—require the least preparation. With tape recorder or journal in hand, the observer tries to record all interesting and relevant occurrences. However, this unstructured approach makes narrative data unreliable. Different observers may focus on different aspects of a situation or give different interpretations to the same aspect. Although the information may be useful during the early stages of a research project, more reliable techniques should be employed as soon as possible.

BOX 10.1

Illusory Correlations

Chapman and Chapman (1967, 1969) asked subjects to sit in a chair and watch pairs of words projected to them on a screen. Four words appeared randomly on the left side and were paired with three words that appeared randomly on the right. Each word on the left (bacon, lion, blossoms, boat) appeared equally often, as did each word on the right (eggs, tiger, notebook). But when the subjects were asked, "When bacon was on the left, what percentage of the time were eggs, tiger, and notebook on the right?" the average of their estimates for eggs was 47 percent. A similar illusory correlation held for estimates of lion-tiger pairings.

The Chapmans studied the phenomenon with the Draw-a-Person test (DAP), at one time widely used as a diagnostic aid by clinical psychologists. College students with no prior DAP experience looked at a series of 45 arbitrarily labelled drawings, for example, the man who drew this is homosexual; the woman who drew this is paranoid. There was no objective relationship between any features of the drawings and any diagnostic category. But relationships were seen anyway. They were of the same type that clinicians claim they see, for example, that people who worry about their intelligence draw figures with big heads. Even when the students looked at the pictures for three consecutive days, were given rulers to measure the drawings, were told what questions they would be asked about the pictures, and were offered a reward for accuracy, some illusory correlations remained.

- Ratings. Ratings require observers to evaluate what they see. For observation of complex behaviors, such as human social interactions, a rating scale may be the most effective procedure. But because judgments are required, reliability is often low. Some raters give consistently high evaluations, some consistently low, and some consistently in the middle range. Also, some raters tend to see others as like themselves, some tend to see others as different; raters who score a person high on one trait tend to score her high on others. More generally, certain traits are rated as going together even when the data do not show any association between them. See Box 10.1 for research on these illusory correlations.
- Checklists. Checklists are sheets of paper that list all the behaviors likely to be observed and have space for indicating the frequency and duration of occurrence of the behaviors. Medley and Mitzel (1963) distinguished between two types of checklist: In one, only behaviors of interest are on the list, and the observer notes their occurrence only. In the other, all behaviors are assigned to some category, and all behaviors are scored; this often requires a 'miscellaneous' category which, if made large enough, makes the two types of checklist equivalent.

Whether or not a checklist is used may greatly affect results. In one study, Levine (1977) supervised two groups of observers of patients hospitalized with spinal cord injuries. Observers with a checklist of 45 preselected behaviors recorded a frequency of occurrence of the behaviors approximately four times greater than what was recorded by the other group.

If you are going to use a checklist, observe enough beforehand so that you can include and unambiguously define all the behaviors likely to occur. Still, have a miscellaneous category and room for notes. If you are interested in all behaviors, and several "miscellaneous" are checked, add to the list. If several behaviors occur together in unvarying sequence, and records of the separate occurrences are not crucial to the research question, you may simplify your task by assigning a single behavioral label to the entire sequence. You need not record the intensity or speed of individual behaviors unless they are important to the research question.

The categories need not be mutually exclusive. In fact, the observer might be interested in seeing which behaviors co-occur. For example, she might test for associations between specific movements and specific vocalizations. Each combination of movement and vocalization could be put in a separate category, but the complexity of the checklist would be increased accordingly.

Develop a List of Behaviors

The researcher must make several decisions about the behaviors to be recorded. This should be done early in the research process.

1. Should the behaviors be classified according to their purpose or by specific movements? If by purpose, then calling for a nurse and pressing a button to get a nurse would be classified in the same way. If by specific movements, then talking with a nurse, doctor, or family member would be classified the same.

    Rosenblum (1978) offered two important reasons for classifying by purpose. First, members of different species may perform different behaviors to achieve the same purpose; but both sets of behaviors may respond similarly to the same experimental manipulations, as for example, sexual behaviors of both rats and humans respond to changes in male and female hormone levels. These parallel responses would be undetectable without classification of behaviors by purpose. Second, individuals vary considerably in how they perform specific behaviors, and this is especially true of people. For example, human mothers pick up and carry their babies in many different ways. Organizing a set of behaviors under a category such as "maternal retrieval" allows comparisons between women.

    Two other methods of classifying behaviors are by their evolutionary origin and according to events that cause the behavior (Hinde 1970). Whichever system is chosen, the researcher should apply it consistently; do not classify some behaviors by movement and some by evolutionary origin.

A major consideration in observing is that the data be unambiguous and uncontaminated by the sophistication and biases of the observer. One way to minimize ambiguity is to avoid making inferences at the stage of data collection, since even behaviors that seem straightforward may be interpreted differently by others. For example, when the 8-year-old English boy peeped into a darkened house, he reported that the couple within were fighting. His American friend, seeing the same tableau, said knowingly that they were making love. At which his French companion sadly nodded his head and said, "Oui, they are making love. Badly."

2. Should the behaviors be molecular or molar? Molecular behaviors are finer, such as muscle twitches, whereas molar behaviors involve larger movements. Two major disadvantages of molecular categories are that (a) they are complex, with many behavioral categories and sometimes difficult distinctions between behaviors; and (b) a given molecular behavior may be a component of two or more molar behaviors.

On the other hand, observations recorded at the molecular level require fewer interpretive judgments and may reveal relationships that would not otherwise be seen. During data analysis, the researcher may decide that several molecular behaviors can safely be lumped into a single larger category; but molar behaviors can not be decomposed after the fact into molecular ones. Bakeman and Gottman (1986) noted that there are generally many levels between the extremes of molecular and molar; they suggested that behaviors should be coded at a level somewhat more molecular than the level at which the research question has been formulated. Whichever level you choose, apply it consistently to all behaviors.

3. Should published taxonomic lists be used? Bakeman and Gottman gave a cautious "yes" to that question, but they warned that the research question should dictate the coding system. A system adopted from someone else might not be sufficiently sensitive to the question. Many books and journal articles (Blurton-Jones, 1972; Grant, 1965, 1968, 1969; Hallberg, 1976; Hutt & Hutt, 1970; Jones et al., 1975; Polsky & McGuire, 1979; Ray & Ray, 1976; Sackett, 1978; White, 1974) contain observational systems used with various human populations. Many journals, among them *Animal Behaviour, Animal Behaviour Monographs, Behaviour, Journal of Animal Behaviour,* and *Primate Behavior,* publish observational studies on animals.

Decide How to Record the Behaviors

Scoring can be based on either events (behaviors of interest are recorded as they occur) or time intervals (behaviors are recorded only if they occur during predetermined times, for instance at the onset of each 20 second interval). Although interval recording has been widely used, both Altmann (1974) and Bakeman and Gottman argued that it is generally inferior to event recording. One problem is that several behaviors may occur in rapid succession near the start of the interval, with the observer able to record only one. Researcher bias, that is, the tendency to make errors that support the researcher's hypothesis, is maximized under such circumstances. A second problem with interval recording is that it doesn't allow for the analysis of behavioral sequences, since behaviors that occur in the middle of each time interval are not recorded; but the detection of reliable sequences is often of great interest to researchers.

Three types of research question are given below; the format for event recording is different in each case.

1. The question concerns frequency of events only; neither duration of events nor their sequence are important. A behavior therapist might wish to record how often a smoker lights up a cigarette, how often a cigar, and how often he sucks on hard candies. A three-item checklist with a place for tallying each occurrence would suffice.
2. Sequential information is important. The therapist of the previous example could give each event a code: cigarettes= 1, cigars= 2, hard candy= 3; the data sheet for one session might read 13333132.
3. The duration of events is important. The observer should record not just the occurrence but also the onset and offset of each event. If the events are mutually exclusive, there is no need to record offsets; the onset of event 2 indicates when event 1 ended.

Use Technology

Detailed, informative observations can be made with nothing more than pencil and paper (see Darwin, Charles: *Origin of Species*). But technology has made available attractive alternatives. Dabbs (1975) mentioned several, including devices for measuring subtle changes in facial expression and body positions, and others claimed to detect stress and lying. Tiny cameras and directional microphones make possible long-range unobtrusive observations (and create serious potential ethical problems); other cameras permit night viewing. Lehner (1979) comprehensively discussed both uses and sources of equipment for observing.

Two types of devices are especially valuable to observers. Video cameras allow observers to preserve complete records that can be scored at leisure. Hand-held electronic recorders permit observers to record rapidly with a minimum of distractions.

Be Attentive to Setting

Behaviors often differ dramatically from one setting to another. One of the arguments against laboratory research is that it often fails to accurately predict events outside the lab (cf. Chapanis, 1967). Human participants in laboratory research are generally aware of their roles and are motivated to do well. In more natural settings, neither may be the case. Captive and wild animals often behave differently. For example, caged baboons frequently fight among themselves, and males may literally tear apart a sexually receptive female. Yet free-ranging baboons rarely fight. Kortlandt (1962) wrote: "Zoo observations, incomplete field evidence, and comparisons with other primates have led most writers to assume that wild chimpanzees live in small closed harem groups of 5–15 members. I saw nothing of the kind."

A given behavior may be influenced more by a setting than by traits of particular subjects. An anthropologist who confined his observations to boxing rings would surely conclude that his subjects were highly aggressive. Therefore, be sensitive to setting, and make sure that your behavioral checklist is suitable for it. As Rosenblum (1978) noted, it is difficult to record small facial gestures or limb movements of a large group dispersed over a wide area; and though eye contact between a single mother and child is easily observed, eye contact between a group of mothers and their children is not.

If you observe in other than a natural setting, let your subjects adapt to it. Thus, for example, sleep researchers give their subjects at least one night to adapt to sleeping in a laboratory; dreams during the first few nights are different from later dreams (Hartmann & Cravens, 1973).

Use Unobtrusive Measurements

The benefits of unobtrusive, nonreactive measures are discussed on page 92. In some cases, obtrusiveness declines over time. Therefore, researchers who study animals in the field and children in classrooms often sit quietly for long periods of time, allowing the subjects to habituate to them, before they make their important observations.

Practice

The naturalist Louis Agassiz, renowned as both observer and teacher, required that his students spend long hours learning how to observe. His student, Scudder, described Agassiz' teaching methods (1874). Agassiz gave Scudder a preserved fish called a haemulon and asked him to study it. Scudder returned in 10 minutes to report that the investigation was complete. But Agassiz was gone, so he went back to the fish. "Half an hour passed—an hour—another hour; the fish began to look loathsome. I turned it over and around; looked it in the face—ghastly; from behind, beneath, above, sideways, at a three-quarters' view—just as ghastly." Scudder had been told to use only his hands and eyes, no instruments. He drew the fish and Agassiz, during a brief return, expressed approval and urged Scudder to keep observing. This continued for three days. On the fourth day a second fish of the same group was placed beside the first, and Scudder was told to find the resemblances and differences between the two. More fish followed until the entire family was represented.

Scudder was interested in insects, not fish, but he made continual, surprising discoveries throughout his ordeal; he called it the best lesson he had ever had for the study of insects: ". . . a lesson whose influence has extended to the details of every subsequent study . . ." A year later, he and some fellow students were amusing themselves by filling a blackboard with drawings of grotesque and bizarre creatures. Agassiz entered, noted the fish, and saw they were all haemulons. He knew immediately that Scudder had drawn them. Said Scudder, "True; and to this day, if I attempt a fish, I can draw nothing but haemulons."

Practice can also help when the subjects are alive and moving about. Argyle, Bryant, & Trower (1974) had trainees observe films of children sitting through a classroom lesson and being questioned about it. The trainees were asked to judge which children had understood, and they were then given the correct answers. When they were shown a new film of a second lesson, they showed marked improvement in their judgments.

Dancer, Braukmann, Schumaker, Kirigin, Willner, & Wolf (1978) used a variety of techniques, including lectures, discussions, modeling, video tapes, behavior rehearsal to criterion, and constructive feedback, to teach the skills of observing. Written materials instructed trainees to attend to three broad aspects of behavior:

1. What is the person doing?
    a. Listen to what the person says.
    b. Watch the person's facial expressions.
    c. Watch the person's body.
    d. Note the intensity of these behaviors.

2. What are the circumstances surrounding the behavior?
    a. Note the activity the person is involved in.
    b. Note when the behavior occurred.
    c. Note where the behavior occurred.
    d. Note the conditions immediately preceding the behavior.

3. What is the outcome of the behavior?
    a. Note how the behavior affects others around the behaver.
    b. Note the final physical conditions of the situation.
    c. If relevant, note whether the task was completed.

The trainees watched brief videotaped interactions of youths and wrote descriptions of what they'd seen. They were asked to try to provide descriptions that would enable another person to substantially replicate the observed events; and to eliminate all adjectives and adverbs that did not increase the behavioral specificity of the description. Their trainers rated the quality of each description on a four-point scale, with the primary criterion being how specific it was. Below is an example of ratings of descriptions of verbal statements. (Descriptions of voice tone and action behavior were scored in a similar manner.)

Description	Rating
What he did was inappropriate.	1 (vague and general)
Dale told Steve he was going to hit him.	2 (summary of what was said)
Dale said to Steve, "I'm going to hit you if you don't shut up."	3 (direct quote)
Dale said with fists clenched, "Steve, I'm going to hit you if you don't shut up."	4 (qualifier accompanies quote)

The results were impressive. In each of three separate workshops, trainees increased their average scores by over 100 percent. (Note: Researchers who develop checklists and consider issues like molecular versus molar behavioral units will almost certainly describe behaviors in specific rather than general terms.)

Measure the Reliability of Observations

As discussed in chapter 6, unreliable measures are of little value. A large section of the Journal of Applied Behavior Analysis (1977, Volume 10) was devoted to measuring and improving the reliability of observational data. Hopkins and Hermann (1977) gave a procedure for estimating the reliability of two observers who independently record the occurrence or nonoccurrence of responses during short time intervals: Reliability = the number of intervals in which observers agree either that the response has occurred or that it hasn't, divided by the total number of intervals; the result, multiplied by 100, yields a percentage of agreement. See Box 10.2.

Herbert and Attridge (1975) made the important point that observer agreement is different from reliability, which refers to stability over time and to accuracy in comparison with an acceptable standard. Two observers who agree perfectly may nevertheless be unreliable. However, it is common practice to apply the term reliability to measures of agreement.

Following a series of experiments, O'Leary, Kent, and Kanowitz (1975) concluded that estimates of agreement are often spuriously high. When observers knew they were being checked, the reliabilities were considerably higher than when they were unaware. When aware, they tended to adjust their scoring to match that of the person with whom they were being compared. And their computational errors consistently inflated reliability scores. A partial solution is to monitor observers without telling them; this would not directly improve observer accuracy but would give better estimates of reliability.

Estimates of reliability are sensitive to the number of behaviors in a scoring system and the number of categories actually scored in a given session. The more categories, the lower the reliability. Jones, Reid, & Patterson (1975) found that fewer categories were scored when reliability was being checked than when it wasn't, probably because the observers became more conservative. Also, molar behaviors are less reliable than molecular ones.

BOX 10.2

Estimating the Reliability of Observations

In Table 10.1, each X indicates that a response has been recorded.

The observers agreed for 6 of the intervals (1, 4, 5, 6, 7, and 10), so reliability = 6/10 × 100 = 60%. Unfortunately, the computation is easier to do than to interpret, primarily because two observers who randomly marked Xs on their data sheets would also agree occasionally. So, Hopkins and Herman gave a formula for estimating chance reliability (CR):

Let $pO1$ be the proportion of intervals in which observer 1 records a response.

Let $pN1$ be the proportion of intervals in which observer 1 records no response.

Let $pO2$ and $pN2$ be the corresponding proportions for the second observer.

Then, $CR = [(pO1 \times pO2) + (pN1 \times pN2)] \times 100$.

Table 10.1

	Interval									
	1	2	3	4	5	6	7	8	9	10
Observer 1	X	X		X	X		X	X	X	
Observer 2	X		X	X	X		X			

For the data of Table 10.1,

$CR = [(.7 \times .5) + (.3 \times .5)] \times 100 = 50\%$.

So, the observers did not do much better than chance. Many other methods are available for computing observer agreement. They are applicable to different types of data, and only some of them take into account chance agreements. So, the common practice of publishing coefficients of agreement without indicating how they were derived should be discouraged.

Observers tend to "drift" in their scoring tendencies, that is, over time they change their criteria for scoring. Observers who work together over a long period of time may drift in the same way, thereby earning high reliability scores; but the scores will not be accurate. Kazdin (1977) suggested three methods for controlling drift:

1. Have observers meet periodically as a group, rate behavior, and receive immediate feedback on the accuracy of their observations.
2. If possible, videotape the behaviors to be observed and score the tapes in random order at the end of the study.
3. Periodically bring in newly trained observers and compare their scores with those of old-hands.

One additional strategy for improving the reliability of observations (and of all other research procedures) is to pick suitable behaviors. If the behavior of interest is not easily scored, search for a related one that is. For example, a researcher interested in people's postures during an interview situation might measure the height of their eyes as they sit. With closed circuit television, this is a highly reliable indicator of posture.

PARTICIPANT OBSERVATION: WATCHING THE CORN GROW

Participant Observation

Participant observation, a technique favored by many anthropologists and sociologists, was defined by Taylor and Bogdan (1984) as "research that involves social interaction between the researcher and informants in the milieu of the latter, during which data are systematically and unobtrusively collected." The researcher works to gain entry into a group and then collects data. His goal is to see the world from the perspective of the group members and to communicate his vision. According to Feldman (1980), two participant observation studies of the 1960s (Preble & Casey, 1969; Sutter, 1966) permanently changed how public policymakers view heroin addicts. Taylor and Bogdan gave examples from many other fields.

Because the participant observer wants the group members to act naturally, he tries to be unobtrusive in his data collection procedures. This often requires deception. Deception is also often necessary to gain entry into the groups, especially if they are considered deviant (inmates in prisons and mental hospitals, users of illicit drugs, people claiming contact with extraterrestrial beings). The ethical problems are serious.

Participant observers are not always unobtrusive. Anthropologists, for example, are often so different from the people they study that they can't possibly blend in unnoticed. In such cases, their observations should be regarded with even more caution than usual. Margaret Mead's observations of Samoans, which formed the basis for her *Coming of Age in Samoa,* is in this category. Derek Freeman (1983) argued that many of her conclusions were considerably off the mark, partly because Samoans enjoy making up stories, especially about sex, for visitors.

Robert Mowry Zingg was one of the foremost students of the culture of Mexico's Huichol Indians. Among the artifacts he collected for a museum in New Mexico was a stuffed rat with sewn bits of red cloth for eyes. Zingg described in detail how present-day peyote hunters keep the stuffed animal to commemorate "the rat that stole the fire in the peyote." But according to Berrin (1979), the rat probably was a joke played upon Zingg by the Huichol; it plays no role in the culture.

Some Problems to Be Aware of

Confounding

A confound is a variable not manipulated by an investigator that systematically biases results. To compare frequencies of various responses in different individuals or groups, an investigator might observe them on separate occasions. But if one group is observed in the morning and the other at night, the time of testing is a confound and may be a major cause of differences between them. Place and context of observations may also exert major effects. The best solution is to observe subjects more than once, and at different times, with the order of observation chosen randomly. And to be aware that conditions of observation may profoundly affect results.

Computing Relative Frequencies

Altmann (1974) called attention to another problem with comparative data: some behaviors are easier than others to observe. A report that young children cry more than they smile may rest on the tenuous assumption that crying and smiling are equally likely to be detected; a report that juvenile monkeys fight more than do adults, on the assumption that fighting is equally likely to be recorded in juveniles and adults.

Feedback May Lead to Bias

O'Leary et al. (1975) told observers that they would see how an experimental intervention had modified the behavior of children in a classroom. The children were shown on videotape, apparently before and after the intervention. Actually, however, the tapes were matched so that the crucial behaviors occurred with equal frequency in all cases. After the initial observation sessions, feedback was provided: positive, if an observer scored behavior as changing in the right direction and negative if he reported no change or a change in the wrong direction. The combined effects of expectancies and feedback led to biases in the second round of observing, so that most behaviors were reported as changing in the right direction. Kazdin (1977) recommended that observers should be given feedback only about the accuracy of their observations, not whether they support the research hypothesis.

Sloppy Observing Can Be Dangerous

Beveridge (1950) told the following story: A Manchester physician, while teaching a ward class of students, took a sample of diabetic urine and dipped a finger in to taste it. He then asked all the students to repeat his action. This they reluctantly did, making grimaces, but agreeing that it tasted sweet. "I did this," said the physician with a smile, "to teach you the importance of observing detail. If you had watched me carefully you would have noticed that I put my first finger in the urine but licked my second finger."

Summary

There are several reasons for doing observational studies: to describe nature, generate and test hypotheses, assess behavior, and when other methods are impractical. Prepare for observing by making up a checklist of behaviors that are classified according to a single principle. Classification by purpose is generally best, but behaviors can also be classified by function, evolutionary origin, or causes. Try to pick behaviors that can be easily scored. Use event rather than interval recording, and be as unobtrusive as possible. Practice and seek feedback on the accuracy of your observations.

Key Terms

Confound A variable not deliberately manipulated by a researcher that systematically biases results.

Illusory Correlation People who expect certain associations often report seeing them even if they have not occurred. The inaccurate perceptions are called illusory correlations.

Participant Observation The observer joins in the activities of the individual or group being observed.

Unobtrusive Observation Subjects are unaware that they are being observed.

CHAPTER 11

Surveying

By the time you finish reading this chapter, you should be able to answer the following questions:

How do survey researchers decide on the number of subjects to use?

How are subjects picked to be in surveys?

What are the relative merits of mailed questionnaires, personal interviews, and telephone interviews?

What special techniques are used for asking questions about behavior, knowledge, and attitudes?

Why do different surveyors often report discrepant results?

A college woman became tired of filling out questionnaires about her sex life. Summoned one day to an interview with a visiting psychiatrist, she patiently answered his lengthy questions. Upon leaving, she quietly asked, "Oh doctor, tell me. Does it mean anything that I am so passionately fond of pancakes?" The doctor laughed affably and said, "Why, dear girl, of course not. I'm very fond of them myself."

"Oh, I'm so glad," she said. "I've got a whole drawer full of them in my dorm room and I love to take them out and stroke them."

Interviewers stopped pestering her. But, she reported, the housemother dropped by periodically and went through her dresser drawers.

Surveys are inexpensive; superficially easy to design, implement, and interpret; and require no elaborate equipment. No wonder they are the most widely used method of data collection in the social sciences. Wahlke (1979) reported that about half of 180 articles on political behavior in *The American Political Science Review* were based solely on surveys, and an additional 20 percent on survey as well as other data. Moreover, as can be seen from examples given by Converse and Traugott (1986), surveys impact the lives of most Americans:

- In the 1980 presidential campaign, the League of Women Voters set a threshold of 15 percent popular support in the polls for aspirants to qualify as participants in its televised presidential debates.
- Campaign contributions to political candidates rise and fall with their support in the polls.
- Major national issues are debated with reference to public opinion surveys, and the nation's decision-makers watch them closely.
- Many interest groups use surveys in an attempt to create an aura of popular support for their positions. Because their goal is to promote their positions rather than to seek truth, they often pick unrepresentative samples, ask leading questions, and report results selectively.

The most familiar use of surveys is to answer questions about the members of a population, including objective facts about them and their attitudes, knowledge, and behavior patterns. Questions may focus on differences between subgroups within the population. Surveys of this type, as in the examples above, are usually atheoretical and not part of a comprehensive program to learn why and how phenomena occur. As such, and despite their often tremendous practical significance, they barely qualify as science. But used in other ways, they can be part of scientific programs. They may unearth interesting relationships, as between left-handedness and learning disabilities (see chap. 4), that open up meaningful avenues of exploration. Survey results can be the dependent variable of an experiment; for example, subjects assigned to watch one of two different television programs might be given a post-viewing questionnaire to see whether their attitudes have been changed by the programs. Surveys can be used to classify, with people being assigned to one of two or more conditions on the basis of their answers. Repeated surveys on the same topic may indicate how people change over time. Exploratory surveys, even with nonrepresentative samples, may help investigators learn the key issues about a topic. And surveys may themselves be studied scientifically; for example, Schuman and Presser (1981) systematically varied the order of items on a questionnaire and recorded differences in responses.

The prospective surveyor must make three important decisions: how to select the sample; whether to use mailed questionnaires, personal interviews, or telephone interviews; and how to formulate the questions.

Selecting the Sample

Surveyors, like all researchers (except sequential samplers—see chapter 16), should decide before collecting data how many subjects they will use. To an even greater extent than other researchers, they must use carefully prescribed procedures in choosing specific subjects. And they must often make inferences about people who refuse to participate.

Size of Sample

Before collecting any data, researchers should figure out how many subjects they will need. If too few subjects are used, the data will be unreliable; but using too many is wasteful of time and money. Decisions about sample size affect both confidence in results and sampling error (a measure of expected accuracy when characteristics of a population are estimated from those of a sample). The relationships between sample size, sampling error, and confidence in results apply not just to questionnaires and interviews but also to survey estimates of such diverse things as the number of defective parts in a batch, the frequency of twin births, and the proportion of tomatoes that have worms in them.

A researcher who wanted to find the average height of women in an introductory psychology class could measure the height of each woman; the procedure would be simple and accurate. But if she wanted to know the average height of all women in the United States, she'd have to take a sample and generalize from it to the population. It seems amazing that samples of only 1,500 people are enough for such estimates; yet 1,500 is the standard sample size from which Gallup and other surveyors predict voting patterns. See Table 11.1 for their accuracy since 1936.

For many traits, the scores of individuals distribute symmetrically about the mean for the trait. The distribution is called normal and described by an exact mathematical equation. The method given below for computing sample size entails the assumption that the characteristic being measured is normally distributed, but even fairly large departures from normality can be tolerated. Alternative statistics are available if the normality condition is not met, but the required sample size is then much greater.

Table 11.1 *Record of Gallup Poll Accuracy*

Year	Gallup Final Survey*		Election Result*	
1984	59.0%	Reagan	59.2%	Reagan
1982	55.0	Democratic	55.8	Democratic
1980	47.0	Reagan	50.8	Reagan
1978	55.0	Democratic	54.0	Democratic
1976	48.0	Carter	50.0	Carter
1974	60.0	Democratic	58.9	Democratic
1972	62.0	Nixon	61.8	Nixon
1970	53.0	Democratic	54.3	Democratic
1968	43.0	Nixon	43.5	Nixon
1966	52.5	Democratic	51.9	Democratic
1964	64.0	Johnson	61.3	Johnson
1962	55.5	Democratic	52.7	Democratic
1960	51.0	Kennedy	50.1	Kennedy
1958	57.0	Democratic	56.5	Democratic
1956	59.5	Eisenhower	57.8	Eisenhower
1954	51.5	Democratic	52.7	Democratic
1952	51.0	Eisenhower	55.4	Eisenhower
1950	51.0	Democratic	50.3	Democratic
1948	44.5	Truman	49.9	Truman
1946	58.0	Republican	54.3	Republican
1944	51.5	Roosevelt	53.3**	Roosevelt
1942	52.0	Democratic	48.0	Democratic
1940	52.0	Roosevelt	55.0	Roosevelt
1938	54.0	Democratic	50.8	Democratic
1936	55.7	Roosevelt	62.5	Roosevelt

*The figure shown is the winner's percentage of the Democratic-Republican vote except in the elections of 1948, 1968, and 1976. Because the Thurmond and Wallace voters in 1948 were largely split-offs from the normally Democratic vote, they were made a part of the final Gallup Poll preelection estimate of the division of the vote. In 1968 Wallace's candidacy was supported by such a large minority that he was clearly a major candidate, and the 1968 percentages are based on the total Nixon-Humphrey-Wallace vote. In 1976, because of interest in McCarthy's candidacy and its potential effect on the Carter vote, the final Gallup Poll estimate included Carter, Ford, McCarthy, and all other candidates as a group.

**Civilian vote 53.3, Roosevelt soldier vote 0.5 = 53.8 percent Roosevelt. Gallup final survey based on civilian vote.

From "The Gallup Poll" in *Public Opinion*. Copyright © 1986 Scholarly Resources, Inc., Wilmington, DE. Reprinted by permission.

A good sample reflects the population from which it comes; otherwise, there would be no point in generalizing. But because of chance factors, sample and population are rarely identical. The extent to which a properly chosen sample can be expected to deviate from the population from which it comes is called sampling error. As the number of subjects increases, sampling error is reduced. So, the first step in figuring out sample size is to decide how much error is acceptable. Only by sampling the entire population will sampling error equal zero. (Even then, other errors would still be likely. In lining up to have their heights measured, some women might stand erect, others might slouch; the surveyor might misread the tape slightly, record the measurement incorrectly, or misanalyze the data. These would not be affected by changes in sample size.)

BOX 11.1

Calculating Sample Sizes for Surveys

Suppose that, either from previous experience or a small study designed specifically for the purpose, it is determined that the time to learn a particular task has a standard deviation of 30 seconds. A researcher wishing to estimate time to learn the task, and willing to accept a sampling error of 5 seconds, would compute $N = 30^2/5^2 = 900/25 = 36$. The result tells him that repeated measures of learning times on random samples of 36 subjects would, in 68 percent of cases, yield means within 5 seconds of the population mean. To be confident of the 5 second margin in 95 percent of cases, he would collect data on $36 \times 3.84 = 138$ (always round upwards) subjects.

Intuitively, it might seem that sample size should depend on population size, that a sample adequate for a population of, say, 1,000, might be too small for one of 1,000,000. To some extent, this is true, and a correction factor can be applied. If N is the initially calculated sample size and P the population size, then the final sample size is given by $(N \times P)/(N + P)$. In the above example, $N = 138$. With a population of 1,000, $NP/(N + P) = 138,000/1,138 = 122$; so the sample size need be only 122 instead of 138. With a population of 1,000,000, $NP/(N + P) = 128$. As can be seen, the effect is small and rarely important unless the sample size is a substantial proportion of the population size.

Surveyors often express results in proportions. For example, 53 percent of the population supports Proposition 12. In such cases, use the formula $N = [p(1 - p)]/SE^2$ where $p =$ the proportion expected to vote yes and $1 - p =$ the proportion expected to vote no. If there is no basis for estimating, let $p = 1 - p = .5$; this is the estimate that requires the most subjects. Once again, multiply by either 3.84 or 6.64 to get the desired confidence levels.

Example:

If the acceptable error due to sampling is 0.05 and $p = .5$, then $N = .25/.0025 = 100$. To be 99 percent certain that the observed proportion is within 0.05 of the true proportion, multiply

$$100 \times 6.64 = 664 \text{ subjects.}$$

If the population were 1,000,000, then the corrected sample size would be $NP/(N + P) = 664,000,000/(1,000,664) = 664$.

To find the appropriate sample size n for a given sampling error, SE, use the formula $n = s^2/SE^2$, where s is the standard deviation of the population. The mean of a sample of size n can be expected to fall within 1 SE of the true population mean in 68 percent of cases. To be 95 percent or 99 percent confident, multiply n by 3.84 or 6.64 respectively. (From Appendix E, you can see that 95 percent of cases fall within 1.96 standard deviations of the mean of a normally distributed population, and $1.96^2 = 3.84$; 99 percent of cases fall within 2.58 standard deviations, and $2.58^2 = 6.64$.) See Box 11.1. Although a researcher might be satisfied with some other level of confidence, in practice the 95 percent and 99 percent levels are used almost exclusively.

Shown in Table 11.2 are the appropriate sample sizes for various levels of p and acceptable sampling errors. Pick a few values from the table and see if you can reproduce them.

Table 11.2 *Sample Sizes for Various Levels of p and Acceptable Sampling Errors**

	p				
Sampling Error	.1	.2	.3	.4	.5
.01	900	1600	2100	2400	2500
.02	225	400	525	600	625
.03	100	178	233	267	278
.04	57	100	131	150	156
.05	36	64	84	96	100
.06	25	45	59	67	70
.07	19	33	43	49	51
.08	15	25	33	38	40
.09	12	20	26	30	31
.10	9	16	21	24	25

*Randomly chosen samples of the proper size can be expected to give results within the acceptable sampling error in about 68 percent of cases. To be 95 percent sure, multiply the listed value by 3.84. To be 99 percent sure, multiply by 6.64.

Note from Table 11.2 that, as sampling error is reduced, proportionately more subjects must be sampled to reduce it even more. Thus, to reduce error from 0.1 to 0.04 when $p = .1$, N must be increased from 9 to 57; but to reduce error from 0.04 to 0.02 for $p = .1$, N must be increased from 57 to 225. Note, too, that the closer p is to .5, the larger the sample size must be.

Choosing Individual Subjects

Inferences from samples to populations are justified only if the samples have been chosen properly—no matter how many subjects are used. If I asked everybody in my psychology class to fill out a questionnaire about his or her political preferences, even if it were a very large class I could not properly generalize to all Americans. My sample would consist of students, mostly in the 18–25 year age range, at a single college, Cal State University; and however brilliant, good-looking, and noble they are, they are not representative of all Americans.

I could not generalize to all college students. Students who attend expensive, small, private colleges in the east differ from those at large, western, public institutions.

I could not generalize to all students at Cal State. Psychology, art, and business students differ.

I could not generalize to all psychology students. Freshmen and seniors differ. So do students who intend to specialize in clinical psychology, work in industry, pursue careers in research, and stay in school until they can draw on the trust fund.

I could not even generalize to the other classes I teach. Some are required for many majors within the university, some are not. Classes taught at different times of day may attract different types of students: full-time versus part-time, morning people versus night people, and so forth.

In short, any two self-selected groups are likely to differ, often in subtle but important ways. Still, proper samples can be drawn. The key is that they be drawn randomly. Stuart (1984) defined a simple random sample as "one selected by a process which gives every possible sample (of that size from that population) the same chance of selection." Both simple and complex random samples are described below.

Simple Random Sampling The first step in getting a random sample is to carefully define the population. If I wanted to generalize to the student population at Cal State, I might get a print-out of all students from the registrar. I'd have to decide whether to include them all or only full-time, currently-enrolled students taking courses on the Hayward campus. (We teach extension courses in other cities.) Then, I could write each name on a slip, put the slips in an urn, and pull out as many as needed. Or, I could assign a number to each student and use a table of random numbers to get my sample.

It is relatively easy to get random samples from small populations such as apartment buildings or even universities. Randomly sampling the population of a city is much more difficult, national surveys harder still. Always, the population and hence the criteria for inclusion in the sample, must be carefully specified. Surveyors must decide if there will be age limitations. If, in a city survey, everybody in the city at a specified time will be eligible or if the population will be restricted to permanent residents. Will it include people confined to mental hospitals? Prisons? Will the surveyors try to reach residents traveling abroad? Will their national surveys include (as few do) Alaska or Hawaii?

Few cities have inclusive lists of residents comparable to the registrar's list of students. Researchers have used telephone directories—but some people don't have phones or have unlisted numbers, some have more than one number, and many numbers in the directory belong to commercial or government establishments. (If, instead of using directories, numbers are generated randomly by computer, people with unlisted numbers can be reached and the representativeness of the sample improved; still, some 7 percent to 8 percent of household populations are inaccessible by phone (Converse & Traugott, 1986). Lists of voters and property owners have obvious limitations. At present and for the foreseeable future, Stuart's definition of random sampling is an unattainable ideal. And deviations from randomness reduce the accuracy of generalizations from a sample.

Systematic Sampling Given an inclusive list of a population, the first name for a sample might be chosen randomly; then, in the procedure known as systematic sampling, every nth name would be taken until the full complement was reached. This variant of simple random sampling is easy to do, but it is flawed if the list has a pattern to it. For example, starting at an apartment picked at random, a researcher might systematically interview tenants in every fifth apartment of a tall building. But if the apartments differ in desirability according to their location on the floor, and there are 10 apartments per floor, she will end up with a biased sample.

Stratified Sampling Recall the formula $n = s^2/SE^2$; transposed, it shows that both the necessary sample size and the standard error vary with the standard deviation of a trait in a population. If subgroups within the population are very different with respect to the trait, s^2 will be large; it can be reduced by sampling each subgroup separately in proportion to their population ratios. Thus, to estimate the mean weight of fruit from a bowl containing 50 grapes, 40 apples, and 10 watermelons, the researcher would randomly sample each stratum of fruit independently, to ensure that the sample contained grapes, apples, and watermelons in the ratio 5 to 4 to 1; if their squared standard deviations are symbolized by *g, a,* and *w,* and *n* stands for the total number of fruits sampled, then $(5g + 4a + w)/n$ equals s^2. In this example, because the strata differ considerably, s^2 (and hence, both standard error and necessary sample size) would be decreased. An additional benefit of stratification is that it allows for collection of data on the separate subgroups.

Cluster Sampling Suppose a pollster wanted to interview 500 people in a city. Traveling to 500 scattered addresses would be time-consuming and inefficient, so she might use an alternative procedure. She would divide the city map into clusters—blocks, tracts, or similar large units—and sample them by a random (often stratified) procedure. She would attempt to interview everybody within the cluster. This method, although cost-effective, increases the sampling error for a given sample size whenever people living within a cluster are similar in ways relevant to the purposes of the survey.

Nonresponse

Selecting subjects is one matter. Getting adequate data from them is another. Inevitably, in any large-scale survey, some subjects will not be located; some will refuse or be unable to participate; and some will misunderstand instructions or otherwise give data that can't be used. The response rate—proportion of people selected for a survey who give usable data—is important to consider when trying to evaluate the probable accuracy of a survey. Respondents and nonrespondents differ, often in ways relevant to the purposes of a study. For example, prisoners, soldiers, hospital patients, and children are often excluded from surveys; and they probably differ on average from respondents in attitudes about prison and military reform, and hospital and child care. In actual studies (Aneshensel, Frerichs, Clark, & Vokopenic, 1982; DeMaio, 1980), blacks, older people, city-dwellers, and westerners were less willing than others to grant telephone interviews. Both the proportion of nonrespondents and the reason for their exclusion from the final tabulations (unavailability, refusal, or unusable data) affect the degree of bias.

If the proportion of nonrespondents is small, their absence will make little difference. But the return rate with mail surveys is often less than 50 percent, and nonresponse rates for both telephone and personal interviews have increased in recent years. Several reasons have been suggested: (a) in many families, everybody works, so interviewers are less likely to find anybody home during the day; (b) the best time to find people home is early evening (Vigderhous, 1981), but interviewers are reluctant to go into certain neighborhoods after dark; (c) for many people, especially in large cities, English is not the native language; (d) people fear opening their doors to strangers; and (e) participants in early surveys may have believed that their answers would make a difference, but the predominant mood today is that respondents have no influence over either the surveyors or the government (Sharp & Frankel, 1983).

If the proportion of nonrespondents is x, and the difference between respondents and nonrespondents with respect to an issue is y, then the biasing effect of nonresponse is xy. For example, if 35 percent of a sample respond, and 85 percent of them (but only 60 percent of nonrespondents) favor the issue, then the biasing effect is $(.65) \times (.85 - .60) = .1625 = 16.25$ percent. A surveyor who ignored the problem of nonresponse would conclude that 85 percent of the population favored the issue; the actual percentage would be 68.75 percent.

Of course, the preferences of nonrespondents cannot be known with accuracy. The best strategy for dealing with the problem is to reduce their number. Sending an introductory letter before calling for an interview helps (Dillman, 1978). So do small rewards, especially when given before participation rather than promised for afterwards (Schewe & Cournoyer, 1976) and follow-up telephone calls or special delivery letters. But people who respond only after several follow-ups are likely to differ from immediate respondents on variables central to the survey. For example, in a telephone survey during the 1984 presidential campaign, Reagan edged Mondale by three percentage points among those interviewed at one call. After including results from people who responded to a callback, the lead increased to six percentage points, and the final results, which included people who were reached after as many as 30 callbacks, gave Reagan a 13 point lead (Traugott, 1987). Two other problems with follow-ups are (a) they are costly and (b) they are impractical when deadlines are short.

The *Literary Digest* survey of 1936, which led to the strikingly incorrect conclusion that Alf Landon would win the presidential election by a wide margin over Franklin Delano Roosevelt, erred because of a large nonresponse rate. A systematic sampling procedure was used to pick 10 million names from lists of registered voters, and sample ballots were mailed to them. But people who feel strongly about an issue and want the underdog to win are most likely to respond to mail surveys (Bryson, 1976), and only 2.3 million ballots were returned.

Choosing the Method of Data Collection

Three major methods of data collection are used for surveys: mailed questionnaires, personal interviews, and telephone interviews. (A fourth method is to collect data from people who happen to gather in a single place such as an airplane or lecture hall; because of the nonrandom selection of subjects, few valid generalizations can be made from such studies.) O'Keefe (Zimmerman & Clark, 1987) reported that recent costs of professionally conducted surveys have been $15 per respondent for self-administered and mail questionnaires, $25 for telephone interviews, and $40 for personal interviews.

Mailed questionnaires have several advantages over personal interviews:

- Researchers can cover a wide geographic area at low cost, whereas interviewers must often travel long distances and to dangerous places; interviewers are probably often tempted to make up data, and they may not always resist temptation.
- Because interviewers are not used, the survey staff can be kept small; thus, training and payroll costs are reduced.
- There are no interviewer effects (see p. 168).
- Respondents can take their time thinking about difficult questions.
- The processing of data is relatively easy.
- Respondents given assurances of confidentiality are more likely to give honest answers to sensitive questions. For example, Seattle residents were surveyed either by secret ballot or interview about how they would vote on an upcoming bill to give money to war veterans. The first group was less favorable, and their responses predicted the actual outcome much more accurately (Edwards, 1957). (But confidentiality can be achieved even during personal interviews, and mailed questionnaires may have no special advantage in eliciting honest answers to sensitive topics (DeLamater, 1982).

The following are disadvantages of questionnaires as compared with personal interviews:

- The response rate is considerably lower. (Sudman and Bradburn (1984) noted that surveys of special groups, such as members of professional societies, yield comparable response rates for mailed questionnaires and personal interviews.)
- Results come in more slowly.
- Mailed questionnaires are unsuited to certain segments of the population—the poorly educated, the aged, and others who have trouble reading or filling out answers to questions.
- Questions must be kept simple, and unclear questions cannot be clarified.
- Investigators cannot follow up interesting answers with additional questions.
- Someone other than the intended respondent may fill out the questionnaire.
- Later questions may suggest answers to earlier ones, and respondents may then change their answers.

The telephone interview at one time led to underrepresentation of poor people, many of whom did not have telephones. But telephones are now found in more than 90 percent of U.S. households, and many people without them are unlikely to be located by other survey methods. Random dialing techniques let interviewers reach people with unlisted numbers.

Telephone interviews share most of the advantages and few of the disadvantages of the other methods of surveying. Telephone interviewers must be trained, but costs are low because they need not travel and they typically work in a centralized location under close supervision. A wide geographic area can be sampled, and response rates are not much different from those of personal interviews. Results are obtained rapidly. Although less suited than personal interviews for detailed, probing questioning, telephone interviews are superior in most other respects. (On the other hand, as one who has often received calls at inopportune times, I urge surveyors to consider whether such obtrusive behavior is warranted by their projects.)

Making up the Questionnaire

Sudman and Bradburn (1982) wrote that analysis of the concepts involved in a research question should be the first step in designing a questionnaire. Step 2 should be to formulate specific questions to measure the concepts. Specificity is crucial. Thus, for example, it is not enough to decide to measure "attitudes" of potential voters toward a particular candidate; specific attitudes should be indicated, such as toward the candidate's personality or platform. As questions are made increasingly specific, the research question is clarified.

Four classes of questions appear on surveys: demographic, behavior, knowledge, and attitude. Sudman and Bradburn gave their choices of specific wordings for standard demographic questions (Social Science Research Council, 1975; Sudman & Bradburn, 1982; chapter 7) and suggested an interesting strategy for formulating all questions: Review the literature to see what is available on the topic of interest, and then plagiarize. They pointed out that survey researchers encourage the repetition of questions, and they listed several valuable sources: the Gallup Poll, CBS-New York Times Poll, and Polls Section of Public Opinion Quarterly. They also listed the following American archives of survey research data:

Behavioral Sciences Laboratory, University of Cincinnati, Cincinnati, OH 45221

Data and Program Library Service, University of Wisconsin–Madison, 4451 Social Science Building, Madison, WI 53706

Institute for Research in Social Science, Manning Hall, University of North Carolina, Chapel Hill, NC 27514

Inter-university Consortium for Political and Social Research, University of Michigan, Ann Arbor, MI 48106 (Institute for Social Research Archives are at same address)

National Opinion Research Center, University of Chicago, 6030 South Ellis Avenue, Chicago, IL 60637

Roper Center, University of Connecticut, Storrs, CT 06268
Survey Research Center, University of California, Berkeley, CA 94720

Always pretest questions; and if, following a suggestion by Belson (1968), you ask pretest respondents to tell you in their own words what each one means, you may be able to identify and modify any subject to misinterpretation. These may surprise you. For example, a respondent was asked about the item "It is better not to try to plan when to have children, but just to accept them when they come;" she said "Of course you accept them when they come—you can't just leave them in the hospital." (Sudman & Bradburn, 1982)

If your questions put demands on the respondents' memories, let them use household records (canceled checks, insurance policies, bills for home repairs, etc.); also, give memory cues. Thus, instead of asking them to "Name all the major appliances you own," provide a list of major appliances and ask them to check the appropriate boxes. An exhaustive list may be impractical ("Name all the magazines you have read in the past five years."), so you will need an "Other" category. But be as complete as possible, since items not mentioned on the list are likely to be underreported.

Do not force respondents to give a single answer to what is actually two separate questions. Do not, for example, ask "Do you support the President and his proposed tax increase, or do you support the challenger and her plan to reduce taxes?" By subtle use of such double-barreled questions, as in the following example from Sudman and Bradburn, dishonest surveyors can manipulate answers: "Are you in favor of building more nuclear power plants so that we can have enough electricity to meet the country's needs, or are you opposed to more nuclear power plants even though this would mean less electricity?" People with strong feelings about having enough electricity might answer "Yes" even though preferring various alternatives to nuclear power; but their answers would be tabulated on the side of nuclear power. Avoid such problems by making sure that each of your questions measures one attitude only.

Writing Questions About Behavior

Some questions may be threatening to respondents, whether about behaviors generally perceived as socially desirable (Do you have a library card? Did you vote in the last election?) or undesirable (Have you ever been convicted of driving while under the influence of liquor? How often do you masturbate?). Not surprisingly, people often overreport desirable and underreport undesirable behaviors (cf. Parry & Crossley, 1950; Warner, 1977). Sudman and Bradburn offered advice for increasing the honesty of responses to threatening questions.

1. Questions may be closed (the respondent is given a specific set of answers from which to choose) or open (the respondent answers in her own words). For threatening material, open questions are better. This is especially true when asking about frequencies.

Example:
Question A is preferable to question B.
A. On the average, how many cigarettes do you smoke per day?
B. On the average, how many cigarettes do you smoke per day?

   0   10   20   30   40   50   60

2. Long questions increased the reported frequencies of socially undesirable behaviors about 25–30 percent over the standard short questions (Sudman & Bradburn, 1982). Lengthening a question is especially useful if the additional material makes it less threatening.
Example:
Question A is preferable to question B.
A. An estimated 60 million Americans have tried marijuana. Many of them report that it enhances their appreciation of music and increases creativity. Have you ever smoked marijuana?
B. Have you ever smoked marijuana?

3. If questions are worded identically for everybody, some respondents may be inhibited by unfamiliar words; but changing the words may change the meaning of the questions. Sudman and Bradburn suggested, as a compromise position, that respondents be asked which words they prefer to use to describe specific activities. For example, questions about sexual intercourse might be prefaced by the statement: "Different people use different words for sexual intercourse. What word do you think we should use?" Then, whichever word the respondent picked would be substituted for "sexual intercourse" for the remainder of the interview.

4. People are more likely to report undesirable behaviors if questions about them are carefully worded:
A. Suggest that the behavior is very common. For example, "Everybody tells a lie once in awhile. Have you lied to anyone during the past seven days?"
B. Assume the behavior and ask about details. For example, "How often do you masturbate?"
C. Use an authority to justify the behavior. For example, "Many doctors believe that drinking wine reduces heart attacks. Have you drunk any wine in the past month?"
D. Ask if the respondent ever behaved in the undesirable way, even once, rather than about current behavior. (For desirable behaviors, the reverse phrasing is preferable. Thus, question A is better than B, and C is better than D.)
Example:
A. Have you ever, even once, driven faster than the posted speed limit?

B. During the past week, did you drive faster than the posted speed limit?
C. Did you use a seat belt the last time you got in a car?
D. Have you ever worn a seat belt?

5. If questions about socially desirable behaviors are introduced with the casual phrase "Did you happen to . . . ," respondents are less likely to overreport them. For example, "Did you happen to attend any concerts last month?"

An alternative to traditional survey methods for finding answers to threatening questions is to use informants—selected individuals who report on behaviors of groups familiar to them rather than on their own behaviors. In one study of alcohol consumption (Smart & Liban, 1982), informants were picked who represented various subgroups within the population being surveyed; they were not randomly selected. They were given questionnaires and told to answer, not for themselves or their families, but for their occupational groups as a whole. Compared with standard survey methods, the informant method reported higher rates of drinking, heavy drinking, and problem drinking, and smaller sex differences. These estimates were closer to those based on alcoholic beverage sales.

Writing Questions About Knowledge

Respondents may worry about appearing ignorant. Reduce their concerns by introducing questions with phrases such as "Do you happen to know . . ." and "Can you recall, offhand. . . ." Also, include an "I don't know" among the possible answers, so they realize that this answer is acceptable.

Respondents shown a list of names and asked to indicate those they know, are likely to overstate their familiarity. To estimate the extent of the overstatement, add one or more fictitious but plausible sounding names to the list. If the fictitious names are picked by 2 percent of the people, then the best estimate is that familiarity with the other names has been overstated by 2 percent. In actual studies, about 15 percent of respondents reported that they had heard of a fictitious civil rights leader (Sudman & Bradburn, 1982), and 5 percent of high school students that they had used a fictitious drug (Petzel, Johnson, & McKillip, 1973).

For questions that have a numerical answer, do not give a list of alternatives. Make such questions open ended.

Writing Questions About Attitudes

Many instruments for measuring attitudes have been standardized, published, and evaluated. The sources listed on page 85 give descriptions and acquisition information and may eliminate the need to write a new questionnaire. If you must construct your own, write questions to assess knowledge before asking about attitudes; this screens out respondents who know little about the topics but might nevertheless express attitudes for fear of appearing ignorant.

Attitudes have three components: evaluative (favorable or unfavorable, as in "Do you approve of the President's position on a tax increase?"), cognitive (knowledge, thinking, as in "Can you name the President's Supreme Court appointees?"), and action (willingness or intention to do something, as in "Are you willing to donate at least $50 to the Committee to Help Re-Elect the President?"). Make sure that the component(s) measured by your questionnaire is both necessary and sufficient to answer your research question.

The strength of each component can be measured in three different ways: by asking a separate question about strength; by building a strength dimension directly into the question ("Indicate by picking a number from 1 to 7, with 1 representing strong disapproval and 7 strong approval, your feelings about the Supreme Court's decision on abortion?"); and, as discussed in chapter 6, by asking a series of independent questions that can be combined to give a single strength score.

Putting the Questions in Order

Three goals should direct the ordering of questions: to make the respondents' task as pleasant as possible, to minimize the time to complete the task, and to maximize the accuracy of responses. With those in mind, the following suggestions are offered:

- Start with nonthreatening questions likely to interest respondents and that can be answered with little effort.
- Introduce topics with general questions, then move to specifics. Keep all questions on a single topic together, and introduce new topics with transitional phrases.
- Use several kinds of questions and vary their lengths. Let a "Yes" be the conservative answer to some questions and the liberal answer to others. (Some people have response biases; they tend to answer all questions with "Yes" or all with "No." Some people tend to agree with statements, others to disagree, regardless of their content (Cronbach, 1950).
- When appropriate, ask screening questions. For example, instead of asking a series of questions on whether the respondent has recently read any of a long list of magazines, ask first, "Did you happen to read any magazines in the past two weeks, or not?" Sudman and Bradburn cautioned against overuse of screening questions, because respondents may learn to say "No," in order to save themselves from the follow-ups.
- Ask as few questions as possible to achieve the research objectives. Babst (1978) showed that answers to a single question on friends' use of marijuana accurately predicted answers to 100 other questions.

Assembling the Questionnaire

Strive to make the questionnaire neat and easy to read. Number all questions and leave spaces between them. Precode answers to closed questions. That is, assign every possible answer a one-digit code (two digits if there are more than 10 possible answers) as precoding makes data processing and analysis much easier. Write a cover letter that gives the name of the organization conducting the study and describes its purpose. Provide guidelines for answering and, for mail questionnaires, give the return deadline. End with a thank you.

Surveyors should pretest questionnaires on people similar to those who will be in the final sample, and they should rewrite or discard items subject to misinterpretation. Oppenheim (1966) wrote that frequently omitted, crossed out, or modified items are suspect, and also those that yield many "Don't Know" responses. Questionnaires designed to measure a single dimension, such as a particular attitude or knowledge of a single subject, should be scaled and item analyzed.

Why Survey Houses Disagree

During the 1984 American presidential campaign, commercial survey houses repeatedly reported widely different results, even when they asked virtually identically-worded questions at virtually the same times. Moreover, houses were consistent in over- or underestimating Reagan's victory margin over Mondale (Kiewiet & Rivers, 1985). Converse and Traugott (1986) offered many reasons why such discrepancies are not rare. Their analysis can help people to both evaluate and plan surveys.

Sampling Error

As indicated above, unless everybody in a population is surveyed, sampling error is unavoidable. For national surveys, most reputable houses sample 1500 people. For $N = 1500$, with a 95 percent confidence level and p estimated at 0.5, sampling error equals $\sqrt{(3.84 \times .5 \times .5)/1500} = .0253 = 2.53$ percent. But the calculation presumes simple random sampling which, as Converse and Traugott pointed out, is prohibitively expensive to implement for large surveys. Cluster sampling is the normal procedure, and sampling error increases with the degree of homogeneity within clusters.

Variability in Sample Composition

Surveyors may select samples from lists of all adults in an area, lists of all registered voters, or lists of all likely voters. Telephone interviews, personal interviews, and mailed questionnaires reach different samples, and the nonresponders to each may differ. Survey houses do not have uniform follow-up efforts to reduce nonresponse rates, and their final samples have different proportions of early responders, late responders, and nonresponders.

Different Interviewers

When people were asked questions about race, the race of the interviewer influenced their responses. Blacks were more militant and whites gave more pro-black answers to black than to white interviewers (Hatchett & Schuman, 1975; Sudman & Bradburn, 1974). Similar interactions seem likely on other sensitive issues, although they have not been clearly demonstrated (DeLamater, 1982). Each survey house has unique training procedures and standards for interviewers, and these may affect responses. In one study (National Center for Health Statistics, 1977), interviewers were required to find out if respondents had been hospitalized during the preceding year. The project supervisors checked the data against hospital records, and they reported a strong negative correlation between interviewers' accuracy and the size of their case loads.

Brenner (1982) received permission from six experienced interviewers conducting a large-scale survey to tape-record several of their sessions. Upon analyzing the tapes, he found that only two-thirds of the questions had been asked as required. The interviewers had often asked probing questions, and the majority of these were inappropriate; in one-fifth of all instances, probes requested information unrelated to the purpose of the study. Deviations from proper procedures were associated with large, consistent differences in the answers received.

Wording, Sequence, and Context of Questions

Surveyors from different houses rarely ask identically worded questions, and slight changes may yield quite different results. Consider the following questions regarding Gerald Ford's first two months as president. The first question is from a Gallup Poll, the second is from a Harris poll.

A. "Do you approve or disapprove of the way President Ford is handling his job as president?"
B. "How would you rate the job Ford is doing as president—excellent, pretty good, only fair, or poor?"

Gallup reported a 45 percent approval rating; Harris scored "excellent" and "pretty good" as positive responses and reported that only 38 percent of Americans approved of Ford.

As the specificity of questions is increased, response rates change. Bradburn (1982) contrasted responses to three questions from a Gallup Poll of 1945:

	Yes	No	Don't Know
A. Do you think the government should give money to workers who are unemployed for a limited length of time until they can find another job?	63%	32%	5%

	Yes	No	Don't Know
B. It has been proposed that unemployed workers with dependents be given up to $25 per week by the government for as many as 26 weeks during one year while they are out of work and looking for a job. Do you favor or oppose this plan?	46	42	12
C. Would you be willing to pay higher taxes to give unemployed persons up to $25 a week for 26 weeks if they fail to find satisfactory jobs?	34	54	12

Willick and Ashley (1971) interviewed college students about their political party preferences. They were much more likely to give an indefinite answer if first asked about their parents' party preferences. In a 1979 survey, Schuman and Presser (1981) asked matched samples of respondents whether married women who do not want more children should have the right to legal abortions; 60.7 percent of those who heard this as their first question about abortion said "Yes;" but of people who were asked first about the right to abortion in the case of a defective fetus, only 48.1 percent said "Yes."

Items appearing first on a list receive more favorable comments than items further down (Sudman & Bradburn, 1982), but early questions about presidential popularity are more likely than later ones to elicit a "No opinion" answer (Sigelman, 1981). See Box 6.4 for other examples of wording and context effects.

The context of a question can refer to both its place in a questionnaire and the setting in which it is asked. The previous studies showed that question order is important, and the potential relevance of setting is shown in two parallel surveys conducted in 1983 by Rootman and Smart (1985). They asked similar (although not identical) questions to comparable samples of teenagers in Ontario. For the most part, respondents gave more socially desirable answers to questions asked at home than at school.

Other Sources of Inaccuracy in Surveys

As shown in Table 11.1, pollsters can be quite accurate in predicting voting behavior from answers to survey questions. But that offers only weak cause for optimism about the accuracy of other types of surveys. In many instances, people's intentions as told to surveyors have been grossly discrepant from their subsequent behaviors. To take just one example, Schwarz (1972) reported that of 72 people who said they planned to buy a car within the next six months, only 33 actually did so. At least four differences between voting surveys and surveys on other matters contribute to error in the latter:

1. For well over 50 years, professional surveyors have carefully refined questions about voting. Few other types of survey questions have received such scrutiny.

2. It seems safe to assume that most Americans are unashamed of their political preferences and try to answer questions about them honestly. But, as previously indicated, many questions are threatening and lead to overreporting of socially desirable, and underreporting of socially undesirable, behaviors. Crowne and Marlowe (1964) developed a scale that distinguishes between people according to their need to "respond in culturally sanctioned ways," and they found that high and low scorers answer differently to questions about number of close friends, marital happiness, general happiness, and other topics. The differences may depend less on their actual status with respect to friends, happiness, and so forth, than with their attempts to answer in ways that will bring social approval.

3. The shorter the time interval between voting surveys and elections, the more accurate the predictions. Time intervals are long between some surveys and the behaviors they predict.

4. The question "Which candidate do you intend to vote for?" places no burden on respondents' memories. In other surveys, questions about the past tend to be underreported, and the magnitude of the effect increases with distance from the asked-about events. For example, the National Center for Health Statistics sponsored a series of surveys, with the following results:

    a. Reports of visits to a physician during the week prior to the survey averaged 0.21; reports of visits during the week before that averaged 0.11 (National Center for Health Statistics, 1971).

    b. Hospitalizations that occurred 1–2 months previously were underreported 4 percent; hospitalizations of 10–11 months previously were underreported 50 percent (National Center for Health Statistics, 1977).

    c. Treatments for chronic illness received during the previous week were underreported 8.6 percent; treatments received 225 to 280 days previously were underreported 51.8 percent (National Center for Health Statistics, 1967).

    d. Respondents' involvement in automobile accidents during the previous three months were underreported 3.4 percent; accidents that had occurred 7–12 months previously were underreported 27.3 percent (National Center for Health Statistics, 1972).

In 1964, Powers, Goudy, and Keith (1978) interviewed a sample of 1,870 men on various aspects of their lives. In 1974, all who could be relocated (a total of 1,332) were asked to recall their situations and attitudes of 1964. Their memories of a decade earlier were strikingly different from what they had reported back then. For example, to a question about their most important value—recreation, comfort, friends, or work—only 42 percent gave the same answer both times. Only 40 percent gave the same answer when asked to identify their 1964 yearly family income from a set of broad categories. Only 75 percent answered consistently to a question about the number of persons in the household. In most cases, the 1974 answers presented respondents in a more favorable light than did those of the initial interview.

The Kinsey Report: A Case Study

In 1948, Alfred Kinsey and his associates published a landmark book, *Sexual Behavior in the Human Male* (1948). Five years later they published a companion volume, *Sexual Behavior in the Human Female* (1953). The two books, based on interviews with nearly 18,000 people, described to a generally shocked public the sexual practices of Americans. Behaviors such as masturbation, homosexuality, and oral-genital sex, previously regarded by most Americans as abnormal, were found to be common. Kinsey's work opened the way for others to study sexual behavior scientifically.

Terman (1948) praised the magnitude of Kinsey's achievements and the significance of his research, but condemned many aspects of the methodology. Unfortunately, Terman's criticisms of 40 years ago have not been learned by many current surveyors of sexual attitudes and practices (cf. Hite, 1987). So, it seems useful to conclude this chapter with some of Terman's main criticisms.

1. Kinsey said almost nothing about how his questions were worded, other than that the manner of wording varied with the intelligence and personality of the subject being interviewed. But without standardized wording, generalizations from a sample are invalid.
2. There were strong interviewer effects. Kinsey's two primary assistants obtained results that differed significantly from each other and from him.
3. Kinsey warned that "In his tone of voice and in his choice of words the interviewer must avoid giving the subject any clue as to the answer he expects." Yet, a standard technique of his for dealing with uneducated and feeble-minded interviewees was to "pretend that one has misunderstood the negative replies and ask additional questions, just as though the original answers were affirmatives." For example, "Yes, I know you have never done that, but how old were you the *first* time that you did it?" That type of suggestive questioning is likely to generate many false responses.

4. Kinsey's interviewees gave long-distance memory reports, and these received the same weight as reports of current practices. For example, in computing mean frequency of masturbation of 15-year-olds, he gave equal weight to reports of 15-year-olds and of 50-year-olds about when they had been 15. As checks on the accuracy of the memories, he reinterviewed 162 subjects after intervals ranging between 18 months and 7 years; and he analyzed husband-wife agreements of 231 married couples who answered the same memory questions. On the basis of correlation coefficients, he reported "an amazing agreement" between statements of husbands and wives. But Terman disagreed with Kinsey's conclusion, because coefficients were high primarily for vital statistics such as number of years married and number of children; they were only .50 for frequency of sexual intercourse in early marriage, .54 for maximum frequency, and .60 for current frequency. The reinterview data, too, failed to support Kinsey's belief that memory distortions were small.
5. Kinsey did not randomly choose his sample, but instead recruited from many sources and took whoever was available. He had 32 groups of contact people, including groups of male and female prostitutes, pimps, bootleggers, gamblers, prison inmates, thieves, and hold-up men, to help him get volunteers. He gave little information on the sources of many of his subgroups, and it is impossible to judge their representativeness.
6. Kinsey required judgments beyond the capabilities of most people. For example, men were asked to fix the onset of their adolescence at the date of first ejaculation, "unless there has been evidence that ejaculation would have been possible at an earlier age if the individual had been stimulated to the point of orgasm."
7. Terman wrote that many of Kinsey's generalizations went beyond his data. But Terman's statement wasn't strong enough: Because Kinsey did not use random samples, none of his generalizations from samples to broader populations were sound.

Summary

The key steps in designing a survey research project are as follows:

1. Decide what question you want the survey to answer.
2. Search the literature for existing items that you can incorporate into your questionnaire.
3. Write the remaining items, always keeping the research question in mind.
4. Arrange the questions so that all on a topic are together.
5. Make up an answer sheet, with codes for each possible response.

6. Decide on the type of survey—mailed questionnaire, personal interview, or telephone interview.
7. If necessary, train interviewers.
8. Define the population you want surveyed.
9. Pretest the questionnaire on representative subjects from the population, and revise poor questions.
10. Get feedback from the interviewers about the questions.
11. Decide how much sampling error you are willing to tolerate and how confident you want to be that the error in your survey does not exceed that amount. Compute your sample size accordingly.
12. Design a plan for random sampling. If the population has subgroups within it that may differ greatly on the variables of interest, consider using stratified sampling.
13. Contact the people in your sample prior to mailing them a questionnaire or trying to interview them. Explain the purpose of your study and encourage them to participate.
14. Send follow-up letters or make phone calls to nonresponders.
15. Analyze your data and write up the report.
16. Take the rest of the year off.

Key Terms

Cluster Sampling   The units of a population are divided into clusters, such as city blocks or telephone numbers with a particular exchange, and the clusters are randomly sampled.

Demographics   Data about human populations, such as size, growth, density, distribution, and vital statistics.

Response Rate   The proportion of subjects selected to be in a sample who give usable data.

Sample and Population   A population is the group to which we want to generalize. A sample is those members of the group on which we collect data.

Sampling Error   The extent to which a properly chosen sample can be expected to deviate from the population from which it comes.

Simple Random Sample   A sample in which every unit within the population has an equal chance of being selected.

Stratified Random Sample   A sample in which the population is broken up into strata that share one or more characteristics; sampling of the units within each stratum is random.

Systematic Sampling   A sample in which the first unit is picked randomly and the remaining units are picked at predetermined intervals.

CHAPTER 12

Single Subject Research

By the time you finish reading this chapter, you should be able to answer the following questions:

What are the three types of research involving single cases? What characteristic do they have in common?

Under what circumstances can after-the-fact case studies be used as proof?

Do after-the-fact case studies have any scientific merit?

Why should people not draw causal conclusions from after-the-fact case studies?

In 1971, Kolansky and Moore published a highly influential paper in which they concluded that "moderate to heavy use of marijuana in adolescents and young people without predisposition to psychotic illness may lead to ego decompensation ranging from mild ego disturbance to psychosis." (Kolansky & Moore, 1971) They supported the conclusion by citing 38 cases of people who had appeared psychologically healthy until smoking marijuana, but who had then shown such serious symptoms that they required the services of a psychiatrist.

But the support was insufficient for two important reasons. First, the adolescents were not randomly assigned to smoke or not smoke marijuana, so the cause of their "ego decompensation" can't be known. Smokers and nonsmokers probably differ in many ways, for example, in use of other drugs, respect for authority, and general lifestyle. Any of these, rather than the marijuana, may account for differences between them. Second, Kolansky and Moore had only part of the information necessary for an informed decision: the number of marijuana users who later showed psychiatric symptoms. They lacked evidence on the number of users who didn't show subsequent symptoms and the numbers of nonusers who both did and didn't show symptoms. So, they were not in position to assert that marijuana users are more (or less) likely than abstainers to experience psychiatric symptoms.

The Kolansky and Moore paper continued a long medical tradition of analyzing individuals, institutions, or events after many or all of their interesting features have already occurred. These after-the-fact single case studies all share the limitations mentioned, and as a result many methodology textbooks either mention them only so they can be unceremoniously dismissed as pre-scientific or ignore them completely. But single case research comes in three varieties, and even the least rigorous of them, the after-the-fact case study, has virtues. Their distinguishing characteristic is that subjects are studied on an individual basis rather than being assigned to groups for analysis.

After-the-Fact Case Studies

The essence of after-the-fact case studies, such as the one by Kolansky and Moore, is that investigators deduce the cause of a particular event from a single type of evidence. As indicated above, causal conclusions are unwarranted. But despite their limitations, after-the-fact case studies should not be discounted completely. All scientific research begins with the observation of phenomena (cases); scientists try to guess the causes and plan their research accordingly. They should be pleased if given the opportunity to analyze an unusual event. Skinner (1956) wrote, ". . . Here was a first principle not formally recognized by scientific methodologists: when you run onto something interesting, drop everything else and study it." Investigations based on that principle, often called serendipitous, played a major role in Fleming's work that led to the isolation of penicillin, Becquerel's recognition of radioactivity, Pasteur's development of vaccination techniques, recent theories on pulsars, and many other scientific breakthroughs. After-the-fact case studies can lead to serendipitous discoveries, because they provide unique opportunities for problem-solving. Especially valuable in this respect are deviant cases—superficially similar to others in many ways, they nonetheless are somehow strikingly different. And the difference demands explanation.

After-the-fact case studies not only provide problems, they suggest answers. They occur in contexts that limit the range of possible solutions. A person seeking the cause of a severe rash and suspecting something he ate would probably not consider foods eaten beyond 48 hours, a week at most, in the past. And he'd be likely to pay special attention to novel foods and known allergens. Similarly, Kolansky and Moore focused on a plausible link to the psychiatric symptoms of their patients. Their probably erroneous conclusion (see Altman & Evenson, 1973; Leavitt, 1982 for further discussion) came, not from using case study data as a basis for speculation, but because they didn't do additional research. Whereas experiments may provide overwhelmingly strong (although never absolutely conclusive) evidence about why something occurred, after-the-fact case studies only suggest.

Yet after-the-fact case studies can occasionally be used in proofs. A single negative case may sometimes be enough for rejection of a supposedly universal relationship (Dukes, 1965). For example, Lenneberg (1962) wrote of a boy who understood language even though he had never babbled, because he lacked the necessary motor apparatus; this overthrew the belief that babbling is essential to later understanding of language. More recently, documented cases in male and female heterosexuals demolished the view that AIDS is a disease restricted to homosexual men.

After-the-fact case studies can be used for instructional purposes to provide vivid descriptions of abstract principles. This is probably one reason for their popularity in medicine. The Skinner quote, cited previously, was taken from an article aptly entitled "A Case History in Scientific Method."

Complex Analytic After-the-Fact Case Studies

Investigators who lack the means for doing experiments may nevertheless convincingly reconstruct the causes of an event: An astronomer may explain why a star went nova; an evolutionary biologist, why passenger pigeons became extinct; a criminalist, how a crime was committed; a psychotherapist, the origins of a phobic reaction; a historian, the reasons for the fall of the Roman Empire. Their single case research differs considerably from after-the-fact case studies. These investigators seek and are open to many types of evidence. They may exhibit creativity of a very high order in searching for artifacts or other physical traces; interviewing friends, lovers, enemies, witnesses, and specialists in relevant fields; analyzing newspaper reports, hospital records, marriage licenses, and other public documents; and determining the significance of comparable or contrasting cases. They may even conduct surveys or experiments in developing a case study. For example, criminalist Herbert MacDonell conducted laboratory studies on the spattering of blood as part of his analysis of a murder (1984).

Lawyers also seek varied evidence when trying to persuade juries of the innocence of their clients. And the analogy points to a weakness of complex analytic after-the-fact case studies: Whereas lawyers are ethically bound to make the strongest possible cases for their clients, scientists should evaluate evidence dispassionately. Complex analytic after-the-fact case studies offer great latitude, much more than do experiments, in choosing and interpreting materials; this increases the likelihood of bias. A second problem is that memory reports, a major data source for such studies, are notoriously unreliable.

Single Case Experimental Designs

The most popular method for testing the effects of an independent variable is to assign subjects randomly to groups. One group receives the IV, the other a placebo, and differences between the groups are analyzed statistically. Some researchers use a different strategy; they too conduct planned tests, but with only a few subjects. Each subject is exposed at least once and usually many times to each treatment. Changes in the dependent variable are measured repeatedly. Thus, each subject, whether a rat in a Skinner box, a patient receiving behavior modification therapy, or a large institution undergoing a policy change, serves as its own control. This strategy is discussed in chapter 15, where it is argued that single case experimental designs often lead to stronger conclusions than those from experiments with large groups.

Summary

Research in which individual subjects are studied intensively comes in three forms:

After-the-fact case studies, in which a single type of evidence is investigated, should be interpreted cautiously. They offer only weak proof that one event has caused another. But case studies should not be ignored. They often provide interesting problems and may suggest possible causes.

Complex analytic after-the-fact case studies involve analysis of many types of evidence and may yield justifiable causal conclusions.

With single case experimental designs, each subject is exposed at least once and usually many times to each experimental treatment. Such designs are as powerful as those from experiments with large groups.

Key Terms

Serendipity The ability to make discoveries at a tangent from the original line of research.

Single Case Study Research in which data are analyzed at the individual rather than group level.

CHAPTER 13

Using Correlation Strategies to Predict and Assess Relationships

By the time you finish reading this chapter, you should be able to answer the following questions:

Why do scientists do correlational studies?

What are the limitations of correlational studies?

What range of values can correlation coefficients have?

Does a correlation of zero between two variables always mean that the variables are unrelated?

Would you expect a higher correlation between height and weight among basketball players or in the population as a whole?

How are regression equations used for making predictions?

Which is higher, a correlation of $+.3$ or $-.6$?

Besides the size of a correlation coefficient, what determines whether the coefficient is statistically significant?

Why are there several different correlation coefficients?

Which correlation coefficient has the most power?

How have investigators used correlation analyses to try to establish causes?

Correlational studies help answer questions about how variables are related. Although of limited value for establishing causality, they cover a lot of territory. Correlations may be obtained between any variables measured simultaneously or years apart. To take just two examples, McClelland (1961) reported several significant correlations between the need for achievement within different cultures, as measured by a complex scoring system applied to popular songs and stories, and several indicators of economic and social development of those cultures. And Livson (1977), in a study spanning almost 30 years, found that the physical attractiveness of 40-year-old women correlated less highly with their attractiveness at age 12 (correlation of .29) than with their direct expression of hostile feelings at age 12 (correlation of $-.31$).

Interpreting Correlation Coefficients: Part I

Correlation coefficients range from -1.0 through 0 to $+1.0$. The more a coefficient deviates from 0, the stronger the relationship is between the variables. A positive correlation means that increases in one variable are associated with increases in the other, and a negative correlation means the reverse. For example, since tall people tend to outweigh short ones, height and weight are positively correlated; and the correlation is probably negative between hours of television watching per week and college grade point average. Correlations of $+1.0$ or -1.0, which indicate a perfect relationship between variables, are exceedingly rare in social science research. Few are higher than the correlation of 0.88 found for IQ scores of identical twins reared together (Erlenmeyer-Kimling & Jarvik, 1963). The heights of parents and their children show a correlation of about .50.

Do you think the heights of daughters correlate more highly with their mothers' heights and those of sons with their fathers'? Why or why not? (Answer at the end of the chapter.)

The procedures presented in this chapter for calculating correlation coefficients can detect only linear relationships between variables—those described by equations of the form $Y = a + bX$ and represented graphically by straight lines. Shown in Figure 13.1a is a graph in which Celsius scale temperatures are plotted on the X-axis and the corresponding Fahrenheit temperatures on the Y-axis. The equation summarizing the relationship between the two scales is $Y = 32 + 1.8X$. Each data point fits right on the line, because the relationship between Fahrenheit and Celsius is perfect. That is, the correlation between them is $+1.0$.

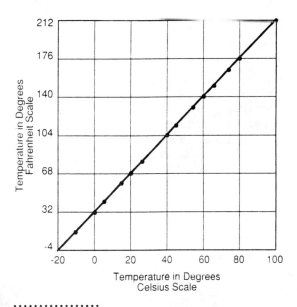

Figure 13.1a
Each dot on the straight line indicates a data point, where temperatures were converted from Celsius to Fahrenheit. Temperatures are plotted in degrees Celsius on the x-axis and degrees Fahrenheit on the y-axis.

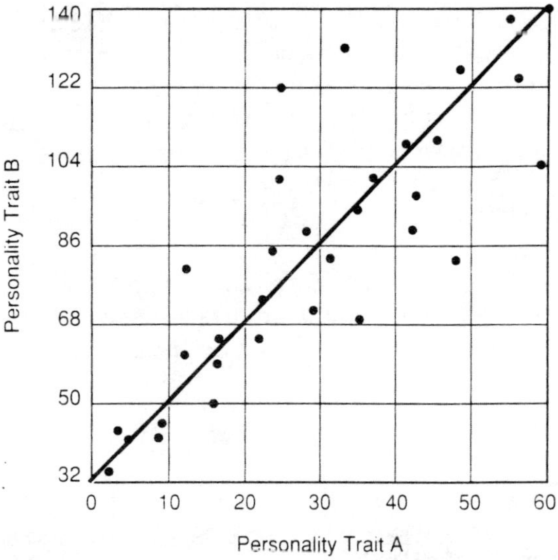

Figure 13.1b
Each dot represents a single person's score on two tests, one to measure personality trait A and one to measure personality trait B.

A fictitious relationship between two personality traits A and B is graphed in Figure 13.1b. The best summary of the linear relationship between the traits is again $Y = 32 + 1.8X$. But because A and B are not perfectly correlated, there are many deviant scores.

A fictitious relationship between two personality traits C and D is graphed in Figure 13.1c. No straight line fits well, although the relationship is described perfectly by the equation $Y = X^4 - 7X^3 + 50$. Many strong relationships, like that between C and D, can't be fit to straight lines. For example, young adults on average run faster than both toddlers and elderly people, so the correlation between running speed and age is best described by a curvilinear function. The procedures for finding such functions are beyond the scope of the present material, but be aware that a low or zero correlation means only that the variables show no linear relationship.

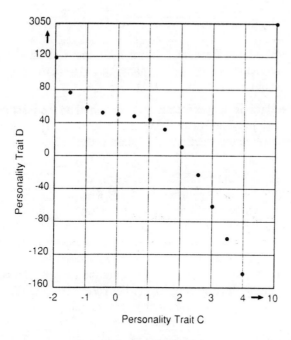

Figure 13.1c
Each dot, including the one at x = 10, y = 3,050, represents a single person's score on two tests, one to measure personality trait C and one to measure personality trait D.

When Should a Correlational Study Be Considered?

1. Correlations are computed to evaluate the reliability of tests. There are three major methods:

   a. split-half: If a test is comprised of many items, the items are divided randomly into two halves and the correlation computed between subjects' scores on both halves. The split-half reliability of the Wechsler Adult Intelligence Scale is about .95, which is about as high as that for any psychological test (Murphy and Davidshofer, 1988).

   b. test-retest: The test is given at two separate times and the correlation between scores computed. The Test Anxiety Questionnaire has 37 items such as:

   26. When you are taking a course examination, to what extent do you feel that your emotional reactions interfere with or lower your performance?

Do not interfere with it at all	Midpoint	Interfere a great deal

   Test-retest reliability over a six-week period was .82 (Mandler & Sarason, 1952).

c. alternate forms: Two different forms of the same test are administered and the correlation between scores computed. The Wonderlic Personnel Test is a 50-item test used as part of the selection procedure for many jobs. Alternate form reliability is about .90 (Murphy, 1984).

2. Correlations between tests indicate whether the tests measure the same thing. The Raven Progressive Matrices Test is a nonverbal intelligence test that correlates .60 with the Stanford-Binet (Keir, 1949).

3. Criterion-related validity, assessed by correlating scores on a test with scores on a criterion, indicates whether the test measures something useful. In one study, the "test" was medical school grade point average of people who had graduated and become physicians. The criteria were 76 measures of doctor performance. Ninety-seven percent of the correlations were nearly zero (Taylor et al., 1965). *[margin note: No relationship]* In another study that calls into question the usefulness of a popular assessment device, Wagner (1949) correlated judgments from interviews with criteria such as grades and performance on standardized tests; the median correlation was .19.

4. Correlations may have theoretical importance, as they did for Sheldon (1942). His personality theory was based on evidence of strong correlations between people's body types and their personalities. He found correlations of about .80 between endomorphy (softness and spherical appearance) and viscerotonia (love of comfort, sociability, and gluttony for food, people, and affection); between mesomorphy (hard, rectangular, muscular body) and somatotonia (love of physical adventure, risk-taking, and vigorous activity); and between ectomorphy (flat chest, linear, fragile) and cerebrotonia (restraint, inhibition, and desire for concealment). Sheldon has been justly criticized on the grounds that he knew the body types of the people whose personalities he judged. Yet his work has many interesting implications and is still widely reported in introductory textbooks. It has led to a great deal of additional research.

5. Correlational studies are used when a variable, for either ethical or practical reasons, can't be manipulated. An experimenter interested in how alcohol affects memory could ask subjects to voluntarily elevate their blood alcohol concentrations; but an investigator wishing to study the relationship between gender and memory could not manipulate genders (unless he had a *very* large budget and the subjects were *very* adventurous). Traits that people carry with them (gender, age, race, country of origin, species, test scores, political attitudes, favorite songs . . .) can't be manipulated

experimentally, but can be studied correlationally. (But recall that correlational studies answer different questions than do experiments.)

6. Correlation does not prove causation, but causes do imply correlations. That is, if one variable causes a reliable change in another, a correlation must exist between them. So, correlational studies may be used as inexpensive pretests of causal hypotheses. For example, an experimental test of the hypothesis that long periods of overcrowding increase human aggression would be enormously expensive. But a researcher could easily compute the correlations between population density of different cities and incidence of violent crimes within the cities. Relevant data are available. Only if the correlations were high enough would the researcher follow them up with rigorous experimental tests.

7. Correlations can be used as a measure of the size of effect of one variable upon another. Reporting effect sizes conveys much more information than mere mention of statistical significance.

8. The technique of regression analysis, closely related to correlation analysis, is used to improve the accuracy of predictions. For example, college admissions officers use regression equations to estimate the likelihood of a freshman's academic success given his SAT scores.

Predicting by Means of Regression Equations

Regression analysis is used for finding the equation that most accurately summarizes the relationship between two variables. (In what follows, I discuss only linear equations, of the form $Y = a + bX$). Once a regression equation is found, new values of X within the range of the original data can be plugged into it and will give the best possible predictions for Y.

The first step in developing a regression equation is to collect data on subjects similar to those about whom predictions will be made. For example, an Army psychologist might give a newly developed test to all enlistees over the course of a year. The next step is to collect data on a criterion; in this example, a reasonable one might be performance ratings given to the enlistees by their superior officers. Third, a regression equation is computed between the two variables. The equation can be used to predict the performance ratings of new recruits from their test scores.

Six subjects are listed in the first column of Table 13.1. The next two columns give their scores on the variables X and Y. To find the regression equation, the values $X \times Y$ and X^2 are also required, so these have been computed and given in columns 4 and 5. Ignore column 6 for now.

Table 13.1 Finding a Regression Equation

Subject	X	Y	X × Y	X²	Y²
1	4	−2	−8	16	4
2	5	10	50	25	100
3	3	4	12	9	16
4	15	12	180	225	144
5	0	−5	0	0	25
6	6	13	78	36	169
Sums	33	32	312	311	458

To find an equation of the form $Y = a + bX$, find values for a and b according to the following formulae:

$$a = \bar{Y} - b\bar{X}$$

$$b = \frac{N\Sigma XY - (\Sigma X)(\Sigma Y)}{N\Sigma X^2 - (\Sigma X)^2}$$

where

$N =$ number of pairs of scores
(in example, $N = 6$)

$\bar{X}$ and $\bar{Y} =$ the means of the X and Y values
and ($\bar{X} = 33/6 = 5.5$; $\bar{Y} = 32/6 = 5.33$)

$\Sigma XY =$ the sum of all the XY products
($-8 + 50 + 12 + 180 + 0 + 78 = 312$)

$\Sigma X \Sigma Y =$ the sum of the X times the sum of the Y
($33 \times 32 = 1056$)

$\Sigma X^2 =$ the sum of the square of each X
($16 + 25 + 9 + 225 + 0 + 36 = 311$)

$(\Sigma X)^2 =$ the square of the sum of the Xs
($33^2 = 1089$)

Then, $b = (6 \times 312 - 1056)/(6 \times 311 - 1089) = 816/777 = 1.05$ and $a = 5.33 - 1.05 \times 5.5 = -.445$ so $Y = -.445 + 1.05X$.

If a seventh subject scored 5 on X, the best guess for her Y score would be $-.445 + 1.05 \times 5 = 4.805$. Yet there was an X score of 5 in the original data, and it was paired with a Y of 10. This shows that the equation does not fit the data perfectly; unless the correlation between X and Y is either $+1.0$

or -1.0, errors in prediction will occur. However, the regression equation minimizes them. (More accurately, it minimizes the sum of the squares of the errors. Squared values are used so that positive and negative errors don't cancel each other out.)

The Correlation Coefficient for Variables Measured with an Interval Scale—The Pearson Product-Moment Correlation

As you will soon learn, there are several different correlation coefficients. If there is a true relationship between variables, the product-moment is most likely to detect it. But the product-moment can be used only with data measured on an interval scale. (Recall that interval scales tell the amount by which two scores differ, in contrast to ordinal scales that indicate only whether one score is better or worse than another.)

The formula for r is

$$r = \frac{N\Sigma XY - (\Sigma X)(\Sigma Y)}{\sqrt{[N\Sigma X^2 - (\Sigma X)^2][N\Sigma Y^2 - (\Sigma Y)^2]}}$$

Note that the numerator and part of the denominator are the same as for b in the regression equation. So, for the data of Table 13.1 for which the regression equation has already been calculated, the only new values needed are $N\Sigma Y^2$ and $(\Sigma Y)^2$. These can be found from the table.

$$r = 816/\sqrt{[(6 \times 311) - 33^2)][(6 \times 458) - 32^2)]}$$
$$= 816/\sqrt{777 \times 1724}$$
$$= .704$$

Even a perfect correlation may occur by chance, so the final computational step is to see if the value is statistically significant. Table 13.2 lists the values of r needed for significance for different numbers of subjects.

Example:

An investigator intends to pursue a line of research only if the product-moment correlation between variables X and Y is at least .70. How many subjects should be run?

The investigator wants a correlation of .70 or more to be significant at the .05 level, and a correlation less than .70 to be nonsignificant. From Table 13.2, a correlation of .71 is required for significance with eight pairs of subjects, and a correlation of .67 is required with nine pairs. So the investigator would need nine pairs.

Table 13.2 Critical Values for Significance at the .05 and .01 Levels (Two-Tailed) of the Product-Moment and Point-Biserial Correlation Coefficients

Number of Pairs	.05 Level	.01 Level	Number of Pairs	.05 Level	.01 Level
3	1	1	21	.43	.55
4	.95	.99	22	.42	.54
5	.88	.96	23	.41	.53
			24	.40	.52
6	.81	.92	25	.40	.51
7	.75	.87			
8	.71	.83	26	.39	.50
9	.67	.80	27	.38	.49
10	.63	.77	28	.37	.48
			29	.37	.47
11	.60	.74	30	.36	.46
12	.58	.71			
13	.55	.68	35	.33	.43
14	.53	.66	40	.31	.40
15	.51	.64	45	.29	.38
			50	.28	.36
16	.50	.62	60	.25	.33
17	.48	.61			
18	.47	.59	70	.24	.31
19	.46	.58	80	.22	.29
20	.44	.56	90	.21	.27
			100	.20	.26

Interpreting Correlation Coefficients: Part II

As can be seen from the computational formulae, correlations and regression equations are closely related. The closer the absolute value of the correlation coefficient is to 1.0, the less error there will be in predicting from X to Y. To understand the relationship better, suppose that somebody had printed each of the Y scores from Table 13.1 on a card, then shuffled the cards and dealt them in random order. Suppose further that, knowing nothing about the X scores, you had to guess the value of each card. If your goal were to minimize errors (for statistical purposes, it's most convenient to minimize the sum of the square of each error), your best strategy would be to guess 5.33 (the mean of the Y scores) for each card. In the first four columns of Table 13.3, I show the X scores, the corresponding Y scores, the values $Y - 5.33$, and $(Y - 5.33)^2$. Column 5 lists each predicted Y from the regression equation $Y = -.445 + 1.05X$, and columns 6 and 7 list $Y -$ predicted Y and $(Y -$ predicted $Y)^2$. The sum of the $(Y - 5.33)^2$ values is 287.34. The sum of the $(Y -$ predicted $Y)^2$ values is 144.51. Now consider the following, recalling that the correlation between X and Y is .704.

$(287.34 - 144.51)/287.34 = .497$; and $r^2 = .704^2 = .497$

Table 13.3 Comparing Errors When Y is Predicted from the Regression Equation and When the Mean of Y is Used

X	Y	Y − 5.33	(Y − 5.33)²	Y'*	Y − Y'	(Y − Y')²
4	−2	−7.33	53.73	3.755	−5.755	33.12
5	10	4.67	21.81	4.805	5.195	26.99
3	4	1.33	1.77	2.705	1.295	1.68
15	12	6.67	44.49	15.305	−3.305	10.92
0	−5	−10.33	106.71	−0.445	−4.555	20.75
6	13	7.67	58.83	5.855	7.145	51.05

*Y' = predicted value for Y

The coefficient of determination, r^2, shows the extent to which the regression equation can improve predictions. If the correlation between X and Y is 0, the best prediction for a new Y value is the mean of Y. If the correlation is not zero and a regression equation is used, errors will be reduced. Let SSY = the sum of the squared errors when predicting from the mean of Y; let SSR = the sum of the squared errors when predicting from the regression equation. Then

$$r^2 = (SSY - SSR)/SSY$$

Variance, a measure of variability, provides another way of understanding the coefficient of determination. The variance of a set of data is given by the formula

$$\text{Variance} = \frac{\Sigma(X - \bar{X})^2}{N}$$

Thus, the variance of the original Y scores is 287.34/6 and the variance of the predicted Y scores is 144.51/6. So, knowing X reduces the variance of Y; the proportion by which variance of Y is reduced is equal to r^2.

Three Potential Problems

1. An investigator who is aware of a subject's score on one variable while measuring a second may allow his biases to intrude and influence the results. (See the comment on Sheldon's research.) Avoid the problem by having different people collect data for each variable.
2. Unless two variables are measured reliably, the correlation between them can't be high. A low reliability means that a variable does not correlate well even with itself. To find the maximum correlation between two variables, multiply their separate reliabilities and find the square root of the product. Suppose, for example, that a scientist is interested in the relationship between activity level and verbal aggression. If activity and aggression can be measured with .72 and .50 reliabilities, respectively, the highest possible correlation between them is the square root of .72 × .50 = .60.

Test-Retest Reliability

One way to estimate reliability is to administer a measure twice and find the correlation between the two sets of scores. Columns 1, 2, and 3 of the table below list 10 people and the scores they earned on separate administrations of a test. Call the first scores X, the second, Y. To compute the product-moment correlation, find, for each pair of scores, XY, X^2, and Y^2. These have been worked out in columns 4, 5, and 6.

Subject	X	Y	XY	X^2	Y^2
1	36	40	1440	1296	1600
2	44	47	2068	1936	2209
3	30	26	780	900	676
4	25	28	700	625	784
5	42	39	1638	1764	1521
6	50	45	2250	2500	2025
7	47	50	2350	2209	2500
8	43	50	2150	1849	2500
9	35	41	1435	1225	1681
10	38	40	1520	1444	1600
	390	406	16331	17684	17096

The additional values to be calculated are $(\Sigma X)^2$ ($= 390 \times 390 = 152100$), $(\Sigma Y)^2$ ($= 406 \times 406 = 164836$), and $\Sigma X \times \Sigma Y$ ($=390 \times 406 = 158340$). The numbers can now be plugged into the formula for the product-moment correlation.

$$r = \frac{(10 \times 16331) - 158340}{\sqrt{[(10 \times 17684) - 152100][(10 \times 17096) - 164836]}}$$

$$= \frac{5070}{\sqrt{24740 - 6124}}$$

$$= 5070/12310 = .412.$$

Notes:

1. A good estimate of test-retest reliability requires many more than 10 subjects.
2. Test-retest reliability is often impractical. When people are tested twice, their second round answers may be affected by memories of the first. More generally, any measure may change what is being measured.

....

3. The strength of a correlation is tied to the specific sample tested. The more homogeneous a sample with respect to one or both variables, the smaller the correlation between the variables. Thus, the correlation between height and weight is lower among basketball players than in the population as a whole; between reading ability and school performance, lower for college than grade school students (because poor readers are likely to drop out

BOX 13.2

Item Analysis

An important part of test construction is elimination of bad items. One way to do this is to correlate scores on each item with scores on the entire test. Suppose, for example, that 20 people assign a number between 1 and 5 to a series of statements about a particular issue; 5 indicates strong agreement and 1 strong disagreement with each statement. The scores on the items are summated to get total scores for each person, as shown in column 2 of the table to the right. To measure the discriminative power of, say, item 5, list each person's score on that item, as in column 3.

Then,

$N\Sigma XY = 45560$; $\Sigma X \Sigma Y = 41122$; $N\Sigma X^2 = 53522$;

$(\Sigma X)^2 = 502681$; $N\Sigma Y^2 = 4080$; and $\Sigma Y^2 = 3364$;

and $r = \dfrac{45560 - 41122}{\sqrt{(535220 - 502681)(4080 - 3364)}}$

$= \dfrac{4438}{4827}$

$= .919$

Item 5 discriminates very well.

Testee	Total Score	Score on Item 5
Ann	20	2
Bob	45	5
Cass	40	3
Dan	38	3
Ed	20	1
Flo	33	2
Gail	44	4
Hy	29	1
Ike	30	2
Joe	50	5
Ken	43	4
Lou	40	3
Meg	47	5
Ned	23	1
Oby	25	2
Pat	40	3
Quinn	35	3
Roy	42	4
Sam	25	1
Tom	40	4

---

of school before reaching college); and between scores on a screening test and job performance, lower when only those who pass the test, rather than all testees, are considered. So, be careful about generalizations from correlational data.

Box 13.1 shows how to compute test-retest reliabilities. Box 13.2 shows a method, widely used in the construction of a type of questionnaire called a Likert scale, for analyzing the effectiveness of individual test items.

When There Are Only Two Categories for Each Variable—Phi Coefficient

Because variables cannot always be measured on an interval scale, the product-moment correlation is not always appropriate. When both variables are measured nominally and have only two possible values, the phi coefficient should be used. The value of phi always falls between 0 and +1.

Table 13.4 *Data for Computing the Phi Coefficient*

		Sensitive Skin	Insensitive Skin
Level of Unresolved Hostility	High	7 (A)	5 (B)
	Low	3 (C)	5 (D)

Suppose an investigator tests the hypothesis that people with sensitive skins, as measured by their reactions to allergens, have a great deal of unresolved hostility. She finds 10 people with sensitive and 10 with nonsensitive skins, gives them a test to measure unresolved hostility, and uses the two sets of scores to assign each person to one of four groups as in Table 13.4.

The formula for phi is

$$\text{phi} = \frac{|AD - BC|}{\sqrt{(A + B)(C + D)(A + C)(B + D)}}$$

$|AD - BC|$ means the absolute value of $AD - BC$, so it is always positive.

In the example,

$$\text{phi} = \frac{35 - 15}{\sqrt{12 \times 8 \times 10 \times 10}} = \frac{20}{\sqrt{9600}} = .20$$

To test for the significance of phi, use the formula $X^2 = N\text{phi}^2$ where $N = A + B + C + D$. So, $X^2 = 20 \times (.20)^2 = .8$. If X^2 is greater than 3.8, the results are significant at the 0.05 level (two-tailed). For the 0.01 level, X^2 must reach 6.6. So the result is not significant.

Note: For simplicity's sake, I chose small numbers to work with. But for a proper X^2 test, each of the four cells must have at least five cases.

Boxes 13.3 to 13.5 show how to calculate correlations when (a) both variables are measured nominally and one or both of them is assorted into more than two categories, (b) both variables are measured on ordinal scales, and (c) one variable is measured on an interval and one on a nominal scale. One additional method for interval data is shown in Box 13.6. It has low power (much less likely than the product-moment correlation to detect a relationship) but can be computed rapidly.

Partial Correlation

Tobacco company spokespeople, recognizing that the evidence for a relationship between cigarette smoking and cancer is almost all correlational, might argue (have argued) that a third factor such as anxiety causes both cancer and smoking. Suppose they do a correlational study and find that level of anxiety correlates .60 ($r_{CA} = .6$), smoking, .50 ($r_{SC} = .5$), with incidence of cancer. If both correlations are statistically significant, would they be justified in claiming that the effects of smoking are largely irrelevant to the genesis of cancer?

BOX 13.3

The Contingency Coefficient

The contingency coefficient, C, is used to find the extent of association between any two variables. Its main application is to variables measured nominally, one or both of which can be assigned to more than two categories. C, like the phi coefficient, can never be negative. The formula for C is

$$\sqrt{\frac{\chi^2}{\chi^2 + N}}$$

Calculating C requires prior calculation of χ^2, a procedure given in most introductory statistics textbooks. The significance of C is found by testing the significance of the χ^2. Although widely used in behavioral research, C has several limitations (Siegel, 1956):

- Even if two variables are perfectly associated, C cannot reach one. Its upper limit depends on the number of categories of X and Y.
- Two contingency coefficients are not comparable unless derived from contingency tables of the same size.
- C is not directly comparable to other measures of correlation.

One type of evidence to answer the question is provided by the technique of partial correlation. To compute the partial correlation, the tobacco advocates would first have to find the correlation between level of smoking and stress. Suppose it were 0.2. Then, the correlations between smoking and anxiety (r_{SA}) and anxiety and cancer (r_{AC}) could be statistically eliminated; and the partial correlation between smoking and cancer ($r_{SC \cdot A}$) would equal

$$\frac{r_{SC} - r_{SA}\, r_{CA}}{\sqrt{1 - r_{SA}^2}\sqrt{1 - r_{CA}^2}} = \frac{.5 - (.2)(.6)}{\sqrt{1 - .2^2}\sqrt{1 - .6^2}} = \frac{.38}{\sqrt{.96}\sqrt{.64}} = .49$$

The formula for the correlation between any variables 1 and 2, with the effects of variable 3 statistically eliminated, is

$$r_{12 \cdot 3} = \frac{r_{12} - r_{13}\, r_{23}}{\sqrt{1 - r_{13}^2}\sqrt{1 - r_{23}^2}}$$

Correlation and Causation—Additional Comments

Correlation does not prove causation.
Correlation does not prove causation.
Correlation . . .

Those words, drummed into the brains of social science majors from their first exposure to methodological issues, are discouraging; scientists generally seek causes yet must often rely exclusively on correlational data. Several recent developments, and some not so recent, have been aimed at resolving this problem. See Blalock, 1984; Blalock, 1985; Heise, 1975 for more extensive discussions.

BOX 13.4

The Spearman Rank-Order Correlation

The Spearman r is used when people or objects are ranked on each of two variables, as in Table 13.5. The last two columns of the table indicate the differences between the rankings on each variable (Ds) and the squares of those differences (D^2s) for each subject. (Since the square of a negative number is positive, the D^2s will always be positive; so the signs of the Ds can be ignored.)

The formula for computing the Spearman r is

$$r = 1 - \frac{6\Sigma D^2}{N(N^2 - 1)}$$

where N is the number of paired scores.

So, $r = 1 - (6 \times 992)/(20 \times 399) = .25$.

If two scores are tied, assign each of them the average of the ranks that would have been assigned had no ties occurred. If the two subjects scoring highest on the first test had gotten identical scores, they would have each received a rank of 1.5. If there are many ties, a correction factor must be incorporated (Rosenthal & Jacobson, 1968).

To test for statistical significance, if there are at least 30 pairs, use the formula

$$z = r\sqrt{N - 1}$$

If z is greater than 1.96 or less than -1.96, the results are significant at the 0.05 level. For the 0.01 level, z must reach ± 2.58. If there are fewer than 30 pairs, use Table 13.6.

Table 13.5 Computing the Spearman r

Rank on First Test	Rank on Second Test	Difference Between Ranks	Difference Squared
X	Y	D	D^2
1	3	2	4
2	7	5	25
3	17	14	196
4	13	9	81
5	2	3	9
6	16	10	100
7	9	2	4
8	5	3	9
9	4	5	25
10	10	0	0
11	12	1	1
12	1	11	121
13	15	2	4
14	20	6	36
15	18	3	9
16	14	2	4
17	19	2	4
18	8	10	100
19	11	8	64
20	6	14	196
			992

Example: The scores of nine people on two variables, X and Y, are given in Table 13.7. Compute the Spearman r.

First, convert the scores to rankings as in Table 13.8. Then, from the formula,

$$r = 1 - (6 \times 112.5)/(9 \times 63)$$
$$= 1 - 1.19 = -.19.$$

The correlation is not statistically significant. Yet X and Y are strongly, in fact perfectly, related by the equation

$$Y = X^4 - 7X^3 + 50.$$

The same equation was given earlier and is repeated to emphasize that the correlational procedures of this chapter are suitable for linear relationships only.

Table 13.6 Table of Critical Values of r_s, the Spearman Rank Correlation Coefficient*

N	Significance level (one-tailed test)	
	.05	.01
4	1.000	
5	.900	1.000
6	.829	.943
7	.714	.893
8	.643	.833
9	.600	.783
10	.564	.746
12	.506	.712
14	.456	.645
16	.425	.601
18	.399	.564
20	.377	.534
22	.359	.508
24	.343	.485
26	.329	.465
28	.317	.448
30	.306	.432

*Adapted from Olds, E. G. 1938. Distributions of sums of squares of rank differences for small numbers of individuals. Ann. Math. Statist., 9, 133–148, and from Olds, E. G. 1949. The 5% significance levels for sums of squares of rank differences and a correction. Ann. Math. Statist., 20, 117–118, with the kind permission of the author and the publisher.
Used by permission of the Institute of Mathematical Statistics.

Table 13.7 Fictitious Scores of Nine People on Two Variables

X	Y
0	50
1	46
2	10
3	58
4	-142
5	-200
6	-156
7	50
8	562

Table 13.8 Data from Table 13.7 Expressed as Rankings

X	Rank of X	Y	Rank of Y	D	D²
0	1	50	6.5	5.5	25
1	2	44	5	3	9
2	3	10	4	1	1
3	4	58	8	4	16
4	5	-142	3	2	4
5	6	-200	1	5	25
6	7	-156	2	5	30.25
7	8	50	6.5	1.5	2.25
8	9	562	9	0	0

BOX 13.5

The Point-Biserial Correlation

The point-biserial correlation is used when one variable is expressed on an interval scale and the other is dichotomous, for example, male or female, Democrat or Republican, above C average or below C average. An important use for the point-biserial is to test for strength of relationship in experiments involving two groups (see p. 331).

Table 13.9 presents a hypothetical set of test scores of students in a small class. The point-biserial can be used to find the relationship between test scores and gender. First, assign all the females to category 1 and all the males to category 0. (One level of the variable must be assigned a 1, the other a 0, but which gets which is arbitrary.)

If the variable measured on the interval scale is called Y, the formula for the point-biserial r is

$$r = \frac{N\Sigma Y_1 - N_1 \Sigma Y}{\sqrt{N_1 N_0 [N\Sigma Y^2 - (\Sigma Y)^2]}}$$

where ΣY = the sum of all the Y values

21 + 20 + 20 + 20 + 18 + 15 + 13 + 12 + 10 + 9 = 158

ΣY^2 = the sum of the squares of each Y

$21^2 + 20^2 + 20^2 + 20^2 + 18^2 + 15^2 + 13^2 + 12^2 + 10^2 + 9^2 = 2684$

$(\Sigma Y)^2$ = the square of the sum of each Y

$(21 + 20 + 20 + 20 + 18 + 15 + 13 + 12 + 10 + 9)^2 = 249664$

Table 13.9 *Data for Computing the Point-Biserial Correlation*

Student	Score	Category
Donna	21	1
Erika	20	1
Anthony	20	0
Marianne	20	1
Gayle	18	1
Phyllis	15	1
Bob C.	13	0
Carlos	12	0
Bob M.	10	0
Corinne	9	1

ΣY_1 = the sum of test scores of people in category 1

21 + 20 + 20 + 18 + 15 + 9 = 103

N_1 = the number of people in category 1 = 6
N_0 = the number of people in category 0 = 4
N = the total number of people = 10

So, $r = \dfrac{(10 \times 103) - (6 \times 158)}{\sqrt{(6 \times 4)[(10 \times 2684) - 158^2]}}$

$= \dfrac{1030 - 948}{\sqrt{24 \times (26840 - 24964)}}$

$= 82/212.2$

$= .39.$

The significance of the point-biserial, like that of the product-moment, can be tested with Table 13.9. For 10 pairs, the correlation must be at least .63 to be significant at the .05 level. So, the results are not significant.

BOX 13.6

A Quick and Dirty Method

This method for testing whether a correlation is statistically significant is simple but has low power. To compute, use the data of Table 13.7 and do the following:

1. Get the means of X and Y.

$\bar{X} = (0 + 1 + 2 + 3 + 4 + 5 + 6 + 7 + 8)$
$= 45/9 = 5.0$
$\bar{Y} = (50 + 46 + 10 + 58 - 142 - 200 - 156 + 50 + 562)/9 = 276/9 = 30.7$

2. For each pair, determine if X and Y are either both above or both below the means for their groups (if so, score the pair +); or if one of the pair is above and the other below (score the pair −); or if one or both of the pair is at the mean for its group (score the pair 0).

The third and fifth pairs (2/10 and 4/−142) are both below the means, and the eighth and ninth (7/50 and 8/562) both above, so there are 4 +s. There are also 4 −s, and pair 6 gets a 0.

3. Subtract the number of minuses from the number of pluses.

$4 - 4 = 0.$

4. Find $2\sqrt{N}$, where N = the number of pairs that were scored either plus or minus.

$2\sqrt{8} = 5.68.$

5. If the result from step 3 is greater than the result from step 4, the correlation is significant at the 0.05 level. To test at the 0.01 level, multiply $2\sqrt{N}$ in step 4 by 2.6.

Table 13.10   Fictitious Scores on Three Tests

Subject	X	Y	Z
1	31	21	1
2	19	23	2
3	18	18	3
4	12	15	4
5	6	20	5

Listed in Table 13.10 are scores on tests X, Y, and Z. Questions 1–3 pertain to the table and should be answered without using pencil and paper.

1. Does test Y or Z correlate higher with test X?

2. Is the Spearman rank order or the Pearson product-moment correlation higher between X and Z?

3. If the first subject's X score was changed from 31 to 100, would the Spearman or Pearson correlation between X and Z be more affected?

Table 13.11 *Fictitious Data on Correlations in IQ Scores Between Children and Both Their Adoptive and Biological Parents*

Adoptee	100	105	110	115	120	122	125	127	130	133
Adoptive Parents	125	120	145	118	133	150	121	111	129	127
Biological Parents	75	80	85	89	94	97	99	103	110	112

The fictitious data of Table 13.11 are based on an actual study, but the results are much more decisive than in the original. Row 1 lists the adult IQs of 10 people who were adopted at birth. Immediately underneath, in row 2, are listed the mean IQs of their adoptive parents. Row 3 lists the mean IQs of their biological parents.

4. Do the IQs of the adoptees correlate more highly with their adoptive or biological parents? Do not use pencil and paper to figure it out.

5. Do the results support the conclusion that a child's environment has little effect on her adult IQ?

If a high, reliable correlation exists between X and Y, then X may cause Y or Y may cause X or another variable or variables may cause both. (In many cases, X and Y are interdependent such that X causes Y and then Y causes a stronger X; for example, a depressed person may sit home and avoid friends, which may deepen the depression.) Simple reflection may eliminate one or more of the possibilities. High school and college grade point averages correlate significantly. It is unlikely that high school grades affect college grades to any great extent, and inconceivable that the causal arrow is in the opposite direction. But why?

The conclusion that high school grades do not appreciably influence college grades comes from the belief that academic performance at all levels is determined primarily by factors such as intelligence, motivation, and work habits. Nevertheless, Rosenthal and Jacobson's research (1968) suggests that a student's transcript may bias teachers' ratings of the student; and both self-esteem and aspiration level are influenced by grades and probably affect subsequent performance (Wattenberg & Clifford, 1964).

The statement that college grades do not affect high school grades is regarded with much more confidence, in fact, with certainty, and would be so regarded even if nothing were known about the roles of intelligence, motivation, and work habits. The reason is that our logic requires causes to precede

Table 13.12 *Fictitious Data on Relationship Between Stress and Drug Abuse*

		1989				
		LS/NA	LS/DA	HS/NA	HS/DA	Total
	LS/NA	24(A)	6(B)	4(C)	15(D)	49
1986	LS/DA	12(E)	16(F)	5(G)	10(H)	43
	HS/NA	11(I)	8(J)	7(K)	17(L)	43
	HS/DA	9(M)	13(N)	2(O)	18(P)	42
	Total	56	43	18	60	177

effects. An extension of the logic forms the basis for techniques in which correlations are analyzed for causes. Both techniques described below require that correlations between variables be studied at two different points in time.

Lazarsfeld's Approach

Heavy drug abusers tend to have high levels of stress (Lawson & Winstead, 1978), but whether drug abuse causes the stress or stressed people seek solace in drugs is uncertain. Table 13.12 presents hypothetical data—information on people who, in both 1986 and 1989, took a stress inventory and answered a questionnaire about their drug use. Both variables were dichotomized at each time period, i.e., the subjects were classified as having either low or high stress (LS or HS) and being either drug abusers or non-abusers (DA or NA). (To use the approach introduced by Lazarsfeld (1948), the variables must be dichotomized.) The numbers in Table 13.12 represent the number of people in each condition. For example, 12 people were both LS/DA in 1986 and LS/NA in 1989. The letters in parentheses name the cells for reference.

The cells along the major diagonal (A, F, K, P) show the number of people who did not change between 1986 and 1989. The cells along the minor diagonal (D, G, J, M) show the ones who changed in both traits. These are not considered further. Four cells (E, L, I, H) move in a direction that increases the correlation between high stress and drug abuse; for example, people in cell E move from low stress and drug abuse in 1986 to low stress and non-abuse in 1989. The remaining four cells (B, O, C, N) move in a direction that reduces the correlation.

One of several formulae suggested by Lazarsfeld (1948) for computing the index of effect of variable A on variable B is

$$I_{AB} = \frac{EL - IH}{(E + H)(I + L)} - \frac{BO - NC}{(B + C)(N + O)}$$

Table 13.13 Fictitious Data Showing the Relationship Between Drug Status in 1986 and Stress Level in 1989

		1989 Stress Level	
		LS	HS
1986 Drug Status	NA	24 + 6 + 11 + 8 = 49	4 + 15 + 7 + 17 = 43
	DA	12 + 16 + 9 + 13 = 50	5 + 10 + 2 + 18 = 35

Table 13.14 Fictitious Data Showing the Relationship Between Stress Level in 1986 and Drug Status in 1989

		1989 Drug Status	
		NA	DA
1986 Stress Level	LS	24 + 4 + 12 + 5 = 45	6 + 15 + 16 + 10 = 47
	HS	11 + 7 + 9 + 2 = 29	8 + 17 + 13 + 18 = 56

The A variable is the one that changes in cells C, H, I, and N from time one to time two. In Table 13.12, stress is A, drug use is B.

$$I_{AB} = \frac{(12 \times 17) - (11 \times 10)}{(12 + 10)(11 + 17)} - \frac{(6 \times 2) - (13 \times 4)}{(6 + 4)(13 + 2)}$$

$$= \frac{204 - 110}{22 \times 28} - \frac{12 - 52}{10 \times 15}$$

$$= +.40$$

The positive sign indicates that in this hypothetical example, level of stress has a greater effect on drug abuse than drug abuse on stress.

Cross-Lagged Panel Correlation

Six phi coefficients can be computed from the data of Table 13.12, and a comparison of two of them forms the basis for a technique developed by Campbell (1963). The first correlation is between variable A at time one and variable B at time two. The second is between B at time one and A at time two. The two correlations are called cross-lagged.

To find the phi coefficient between drug status in 1986 and level of stress in 1989, construct Table 13.13 from Table 13.12.

To find the phi coefficient between level of stress in 1986 and drug status in 1989, construct Table 13.14.

Notice that the cross-lagged correlations use all the data from Table 13.12, whereas Lazarsfeld's technique uses only half of it. In addition, the variables need not be dichotomized for computation of correlations; and precision of measurement is greater for numerical measurements as opposed to dichotomies. The phi coefficient between 1986 drug status and 1989 stress level is

$$\frac{(49 \times 35) - (50 \times 43)}{\sqrt{92 \times 95 \times 99 \times 78}} = -.056$$

The phi coefficient between 1986 stress level and 1989 drug status is

$$\frac{(45 \times 56) - (47 \times 29)}{\sqrt{92 \times 85 \times 74 \times 103}} = .15$$

The conclusion would be the same as that from Lazarsfeld's approach: High levels of stress are more likely to precede abuse of drugs than is drug abuse likely to precede stress. But the conclusion would not be justified that heavy stress causes drug abuse. Lazarsfeld's approach and cross-lagged correlations are useful for indicating probable sequences of events (heavy stress precedes drug abuse) and eliminating possible causes (drug abuse is a major cause of stress), but they share the limitations of all correlational studies for establishing causes.

Summary

Correlational studies answer questions about the extent to which variables are related. Correlation coefficients range between -1 and $+1$. The higher the absolute value, the stronger the relationship. The correlational calculations described in this book can detect only linear relationships.

Correlations are used to compute the reliability and criterion-related validity of measurements, to estimate the size of effect of one variable upon another, to generate causal hypotheses, and when manipulation of variables is impractical. Regression equations are used to make predictions.

There are several different correlation coefficients, because there are different types of measurements.

When both variables are nominal and one or both of them is assorted into more than two categories, use the contingency coefficient.

When both variables are nominal and dichotomous, use the phi coefficient.

When numerical variables are ranked, use the Spearman.

When both variables are measured on an interval scale, use the Pearson product-moment correlation.

When one variable is measured on an interval and one on a nominal scale, use the point-biserial.

The technique of partial correlation is used to measure the relationship between two variables after the effect of a third variable on both of them has been eliminated.

Approaches developed by Lazarsfeld and Campbell are used to learn which of two correlated variables preceded the other in time. Establishing priority of occurrence simplifies causal analysis.

Key Terms

Coefficient of Determination The square of the correlation coefficient. It ranges between zero and one and indicates the degree to which knowledge of one variable accounts for the variance of the other. Thus, for example, since the correlation between heights of parents and their children is 0.50, 25 percent of the variance in the heights of children can be accounted for from knowledge of their parents' heights.

Correlation Coefficient A measure of the relationship between two variables. A coefficient of $+1$ indicates a perfect relationship; a coefficient of -1 also indicates a perfect relationship, but with high values of one variable associated with low values of the other; a zero correlation indicates no relationship.

Dichotomous Measurement Dividing into two parts or categories.

Linear Relationship Relationships that can be fit to equations of the form $Y = ax + b$, where a and b are constant values.

Nominal Measurement Measurement in which objects are assigned to categories. Quantity is not involved, i.e., there is no implication that one category is more or less than another.

Partial Correlation The correlation between two variables when the effects of a third variable on both of them have been eliminated.

Regression Equation The equation that best fits the relationship between X and Y and which can be used to predict the value of Y, given X.

Answers

It might seem that correlations would be higher between mothers and daughters and between fathers and sons, but that is not so; the difference in average height between men and women is irrelevant. If the tallest mothers tend to have the tallest sons and the shortest mothers the shortest sons, then the correlation between them will be high—no matter what the average difference in height between mothers and sons.

1. Test Y, although the correlation is negative.
2. Spearman
3. Pearson
4. Higher, in fact perfect, with biological parents.
5. No. Each adoptee's IQ was higher than that of her biological parents.

CHAPTER 14

Experimenting: Two Groups

By the time you finish reading this chapter, you should be able to answer the following questions:

What are the two key features of experiments?

What makes experimentation such a powerful technique?

Does random assignment ensure that experimental and control groups will be equal?

What plausible explanations of data, other than that the IV has acted on the DV, are ruled out by experiments?

What should be an experimenter's concerns in deciding how many subjects to use?

What are Type I and Type II errors?

What is meant by the power of an experiment?

How can power be increased?

Under what circumstances is random assignment not appropriate?

How is a table of random numbers used for randomization?

What is the reason for claiming that an animal experiment with, for example, 1,000 mice, is actually a small-sample experiment?

Why must surveyors strive for representative samples although experimenters do not?

What are field experiments?

How can experimenters tell almost instantly if results from a two-group experiment are significant at the .05 level?

> "The observer listens to nature; the experimenter questions and forces her to reveal herself."
>
> *George Cuvier, French zoologist (1769–1832)*

Defining Characteristics of Experiments

Two key features distinguish experiments from other types of social science research. First is random assignment. Subjects are assigned randomly to groups that receive either different levels of an independent variable or different IVs; alternatively, subjects may be exposed to each IV but in random order. (Designs of this type are discussed in the next chapter.) Researchers may often be tempted to make assignments on the basis of convenience, for example, people in the front of a room to one group and in the back to another. But people in the two locations might differ in motivation, personality, eyesight, and so forth. Subjects chosen by any nonrandom procedure may differ. Random assignment does not guarantee initial equality, but it greatly improves the odds.

The second key feature is that the groups are treated differently in one and only one respect. If a researcher wanted to study the effects of caffeine on heartrate, she might give a caffeine pill to the experimental group. Control subjects would receive an identically-appearing pill without caffeine. She would test both groups in the same room with the same instructions, and she would measure their heartrates in the same way. Then, if their average heartrates differed, and given that they were presumed equal initially, she could conclude with considerable confidence that caffeine caused the difference.

Reasons for Experimenting

Experimenters Seek Answers to "What Would Happen If . . . ?" Questions

The discovery of interesting phenomena cannot wait on naturalistic observations. Only by carefully arranging circumstances can scientists expect to see phenomena like rats pressing bars for food rewards and people lowering their blood pressures by attending to feedback. In the sixteenth century, Francis Bacon recognized the need for scientists to indulge in experiential play, and he called the invented experiences experiments. Experiments allow scientists to answer "What would happen if . . . ?" questions, and they extend the range of phenomena that can be studied. (Observations extend the range of phenomena in a different way, because they are less limited than experiments by the imagination of the scientist.)

Experimenters Test Logically Deduced Consequences from Theories

Karl Popper, probably the most influential philosopher of science of the twentieth century (see pp. 8–9), argued that scientists must constantly strive to develop bold theories and then test the theories ruthlessly. The experimental method is the most powerful ever devised for testing logically-deduced consequences of scientific theories. Given the prediction that event Y will occur under circumstance X, experimenters produce X and see if Y occurs.

Experimenters Can Control Variables A researcher might be able to test the relationship between stress and aggressiveness merely by observing carefully. But, unable to manipulate subjects' stress levels systematically, he'd have to wait for spontaneous changes. And he'd have no control over the occurrence of other variables that might influence aggressiveness. One of the great virtues of experiments is that they allow control over both the independent variable and other variables.

Experimenters Can Eliminate Otherwise Plausible Alternatives to Their Interpretations of Data Nonexperimental tests of theory are generally inconclusive because they fail to eliminate several plausible alternative explanations of results. By contrast, the logic of experimentation forces the conclusion that the IV has (hasn't) significantly influenced the DV. Contrast *a* and *b* below.

a. During the course of a night, while campers sleep, a tape recorder plays a continuous message telling them to drink milk at breakfast time. The next morning, each camper drinks a large glass of milk.
b. Campers are randomly assigned to be exposed to either continuous music or the message to drink milk. The next morning, campers in the second group drink significantly more milk on average than those in the first.

Study *a* is not an experiment. If each camper drank even two large glasses of milk the next morning, the reason would be unclear. The message might have affected them but, to consider just one plausible alternative, they might each have normally drunk at least two glasses of milk every morning. Study *b* is an experiment. If campers exposed to the message and otherwise treated just like the music group drank significantly more milk, a conclusion that the message was responsible would be on much firmer grounds.

Several non-experiments and faulty conclusions derived from them are listed in Box 14.1. Try to think of an alternative explanation for each.

Sample Size

Types of Errors

Before collecting any data, researchers should figure out how many subjects they will need. If too few are used, the data will probably not be statistically significant. Use of too many subjects is wasteful of time and money. When painful or injurious procedures are involved, as in many animal experiments, the number of subjects should be limited for humane reasons. Proper sample size depends on the research design, statistic used, and tolerable error. Because chance factors can never be eliminated, researchers always risk committing one of two types of error, called Type I and Type II. See Table 14.1. As sample size is increased, the likelihood of making a Type II error is reduced.

BOX 14.1

Alternative Interpretations Exist for Nonexperimental Data

The conclusions drawn from the following studies might be correct, but there are plausible alternatives. Think of one for each study.

1. A teacher tries a new teaching method with her second grade class. The class does better on an achievement test at the end of the year than on a similar test at the beginning of the year. She concludes that the new method works.
2. In the month following a well-publicized campaign to crack down on speeders, car fatalities decline substantially from the previous month; the highway patrol attributes the decline to the campaign.
3. In Prussia, until 1883, suicides were recorded by local police stations. In that year, recording was made the responsibility of a national bureau and 20 percent more suicides were reported. The conclusion was that the suicide rate, for reasons unknown, had increased 20 percent.
4. At 6:00 P.M., volunteers are seated in comfortable stuffed chairs in a dark room. At 11:00 P.M., soft music is piped into the room, and by midnight most of the volunteers are asleep. The researcher concludes that the music is an effective way to put people to sleep.
5. Fifty volunteers receive a series of painful electric shocks while 50 control subjects listen to a motivational tape. Half the people in the shock group get angry and quit before the experiment is over. The remaining 25 do better on average than the control group at a motor coordination task. The experimenter concludes that electric shock, properly applied, improves motor coordination.
6. Fifty volunteers receive a powerful drug, and their scores on a memory test are compared with those of 50 nonvolunteers. The drug takers do better, leading to the conclusion that the drug improves memory.

Answers

1. *Maturation.* Older children do better on average than younger ones, and the children of this study aged nine months during the school year. Maturation effects refer to any changes that occur because of the passage of time (fatigue, increased or reduced motivation, changes in health, etc.).
2. *History.* The weather may have been exceptionally good during the second month. The first month may have been a holiday season when fatalities are normally high. History refers to events besides the independent variable that take place between measurements and may affect them.
3. *Instrumentation.* The local stations and the national bureau used different criteria for suicide. Instrumentation effects refer to any changes in the measuring instrument that affect scores.
4. This one is a check to see if you've been paying attention. It's a maturation effect.
5. *Mortality.* Drop-outs often differ in relevant ways from subjects who remain in experiments, so if groups differ in dropout rates (mortality), conclusions are suspect. (Differential mortality presents problems of interpretation to experimenters as well as to other researchers.)
6. *Selection.* Volunteers to take a powerful drug differ in many ways from non-volunteers. A researcher, observing the differences after administering a drug, might falsely attribute them to the drug. Selection effects are caused by nonrandom assignment to groups.

Table 14.1 *Type I and Type II Errors*

		Does the IV Really Affect the DV?	
		Yes	No
Does the Statistical Analysis Indicate that the IV Affects the DV?	Yes	Correct Conclusion	Type I Error
	No	Type II Error	Correct Conclusion

Suppose somebody, trying to decide on the fairness of a coin, flips it six times and agrees to call it biased if heads comes up each time. A false conclusion is possible, as even fair coins occasionally turn up heads six times in a row. The probability can be calculated precisely and is $(½)^6 = .016$. So, with six flips of a fair coin, the risk of a Type I error $= .016$.

Social scientists must also risk false conclusions. They typically accept results that can be expected to yield Type I errors in no more than 5 percent of cases, which is the standard I've used throughout this book. Different standards are possible. For example, the coin flipper might accept either five or six heads as proof of bias. From the binomial theorem—see page 52—the probability of getting five heads in six tosses of a fair coin is $(6!/5!1!) \times .5^5 \times .5^1 = .094$. Add to that the .016 probability of six of six heads, and the risk of a Type I error would be .11.

Some researchers prefer a more stringent level, .01. The risk can be made even smaller, infinitesimally small, even zero. (A researcher who refuses to take 1,000 or 1,000,000 or any number of consecutive heads as proof that the coin is biased will not make a Type I error.) But the purpose of research is to discover relationships, not to shy away from recognizing them. And that means risking errors.

....

1. Suppose a coin-flipper suspects bias but doesn't know whether it's toward heads or tails. She flips five times and gets all heads. Are the results significant at the .05 level?

2. What happens to the probability of making a Type I error as sample size is increased?

....

Power

The term power refers to the likelihood of getting statistically significant results when a true relationship exists or there are treatment effects. Power is equal to 1 − the probability of making a Type II error. This can't be known in advance. It depends on the strength of the relationship, which is what the experiment is designed to find out.

Table 14.2 *Probability of Getting Exactly X Heads in Six Tosses of a Biased Coin*

X	Bias = .9 Toward Heads Probability	Bias = .6 Toward Heads Probability
0	.000	.004
1	.000	.037
2	.001	.138
3	.015	.277
4	.098	.311
5	.354	.187
6	.531	.047

The first column of Table 14.2 lists the probability of each possible outcome from a set of six tosses of a coin with a .90 bias toward heads. The second column does the same for a coin with a .60 bias. A coin-flipper testing for bias and setting the probability of a Type I error at .05 would, as shown above, require six of six heads. Note from Table 14.2 that such an outcome can be expected .531 of the time if the bias is .90, .047 of the time with a bias of .60. Thus, power would be .531 and .047, respectively. The probability of a Type II error would be .469 in the first case and a whopping .953 in the second. In other words, with the chance of a Type I error set at .05, a researcher who tossed a coin six times, if it were biased .60 toward heads, would have only a .047 chance of getting statistically significant results.

One way to increase power is to accept a greater risk of a Type I error. If the .11 level were used rather than the .05, five heads in six flips would lead to the conclusion of bias. Then, if a coin were actually biased .90 toward heads, power—in this case, the probability of getting at least five heads—would equal .354 + .531 = .885. With a .60 bias, power would equal .187 + .047 = .235. Although convention (and publication requirements in most scientific journals) keeps the risk of a Type I error at .05 or less, a more lenient criterion may sometimes be preferable. A terminally ill person offered an experimental treatment might try it even if the probability of failure were estimated to be much more than .05.

Power can be increased without changing the probability of a Type I error, simply by increasing sample size. The probability of a fair coin turning up heads at least 10 times in 12 flips is .0158, virtually the same as that of six heads out of six. So, a researcher who calls the coin biased on the basis of either outcome runs the same risk of a Type I error. But with 12 flips, if the coin is biased .90, the probability of a Type II error reduces from .469 to .111; and if the bias is .60, from .953 to .917.

It seems pointless to do a study that will probably fail even if the research hypothesis is true. Yet researchers who use too few subjects, just as the person who flips a coin only six times, put themselves in such a position; unless the

Table 14.3 Numbers of Subjects Needed to Detect Small, Medium, and Large Effects at Different Levels of Power and with the Probability of a Type I Error Set at .05

	Power	Small	Effect Size Medium	Large
F test, two groups	.5	193	32	13
	.8	393	64	26
	.95	651	105	42
	.99	920	148	58
Sign test	.5	384	44	17
	.8	783	85	30
	.95	1294	138	49
Correlation coefficient	.5	385	42	15
	.8	783	85	28
	.95	1294	139	46

For the *F* and sign tests, the numbers in the table refer to numbers of subjects per group. For the correlation coefficient, the numbers refer to numbers of paired scores.

data show an extremely powerful relationship, they will be nonsignificant. But samples may also be too large, not only for the reasons listed in the introductory paragraph of this section, but also because they may yield statistically significant results on the basis of trivial effects. Therefore, researchers should decide on the size of their samples only after they have made decisions about power and the minimum effect size that would be scientifically meaningful.

Cohen (1969) suggested that, in the absence of other bases for choice, power should be set at .80. The minimum acceptable effect size varies with each research project; even tiny effects may sometimes have dramatic impacts. Power, probability of a Type I error, sample size, and minimum acceptable effect size are interrelated; when any three of them are set, the fourth is completely determined. Cohen's book contains many tables that incorporate these relationships and can be of immense help in planning research projects. But just a few key numbers, such as those of Table 14.3, will often suffice. In each case, the probability of a Type I error is set at .05.

Contained in Table 14.3 is information about three statistical tests: the *F* test and sign test for two-group experiments (both described later in the chapter) and the correlation coefficient of the previous chapter. For each test, sample sizes needed to detect small, medium, and large effect sizes are given. For example, suppose a researcher sets the Type I error at .05 and plans to analyze results with an *F* test. If she wants to have an 80 percent chance of getting statistically significant results and guesses that the actual difference between groups is small, she should use 393 subjects per group. If she guesses that the difference is large, she should use 26 subjects per group.

For the F test, a small effect size is defined as one in which the difference between population means divided by the standard deviation of either population equals 0.2. Then, if the populations are normally distributed and equally variable, 57.9 percent of the lower scoring population will score less than the mean score of the higher scoring population. For medium and large effect sizes, the difference between population means divided by the standard deviation of either population equals 0.5 and 0.8, respectively. And the percent of the lower scoring population that score less than the mean of the higher scoring population is 69.1 and 78.8 percent, respectively.

Cohen defined effect sizes for different tests to be comparable. For the sign test, if 55 percent of the higher scoring members of each pair come from the same group, effect size is small; proportions of .65 and .75 are called medium and large, respectively. For correlation coefficients, small, medium, and large correspond to coefficients of .1, .3, and .5, respectively.

Note from the table that tests differ in power. To detect medium size effects with power of .5, 44 pairs of subjects are needed for the sign test and only 32 for the F test.

Table 14.3 can help answer four types of questions. Use the table to answer the following questions for the F test. Cohen's book also has tables for Type I errors of .01 and .10, for experiments on other than two independent groups, and for other statistics.

....

3. What significance level must be set to detect a medium effect with 50 subjects and power = .80? (This cannot be answered without the help of additional tables.)

4. For a significance level of .05, a sample size of 50, and a large effect size, what is the power?

5. For a significance level of .05 and power of .80, how many subjects are needed for detecting a small effect?

6. For a significance level of .05 and a sample size of 105, what must the effect size be to have power of .95?

....

How to Randomly Assign Subjects

Assigning subjects randomly to groups is one of the key characteristics of experiments. But random assignment is often impossible (as when comparing characteristics of first and later born children), unsuitable, or infeasible. Cook and Campbell (1979) gave the hypothetical example of a researcher wishing to compare job placements and income supplements as methods for reducing recidivism rates of ex-convicts. A randomized experiment would give the clearest answers but take a long time. Instead, records could be compared of people who had already been given jobs or income supplements upon release from prison. The alternative approach would not conclusively establish cause, but its failure to do so might be outweighed by the advantage of accumulating data rapidly.

Many different procedures are used to randomize, including tables of random numbers, computerized programs, dice, shuffled cards, roulette wheels, and so forth. Cook and Campbell suggested the following method: (a) make up or obtain lists of all the potential subjects (rats, people, families, schools) in a relevant population (much as is done for surveys), (b) assign unique numbers to all of them, each with the same number of digits (if the highest number has three digits, all should have three digits; write 1 as 001, 10 as 010), (c) consult a table of random numbers; these are available as separate publications and in appendices in statistics books. If each potential subject has been given a three-digit number, start at the top of a page picked at random from the table and read down a three-digit wide column. Skip numbers that haven't been assigned to a subject and write down the rest until the desired sample size is reached, and (d) designate the first half of the chosen sample the experimental group, the second half the controls.

Variability in Experiments

Suppose subjects are randomly assigned to experimental and control groups and then, before being exposed to an IV, given a test. Even though nothing has been done to them, their scores will almost certainly vary. The following sections explain why and in so doing prepare the way for learning how to improve two-group experiments and statistically analyze them.

Three sources of variation are inherent in any measurement situation involving people. First, subjects come in assorted sizes, shapes, and ages, and with different motivation and ability levels. Second, because of the physical impossibility of having two objects in the same place at the same time, the subjects can't be treated exactly alike. Some will be tested at 8:15, others at 8:22; some on Tuesday, others on Friday; some by a 60-year-old man, others by a 25-year-old woman; some in complete silence, others while an argument rages outside the laboratory. Third, measuring devices are imperfect and imperfectly used. That is, errors occur. For these three reasons, variation occurs even among subjects assigned to the same group; it is called within-group variability.

The purpose of random assignment is to form equivalent groups with respect to the dependent variable. Then, prior to administration of an IV, the variability between groups will be equivalent to the within-group variability. If an effective IV is given to only one of the groups, or given in different doses, between-group variability will increase; but since all members within a group receive the same level of the IV, within-group variability will not change.

Shown in Table 14.4 are the DV scores of subjects in two fictitious two-group experiments. In the first, subjects were randomly assigned to receive either experimental treatment A or control treatment B; in the second, they

Table 14.4 Results of Two Fictitious Experiments

Experiment 1		Experiment 2	
Group		Group	
A	B	C	D
10	7	20	10
11	6	0	4
9	8	18	14
12	8	2	0
9	6	15	13
9	7	5	12
10	7	19	1
10	7	1	2

Table 14.5 Table of Variance

Source	SS	df	ms	F	p
Between conditions	36	1	36	30.9	.05
Within conditions	12	14	0.86		

received either treatment C or D. In both cases the mean of the treatment group is 10, of the control group, 7. Yet only experiment one yields statistically significant results. The reason is that the within-group variability is a relatively smaller proportion of the total variability in experiment one than two; the between-group variability (that caused by the IV) is correspondingly larger. The expression $\Sigma X^2 - (\Sigma X)^2/N$, called the sum of squares (SS), is a measure of variability and part of several important statistical formulae. The more the scores in a sample deviate from the mean of the sample, the larger will SS be. SS is 0 if all the scores are the same. For example, for the set of scores 5, 5, 5, 5, 5, and 5, $\Sigma X^2 = 25 + 25 + 25 + 25 + 25 + 25 = 150$ and $(\Sigma X)^2 = (5 + 5 + 5 + 5 + 5 + 5)^2 = 900$, $900/6 = 150$, and $150 - 150 = 0$.

If SS is divided by $N - 1$, the result is called the variance; and the square root of the variance, the standard deviation, is the most common measure of variability. Analysis of variance is a statistical technique, based on sums of squares, for comparing within- and between-group variabilities. I give the computatinal procedures in Box 14.2 and apply them to the data of experiment 1. The analysis for experiment 2 is given on pages 226–227.

The results of analyses of variance are summarized in tables like Table 14.5.

BOX 14.2

Computational Procedure for Analysis of Variance

1. Get the sum of the scores for groups A and B.

   $\Sigma A = 10 + 11 + 9 + 12 + 9 + 9 + 10 + 10 = 80$

   $\Sigma B = 7 + 6 + 8 + 8 + 6 + 7 + 7 + 7 = 56$

2. Get the squares of all the A and B scores and sum them.

   $\Sigma A^2 = 100 + 121 + 81 + 144 + 81 + 81 + 100 + 100 = 808$

   $\Sigma B^2 = 49 + 36 + 64 + 64 + 36 + 49 + 49 + 49 = 396$

   $808 + 396 = 1204$

3. Add ΣA and ΣB, square the result, and then divide by the total number of scores.

   $80 + 56 = 136$; $136^2 = 18496$; $18496/16 = 1156$

4. Subtract the result of step 3 from that of step 2. The answer, the sum of squares total (SS_t), reflects the total variability of scores.

   $1204 - 1156 = 48$

5. Square the sum of the scores in group A and divide the answer by the number of subjects in group A. Do the same for group B and add the two answers.

   $80^2/8 = 800$; $56^2/8 = 392$;
   $800 + 392 = 1192$

6. Subtract the result of step 3 from that of step 5. The answer, the sum of squares between groups (SS_b), reflects the variability due to the experimental treatment plus that due to chance factors.

   $1192 - 1156 = 36$

7. Subtract SS_b from SS_t. The answer, the sum of squares within groups (SS_w), reflects variability due only to chance factors.

   $SS_w = 48 - 36 = 12$

8. Calculate the degrees of freedom for SS_t, SS_b, and SS_w.

   df for SS_t = the total number of scores minus 1; $16 - 1 = 15$

   df for SS_b = the number of groups minus 1 (as will be seen in the next chapter, the use of analysis of variance is not restricted to two-group experimental designs); $2 - 1 = 1$

   df for SS_w = the total df minus the between df; $15 - 1 = 14$

9. Divide the between- and within-groups sums of squares by their respective dfs. These give the mean squares (ms) for each.

   $ms_b = SS_b/df = 36/1 = 36$

   $ms_w = SS_w/df = 12/14 = 0.86$

10. The ratio ms_b/ms_w, called the F ratio, is the test of significance. Since ms_b reflects variability due to treatment effects plus chance factors, and ms_w reflects variability due to chance factors alone, if a treatment has no effects the F ratio should be about 1.

    $ms_b/ms_w = 36/0.86 = 30.9$

11. Appendix C lists F values for different degrees of freedom. For 1 and 14 degrees of freedom, a value of 30.9 is significant at less than the .05 level. Note: Many students learn in statistics classes to analyze two-group experiments with the t test. The F test is directly parallel, and results from the two tests are significant at the same level.

To find the effect size associated with a particular F value, additional calculations must be made.

1. Compute eta. Eta = (SS_b/SS_t).

   From calculations on the data of Table 14.4, eta = $36/48 = .75$.
2. Compute $f = (eta^2)/(1 - eta^2)$. For the result from step 1, $f = .5625/.4375 = 1.29$; f can vary between zero and infinity.

In Table 14.3, f values of 0.10, 0.25, and 0.40 can be substituted for "small," "medium," and "large."

Implications for Experimenters of the Analysis of Variance Approach

Analysis of variance is only one of many available statistical procedures for determining if an experimental treatment has had an effect, and it may not be the best one. In addition, many critics argue that overreliance on statistical significance testing has hindered research in the social sciences. But analysis of variance puts the focus on sources of variability, and that is useful regardless of a person's attitude toward significance testing.

Experimenters try to maximize the ratio ms_b/ms_w. Thus, they try to increase the numerator and decrease the denominator of the fraction. The numerator increases with increasing treatment effect, which depends on proper selection of the IV. So experimenters should move carefully from their abstract concepts to operational definitions of variables. Recall from chapter 6 that concepts can be operationalized in many ways, each with a different potential impact on behavior. Creating stress by giving an electric shock will probably produce a stronger effect than by saying "Boo."

Once a concept is operationally defined, it should be given at a level that will maximize the difference between experimental and control groups. Finding the right level may require pretesting.

Decreasing Within-Group Variability   Because there are three independent sources of within-group variability, several strategies can be used for reducing it.

Selecting Subjects

Three important methods exist for reducing the inevitable variability between subjects. One, using them as their own controls by exposing them to two or more levels of the IV, is treated in the next chapter. The other two methods are considered below.

Using a Matched-Pair Design The matched-pairs design often helps reduce ms_{error}. A researcher wishing to compare math scores after exposure to teaching methods X and Y could first give all subjects a preliminary math quiz. Then he'd rank the scores from high to low and pair the two highest, then the next two, and so on. Then he'd randomly assign one member of each pair to method X, the other to method Y.

If scores on the matching test have at least a .30 correlation with the DV scores, the matched-pairs design can result in a substantial savings in subjects. The matching test need not have an obvious relationship with the DV, as long as the correlation is at least .30. If subjects are expensive or otherwise difficult to obtain, the advantage is obvious. The disadvantages are that (a) matching tests take time, (b) subjects may drop out of the study between a matching test and exposure to treatment, and (c) exposure to a matching test may sensitize subjects to the purpose of the experiment and affect scores on the DV.

Analysis of variance can be used to analyze the results of a matched-pair design. A simple alternative, the sign test, is described in Box 14.3. An even simpler method, illustrated with the data of Table 14.7 is as follows:

1. Count the number of pairs excluding ties. Call this number A.
2. Let $B =$ the number of $+$s in the experimental group and C the number in the control group.
3. Let $D =$ the difference between B and C, ignoring the sign.
4. If D is more than twice the square root of A, the result is significant at the .05 level.

From the data of Table 14.7: $A = 26$; $B = 21$; $C = 5$; $D = 21 - 5 = 16$. Twice the square root of $A = 10.2$. Since 16 is more than 10.2, the results are significant.

Using Homogeneous Subjects The experiment can be restricted to only a certain type of subject. Researchers who use human subjects try to make their samples as homogeneous as possible. They may open an experiment only to male volunteers from an introductory psychology class who have not previously participated in a learning study, or to inmates of a particular institution, or to girls in a day care center. Researchers who work with animals are even more demanding. Unless their specific goal is to compare, they typically use only a single species. More than that, they use animals that come from the same supplier, are of the same age and weight, and are housed and fed under identical conditions. The animals are often genetically inbred. Denenberg (1982) argued that they are for all practical purposes the same individual. Experiments with such animals, no matter how many of them are used, are small sample experiments (without, however, the benefits of planned small sample experiments; see pages 242–244 for a discussion).

BOX 14.3

The Sign Test

The sign test is useful for comparing two groups if subjects in each group are matched. It isn't very powerful compared with some other statistical tests. For large samples, power is only about two-thirds; that is, for a given significance level, use of the most powerful test is only two-thirds as likely to lead to a Type II error. But the sign test, simple to compute and versatile, is often used. The steps in computation are as follows:

1. For each matched pair, put a plus by the name of the subject who scored more.
2. Let N = the total number of pluses.
3. Let x = the number of pluses in the group that has the fewer of them.
4. If N is 25 or smaller, turn to Table 14.6. The first column lists N, the next two columns list critical values of x. If the observed x is equal to or less than the number in the table, the results reach significance at the .05 or .01 level (two-tailed).
5. If N is more than 25, use the formula

$$z = \frac{N - 2x - 1}{\sqrt{N}}$$

6. If z is greater than 1.96, the results are significant at the .05 level (two-tailed); if z is greater than 2.58, they are significant at the .01 level.

Examples 1–3 are given to show the versatility of the sign test.

Example 1: One member of each of 27 pairs of identical twins receives special musical training.

Table 14.6 *Critical Values of x for the Sign Test*

N (Total Number of Pluses)	x (Number of Pluses in Group with Fewer Pluses)	
	Significance Level	
	.05	.01
6	0	—
7	0	—
8	0	0
9	1	0
10	1	0
11	1	0
12	2	1
13	2	1
14	2	1
15	3	2
16	3	2
17	4	2
18	4	3
19	4	3
20	5	3
21	5	4
22	5	4
23	6	5
24	6	5
25	7	5

Then the twins are given a test of musical creativity. The twin who does better on the test receives a score of plus. The results are as given in Table 14.7. $N = 26$ and $x = 5$, so

$$z = \frac{26 - 10 - 1}{\sqrt{26}} = 2.94$$

The results are significant at the .01 level.

Table 14.7 Fictitious Data for Matched-Pairs Design

Twin Pair	Exposed to Music	Control Subject
1	+	
2	+	
3	+	
4	+	
5		+
6		+
7	+	
8	+	
9	+	
10	+	
11		+
12	+	
13	+	
14	+	
15		+
16	+	
17	+	
18	+	
19	tie	
20	+	
21	+	
22	+	
23	+	
24		+
25	+	
26	+	
27	+	

Example 2: Subjects are timed twice in a 50-yard dash; half of them listen to rock music before the first dash and a piano concerto before the second, and half listen in the reverse order. Since each subject is his own control, scores from the two conditions are matched and the sign test is appropriate. Let R stand for the times under rock, P for the piano times, and + for the faster scores.

	\multicolumn{14}{c}{Subject}
	1	2	3	4	5	6	7	8	9	10	11	12	13	14
R	+	+		+		+	+	+		+			+	+	+
P			+		+				+		+	+			

From Table 14.6, with $N = 14$, x must be 2 or less to reach significance at the .05 level. So the results are not significant.

Example 3: Over a series of 24 trials, a subject gets the following scores: 43, 37, 40, 43, 35, 44, 40, 54, 42, 47, 46, 43, 49, 46, 61, 51, 50, 47, 54, 48, 52, 59, 46, 45. Do the data show a trend toward improved scores?

The scores can be divided into two halves, with score 1 (43) matched with score 13 (49), 2 with 14, and so on as shown below. The higher score of each pair has been given a plus.

	\multicolumn{12}{c}{Scores}
	1	2	3	4	5	6	7	8	9	10	11	12
first half							+				tie	
second half	+	+	+	+	+	+		+	+			+

From Table 14.6, with $N = 11$, x must be 1 or less for the results to be significant at the .05 level. Since x is 1, the results are significant and there is a trend toward improved scores.

The strategy of reducing within-group variability by using homogeneous samples has a cost attached—researchers may focus on certain types of subjects and exclude others. In psychology, males have been used much more than females in both animal and human research (Carlson, 1971; Hyde & Rosenberg, 1976). This has resulted in the development of theories that apply to only one gender. For example, McClelland, Atkinson, Clark, & Lowell (1953) successfully predicted men's achievement behaviors from various tests. It wasn't realized until much later that predictions from the same tests were inaccurate for women (Alper, 1974). Another danger that comes from having a disproportionate number of male subjects is that some people come to regard male behavior as normal and female behavior, if different, as deviant.

Treating Subjects Uniformly Except for Exposure to the Independent Variable

Social scientists no less than medical researchers, physiologists, and chemists, should try to keep their laboratories free from contaminants. They should strive to make testing conditions exactly the same for all subjects. That is an unattainable ideal, but what can be attained is a uniformity of relevant details. These vary from study to study. For example, uniformity would be lacking if some subjects were tested just before and some just after a brutal and highly publicized boxing match; but within-group variability would probably not be affected much unless the DV were related to aggression—and not at all if the subjects were rats.

As a matter of routine, try to standardize the following factors: lighting conditions, noise level, presence of people other than the subjects, time of testing, both explicit wording and manner of presenting instructions, and the experimenter. (An experimenter's gender, age, attitude, and clothing may all influence behavior.) Other factors, such as the size and color of the laboratory and even the changing odors coming from a nearby restaurant may influence certain behaviors. Be sensitive to the possibilities.

Reducing Measurement Errors

Random errors increase variability. Reduce errors by practicing your scoring systems. If two or more people share the responsibility of scoring, run periodic reliability checks. And double-check all computations.

Is It Bad to Decrease Within-Group Variability?

Critics have argued that experimenters, by demanding rigorous control and minimizing within-group variability, create artificial settings with little relevance to conventional ones. For example, Gergen (1978) wrote, "In the attempt to isolate a given stimulus from the complex in which it is normally embedded, its meaning within the normative cultural framework is often obscured or destroyed. When subjects are exposed to an event out of its normal context they may be forced into reactions that are unique to the situation and have little or no relationship to their behavior in the normal setting."

To study the effects of classical music on memory, an experimenter might ask subjects to memorize lists of words in a soundproof laboratory while listening through headphones to either music or white noise. According to Gergen and others, results from such studies are likely to be inapplicable to meaningful memory tasks in places like homes and schools. At home, for example, telephones may ring, children or siblings may demand attention, and televisions and refrigerators may act as powerful distractors.

But the two key features of experiments—randomly assigning subjects to groups and treating both groups identically except for the IV—put no restrictions on the setting or the values of extraneous variables. Subjects need not be tested in soundless, pitch-black, sterile laboratories. The important requirement is that the testing situation expose all subjects to the same sounds and sights. The subjects of the fictitious music study could be interrupted by a variety of unpredictable (to them only) distractors. They could be studied away from a laboratory. For example, an experimenter might pipe music or white noise into randomly chosen classrooms and then test students on the day's lesson. The advantages and disadvantages of this type of research, field experiments, are discussed later in the chapter.

Isolation of variables serves two distinct purposes. First, it enables specification of the precise conditions under which phenomena occur; this is crucial for scientists, as was pointed out in the discussion of operational definitions. Second, simple systems often provide clues to the workings of more complex ones.

Social scientists are not alone in dealing with the context-specific properties of variables. Organic chemistry is the study of carbon compounds, that is, of the immense number of contexts (molecular configurations, including many found only in laboratories) in which the element carbon is found. Organic chemists owe much of their success to the development of procedures for isolating and purifying these compounds.

Nevertheless, say the critics, experimental variables occur in unusual contexts that distort their normal actions. (For example, subjects in the music study, in common with all human laboratory subjects, would know they were in an experiment.) As a result, laboratory findings often fail to generalize to nonlaboratory settings (cf. Chapanis, 1967; Cronbach, 1975).

But by definition, all distinct settings are unique. Laboratories differ from schools which in turn differ from churches, hospitals, and homes. Generalizations from one to another often don't apply. However, meaningful generalizations are not even possible unless the settings and relevant variables are specified. Then, if the generalizations don't hold up, experimenters can search for reasons in the differences between the settings.

As noted previously, awareness of being an experimental subject affects behavior. But subjects are made aware because of ethical considerations, not because the logic of experimentation requires it. Unscrupulous researchers have performed unnecessary surgery, administered powerful drugs, and withheld effective treatments from unwitting subjects (cf. Beecher, 1966; Jones, 1981).

Nonexperimenters have no advantage here; they must either be equally unscrupulous or risk changing their subjects' behavior by explaining the nature of the research.

Experiments Do Not Elimimate the Need for Judgment

Plausible Alternatives May Remain

Recall from Box 14.1 that experiments eliminate several otherwise plausible explanations for data. One explanation not always eliminated is differential mortality. Thus, in a test of a new surgical technique, more experimental than control animals might die; the survivors would then constitute a biased sample of the original subject population. (Mortality does not refer exclusively to deaths, but to any drop-out from an experiment.) Whether or not differential mortality has occurred must be evaluated from the data.

Experiments render each of the other alternatives of Box 14.1 implausible. A historical event or change in a measuring instrument might affect subjects during the course of an experiment, but is unlikely to affect groups unequally.

In most experiments, one alternative to the explanation that the IV accounted for changes in the DV is that the groups started out unequally. Random assignment to groups minimizes but doesn't eliminate the possibility. If an inequality between experimental and control groups is noticed before administration of the IV, the experimenter can rerandomize; otherwise, the meaning of results is uncertain. Imagine a flawlessly conducted experiment in which, purely by chance, all men had been assigned to one group and all women to the other.

Experimenters Must Balance the Costs and Benefits of Research

A design developed by Richard Solomon rules out the possibility that pretreatment inequality between groups, rather than their exposure to different treatments, accounts for post-treatment differences. The Solomon design requires four groups, two of which are given a pre-test. Then, one of the pretested groups and one of the others is given the experimental treatment. All four groups are given a post-test. The procedure can be represented diagrammatically as in Table 14.8.

Table 14.8 *Solomon Four-Group Design*

Group	Pretest	Experimental Treatment	Posttest
1	Yes	Yes	Yes
2	Yes	No	Yes
3	No	Yes	Yes
4	No	No	Yes

The Solomon design lets investigators evaluate the possibility that pretreatment differences between groups account for posttreatment differences. It has other advantages as well, yet is rarely used. The reason is that it requires twice as many subjects and more than twice as much work (subjects in two of the groups must be tested twice) as the simple two-group design. For most social scientists, the costs exceed the benefits.

Results are Interpreted Within the Framework of Existing Knowledge

Study (c) is analogous in form to nonexperimental study (a), given earlier in the chapter.

> c. During the course of a night, while campers sleep, a tape recorder plays a continuous message telling them to grow taller. In the morning, their average height has increased by six inches.

Such extraordinary results would not be dismissed on the grounds that (c) lacked a control group; and the addition of a control group would make little difference. The plausible alternatives described above would not apply, but others would—specifically, fraud and measurement errors. If the study inspired further research, the goal would be to rule those out as possibilities.

Why scientists accept one interpretation rather than another is not susceptible to simple analysis. An important requirement is that acceptable explanations be compatible with previously established facts; but even well-established facts occasionally prove incorrect. Here, as in all aspects of scientific investigation, rules cannot be applied mechanically; critical reasoning is essential.

Experimenters Must be Sensitive to the Complexity of Variables

When posttreatment differences are found between experimental and control groups, investigators typically attribute them to the actions of a specific IV. But IVs as experimenters conceive them may differ from experimental treatments as subjects receive them. For example, an experimenter who administers a drug is likely to assume that the pharmacological properties of the drug constitute the IV. But differences between drug and placebo groups may occur for any of several reasons. The drug and placebo may have different smells, tastes, shapes, sizes, colors, or temperatures. Or the drug may be dissolved in another substance to increase its stability, and the solvent may be biologically active.

The IV may be entirely responsible for differences between groups, but not in the way interpreted by the experimenter. She may administer a drug because it increases body temperature, which is predicted by a theory to have specific behavioral consequences; the drug may produce the predicted consequences, but only because it induces itching.

Generalizations Cannot be Made Routinely

Only rarely are either subjects or settings for experiments chosen randomly, and generalizations from nonrepresentative samples cannot be made by formula. They require careful judgments and must be considered unproven until tested.

Generalizations to Other Subjects Cannot be Made Routinely   Both surveyors and experimenters generalize from samples to populations, but good surveyors try hard to use representative samples and good experimenters rarely do.

Surveys

The goal of surveys is to estimate the distributions of characteristics in populations. Given a representative sample of a particular size, the probability that the population value differs from it by a given amount can be precisely calculated. But if the sample is not representative, generalizations to the population will be inaccurate. For example, the average height of an all-male sample would be a poor estimator of the average height of the entire population.

Experiments

Experimenters don't ask about the distribution of characteristics in a population, but whether observed actions of an IV on a sample hold in the population as well. If the goal is to discover general laws of behavior, then the population is all of humanity. Representative samples are beyond the reach of even the most dedicated scientists; and the most commonly used samples, regardless of research goal, consist exclusively of either student volunteers (most frequently male) from psychology classes or Norway rats. Valid generalizations from these are possible, but the only way to know for sure is to test them on the appropriate populations. So, whereas surveyors can generalize according to precise formulae, experimenters must make judgments. Recall from above that all interpretations of experiments are open to alternative explanations; this applies to generalizations as well. Contrast (*d*) and (*e*) below:

   d. Student volunteers randomly assigned to an experimental group receive painful electric shocks. Students in a control group have electrodes attached to their wrists, but no current is passed through. The heart rates of experimentals change more than do those of controls.
   e. Student volunteers randomly assigned to an experimental group hear a speech about the need for a doubling of tuition costs. Students in a control group watch a short film on crocheting. Heart rates change more in the experimental group.

It would be reasonable to generalize from the student sample to the general population about the effects of shock on heart rate; and unreasonable to

Table 14.9 *Contrast Between a Typical Small Group Study and a Naturally Occurring Small Group Setting*

Typical Small Group Study	Naturally Occurring Small Group Setting
fixed duration, one hour or less	indefinite duration, typically months or years
all college students	community members
no prior interaction among group members	extensive prior interactions
imposed task	task arising from group needs
casual interactions	meaningful interactions
no enduring local culture	established local culture
no hierarchical relationships among members	hierarchical relationships
closed system: no personnel changes, no network of suppliers and recipients of services	open system

generalize about the effects of speeches on tuition hikes. No plausible explanation comes to mind why students differ from other people in the first case, but a perfectly plausible one is available for the second.

For those interested in pursuing the subject, Rosenthal and Rosnow's book (1975) is worth reading; they gave empirical evidence of several ways in which volunteers and nonvolunteers differ. (Volunteers differ even among themselves; Silverman and Margulis (1973) reported several personality differences between volunteers for a study labeled "Personality Assessment" and another called "Color Preferences.")

Generalizations to Other Situations Cannot be Made Routinely Experimenters generally administer and measure their variables in one particular context, the laboratory. But people's perceptions of the experimenter's purpose affect their behaviors in laboratory settings (Orne, 1962). Other settings also influence behaviors. Bruner (1965) wrote: "I am still struck by Roger Barker's ironic truism that the best way to predict the behavior of a human being is to know where he is: In a post office he behaves post office, at a church he behaves church."

Recognition of the contextual determinants of behavior has inspired development of two fields, ecological and environmental psychology, and led to attempts to classify settings (cf. Barker, 1965; Ittelson, Prohansky, Rivlin, & Winkel, 1974). Behavioral researchers have become sensitive to differences between laboratories and other settings. For example, Wicker (1985) contrasted small groups formed by experimenters with those that occur naturally. Table 14.9 is adapted from his work.

Because the effects of IVs are powerfully dependent upon context, some scientists conduct research in natural settings. Called field experiments, these test the generality of laboratory findings. Underwood et al. (1977) conducted a field experiment to test the proposition that people in a good mood are more likely than unhappy people to help others. Laboratory studies (cf. Aderman, 1972) had shown this to be true. (In the laboratory, mood was manipulated by such devices as having subjects read either cheerful or unpleasant materials; and helping was measured by things like responses to requests for donations of money.) Underwood et al., sought a naturally-occurring situation that affects emotions, so they went to the movies. They asked moviegoers to rate several popular films, and from the results picked three double features playing at the time: one film pair rated as inducing negative emotions and two rated as neutral. They set up collection boxes for a well-known charity outside the theater lobbies and, in correspondence with the laboratory findings, collected less after the sad movies.

7. Was the rating procedure necessary? Couldn't the experimenters have picked the movies on the basis of their own reactions?

8. Subjects were not randomly assigned to watch the sad or neutral movies. Might that have affected results?

9. Can you think of any strategy to overcome the problem of nonrandom assignment?

10. Subjects can be randomly assigned to conditions in field experiments. Suppose you wanted to see if helping behaviors of bystanders toward victims of falls are influenced by the victim's race. Can you design a field experiment with random assignment?

Additional Statistics for Analyzing Results from Two-Group Experiments

When the Dependent Variable is Measured Nominally

Suppose that a DV can have one of two values: pass/fail, agree/disagree, above 15/below 15, etc. Scores can be put in a table, such as Table 14.10, where A, B, C, and D represent frequencies. If the groups are unequal in size, the one with more subjects should be on top (represented by A and B) in the table. If the samples are independent, and neither $A + B$ nor $C + D$ exceeds 15, the Fisher exact probability test can be used. If either $A + B$ or $C + D$ exceeds 15 but neither $A + C$ nor $B + D$ does, the table can be turned around and relabeled. The computational steps, from Siegel (1956), are as follows:

1. Find the values of A, B, C, and D.
2. Find the value of $A + B$ in Appendix D, under the heading "Totals in Right Margin." Find the value of $C + D$ for that $A + B$ value.
3. The column to the right of "Totals in Right Margin" column lists possible B values for the particular $C + D$ value. If the observed B value is not among them, use the observed value of A instead.
4. Look at the entry in the column immediately to the right of the B value. If the value you observed for D is less than that number, the data are significant at the .05 level. (If, in step 3, A is used in place of B, use C in place of D in step 4.)

Table 14.10 *Set-up for the Fisher Exact Probability Test*

	Pass	Fail	Total
Experimentals	A	B	A + B
Controls	C	D	C + D
Total	A + C	B + D	N

Table 14.11 *Fictitious Data for the Fisher Exact Probability Test*

	Pass	Fail	Total
Controls	3	9	12
Took Class	6	4	10

Example: Ten people in a class are selected for special tutoring on study techniques. Their final exam scores are compared with those of the 12 other students in the class. The number of students in each group that passed and failed the class is shown in Table 14.11.

$A + B = 12, C + D = 10.$

From appendix D, with $A + B = 12$ (page 390, with arrow) and $C + D = 10$, the listed B values range from 5 to 12. In the example, $B = 9$. The value to the right of that is 3. Since D is 4, the data are not significant. Note: The Fisher exact probability test can be used even if subjects have not been randomly assigned to groups. For example, instead of having people who were and weren't tutored, the groups could have been male and female, or volunteers for the class and nonvolunteers. In such cases, statistically significant results would be interpreted differently; they would support the conclusion that the groups differ but would not indicate why.

Quick and Dirty Methods

Quick and dirty methods have the virtue of being easy to compute and the disadvantage of low power. They are sometimes handy. The test following is easy to apply but requires two conditions. First, although the groups do not have to be the same size, they can't differ in size by more than 10. Second, the sample that has the highest DV score of the combined set should not also have the lowest score.

Count the number of scores in one sample greater than all scores in the other; add to this the number of scores in the second sample less than all scores in the first. If the sum of these two values is more than 7, the results are significant at the .05 level.

Both experiments of Table 14.4 are based on the same number of subjects, so condition one is satisfied. In experiment one, group A has the highest score (12) and group B the lowest (6); in experiment two, group C has the highest

(20) and group D the lowest (0), so condition two is also satisfied. In experiment one, eight A scores are higher than the highest B score, and eight Bs are lower than the lowest A; $8 + 8 = 16$; since this is greater than 7, the results are significant at the .05 level. In experiment 2, 4 Cs are higher than the highest D and no Ds are lower than the lowest C; $4 + 0$ is less than 7, so the results are not significant.

Summary

Experiments allow researchers to answer "What would happen if . . . ?" questions and draw causal conclusions about data.

The two key features of experiments are (a) random assignment of subjects to groups or random order of treatments to subjects and (b) identical treatment of experimental and control groups except for administration of the IV.

The number of subjects run in an experiment should depend on the experimenter's tolerance for Type I and Type II errors and the minimum size of effect that would be acceptable. Sample size, effect size, and likelihood of making Type I and Type II errors are interrelated.

The greater the difference in dependent variable scores between the experimental and control groups, and the smaller the variability of DV scores of subjects within each group, the more likely it is that data will be statistically significant.

To increase differences between experimentals and controls, pick the IV carefully and give extreme levels to both groups.

To reduce within-group variability, use subjects who are similar in important respects; treat subjects uniformly; make sure measurements are reliable.

Virtually all samples for experiments are nonrepresentative, and the adequacy of generalizations from such samples cannot be made by formula. Each case must be judged on its own merits.

The laboratory is a unique setting. Researchers who wish to generalize to other settings often conduct field experiments.

Key Terms

Analysis of Variance A statistical technique for comparing between-group and within-group variabilities.

Between-Group Variability Variability caused by the independent variable.

Field Experiment These studies meet the criteria for experiments—random assignment of subjects to groups and identical treatment of all subjects except for administration of the IV—but they are conducted in natural settings rather than laboratories.

Fisher Exact Probability Test Used for analyzing two-group data when the groups are independent and small.

F Ratio The ratio of mean squares between groups to mean squares within groups.

History Events besides the IV that occur between measurements of the DV may affect the measurements.

Instrumentation Changes in the measuring instrument during the course of a study may affect scores.

Maturation Changes may occur in subjects during the course of a study because of the passage of time.

Mortality Subjects who drop out of a study may differ in important ways from those who remain. Drop-outs from experimental and control groups may also differ.

Power Power refers to the likelihood of getting statistically significant results when a true relationship exists or there are treatment effects. Power is equal to 1—the probability of making a Type II error.

Selection If subjects are not randomly chosen, the experimental and control groups may differ even before administration of the IV.

Standard Deviation The square root of the variance.

Sum of Squares The expression $\Sigma X^2 - (\Sigma X)^2/N$

Type I Error A Type I error occurs when a researcher concludes that there is a statistically significant relationship between variables, or that an IV significantly modifies a DV, but the variables are actually unrelated or the treatment actually has no effect.

Type II Error A Type II error occurs when variables are related, or a treatment has an effect, but the data don't show it (are not statistically significant).

Variance The sum of squares divided by $N - 1$.

Within-Group Variability Total variability of a set of scores minus between-group variability.

Answers

1. No, the probability of getting five of five heads is $.5^5 = .03$. But the probability of getting five of five tails, which would be an equally strong argument for bias, is also .03. The likelihood of getting either five heads or five tails is .06, which does not quite reach statistical significance. Most researchers, like the coin-flipper in this example, do not specify the direction of outcome beforehand. Their hypotheses take the form, "The experimental and control groups will differ," and are evaluated with *two-tailed* tests of significance. The hypotheses, "The experimental group will do better than the control group," and "The coin is biased toward heads," are directional and evaluated with *one-tailed* tests.
2. Nothing. The experimenter sets the Type I error.

3. An approximate answer can be found from Table 14.3. With power = .8 and medium effect size, 64 subjects are required for a significance level of .05. With 50 subjects, the significance level will be somewhat higher. The exact level, from other tables in Cohen's book, is .10.
4. From the table, 42 subjects are needed for power of .95, 58 subjects for .99. To find the power for 50 subjects, interpolate: power = .97.
5. Read directly from the table. N = 393.
6. Read directly from the table. The effect size is medium.
7. Researchers should directly test, rather than rely on intuition, that their manipulations produce the desired effects. Laboratory researchers use the same manipulations repeatedly and need not test each time; field experimenters should test.
8. Any time subjects are not randomly assigned to groups, there may be a problem. See the next answer for strategies to deal with this.
9. Underwood et al. (1977) exposed subjects to the collection boxes at different times: either as they entered the theaters or as they left. Had the entering subjects donated differently depending on the type of movies they had chosen to see, the results would have been difficult to interpret. But there were no differences. So it was reasonable to conclude that, with respect to giving to charity, people who chose sad movies were no different from those who chose neutral ones. And that differences in giving behavior of people who were leaving the theater were caused by the movies, not by pretreatment differences.

    Random assignment could have been implemented had it been thought essential. For example, people in a large apartment building could have been randomly assigned to receive free tickets to one showing or the other. (If the tickets named the films, differential mortality and trading of tickets would have become problems.)
10. There are many possibilities. Here's a study that was actually done. Piliavin, Rodin, & Piliavin (1970) had accomplices pretend to collapse on New York City subway trains. Some of the "victims" were black, some white, and they pretended to be either crippled or drunk. The investigators found that riders helped cripples irrespective of race, but were more likely to help drunks if they were of the same race.

Answer to analysis of variance problem for groups C and D.

1. Get the sum of the scores for groups C and D.

    $\Sigma C = 20 + 0 + 18 + 2 + 15 + 5 + 19 + 1 = 80$

    $\Sigma D = 10 + 4 + 14 + 0 + 13 + 12 + 1 + 2 = 56$

2. Get the squares of all the C and D scores and sum them.

   $\Sigma C^2 = 400 + 0 + 324 + 4 + 225 + 25 + 361 + 1 = 1340$

   $\Sigma D^2 = 100 + 16 + 196 + 0 + 169 + 144 + 1 + 4 = 630$

   $1340 + 630 = 1970$

3. Add ΣC and ΣD, square the result, and then divide by the total number of scores.

   $80 + 56 = 136$; $136^2 = 18496$; $18496/16 = 1156$

4. Subtract the result of step 3 from that of step 2. The answer is the sum of squares total (SS_t).

   $1970 - 1156 = 814$

5. Square the sum of the scores in group C and divide the answer by the number of subjects in group C. Do the same for group D and add the two answers.

   $80^2/8 = 800$; $56^2/8 = 392$; $800 + 392 = 1192$

6. Subtract the result of step 3 from that of step 5. The answer is the sum of squares between groups (SS_b).

   $1192 - 1156 = 36$

7. Subtract SS_b from SS_t. The answer is the sum of squares within groups (SS_w).

   $SS_w = 814 - 36 = 778$

8. Calculate the degrees of freedom for SS_t, SS_b, and SS_w.

   df for SS_t = the total number of scores minus 1; $16 - 1 = 15$

   df for SS_b = the number of groups minus 1; $2 - 1 = 1$

   df for SS_w = the total df minus the between df; $15 - 1 = 14$

9. Divide the between- and within-groups sums of squares by their respective dfs. These give the mean squares (ms) for each.

   $ms_b = SS_b/df = 36/1 = 36$

   $ms_w = SS_w/df = 778/14 = 55.6$

10. Compute the F ratio: ms_b/ms_w.

    $ms_b/ms_w = 36/55.6$

    Since the F ratio is less than 1, there is no need to compute further; it isn't significant.

CHAPTER 15

Experimenting—Variations on the Two-Group Design

By the time you finish reading this chapter, you should be able to answer the following questions:

What are the advantages of using more than one level of an IV?

In experiments with more than two groups, what is the meaning of a significant F ratio?

What is a factorial design?

What is an interaction?

What is the major advantage and what are the disadvantages of using subjects as their own controls?

Why do some scientists argue that generalizations from carefully chosen small samples are more meaningful than generalizations from much larger samples?

What is dimensional sampling?

Can single subject research designs be analyzed statistically?

What are the advantages when an experimenter serves as her own control?

Textbooks on experimental design (cf. Keppel & Saufley, 1980) describe many variations of the two-group design. Three of them are considered below.

Studying an IV at More Than Two Levels

Recall the problem of chapter 5, where you were asked to test the effectiveness of an alleged hair restorative. The solution is conceptually easy even with an IV of extraterrestrial origin. But a major practical question would remain: If you gave the control group a placebo, what dosage of hair restorative would be best for the experiment group?

If an IV is given at too low a dose, between-group variability will be negligible and results nonsignificant. But high doses may be expensive, painful, or even deadly. Besides, maximum between-group variability often occurs at moderate doses. Figure 15.1 shows the relationship between anxiety and performance. People with no anxiety—completely unconcerned about their performance—don't do as well as those with moderate levels. But too much anxiety is paralyzing, and performance suffers. Probably all treatments that enhance a particular behavior will, if made sufficiently intense, impair the same behavior.

Since the actions of an IV are not known beforehand, an experimenter must estimate its most effective levels in considerable ignorance. He can search the scientific literature for clues and pilot test a few levels. In addition, he can administer more than two levels of the IV, each to a different group of subjects (assigned, of course, randomly). A good strategy is to use levels that are constant multiples of each other, for instance, 4-, 8-, 16-, and 32-hours of food deprivation. This procedure not only increases the likelihood of finding statistically significant differences, but also shows how the DV changes throughout the tested range of the IV.

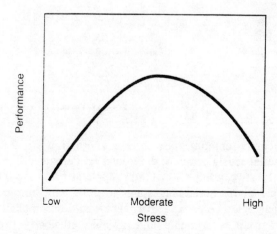

Figure 15.1

BOX 15.1

Analysis of Variance for More Than Two Independent Groups

Arrange the scores in a table. Table 15.1 gives the DV scores from a fictitious study involving 32 subjects who had been randomly assigned to one of four groups. I've made the groups equal in size, which makes computations easier but is not essential.

Table 15.1 Results from a Fictitious Study

Group A	Group B	Group C	Group D
6	5	10	11
3	13	8	4
7	9	15	12
6	10	12	10
5	10	15	6
5	11	13	9
7	10	12	12
5	12	10	5

Computational Procedures

1. Get the sum of the scores for each group.

   $\Sigma A = 6 + 3 + 7 + 6 + 5 + 5 + 7 + 5 = 44$

   $\Sigma B = 5 + 13 + 9 + 10 + 10 + 11 + 10 + 12 = 80$

   $\Sigma C = 10 + 8 + 15 + 12 + 15 + 13 + 12 + 10 = 95$

   $\Sigma D = 11 + 4 + 12 + 10 + 6 + 9 + 12 + 5 = 69$

2. Get the squares of all the scores and sum them.

   $6^2 + 3^2 + 7^2 + \ldots + 9^2 + 12^2 + 5^2 = 2932$

3. Add all the group sums (step 1), square the result, and divide by the total number of scores.

   $44 + 80 + 95 + 69 = 288$; $288^2 = 82944$; $82944/32 = 2592$

4. Subtract the result of step 3 from that of step 2. The answer is the sum of squares total (SS_t).

   $2932 - 2592 = 340$

5. Square the sum of the scores in group A and divide the answer by the number of subjects in group A. Do the same for each group and add the answers.

   $44^2/8 + 80^2/8 + 95^2/8 + 69^2/8 = 2765$

6. Subtract the result of step 3 from that of step 5. The answer is the sum of squares between groups (SS_b).

   $2765 - 2592 = 173$

7. Subtacts SS_b from SS_t. The answer is the sum of squares within groups (SS_w).

   $340 - 173 = 167$

Experimenters sometimes use three or more groups in a single study so they can simultaneously compare different IVs. Perhaps group one would receive hair restorative, group two vitamin supplements, group three hair transplants, group four hypnotic suggestions to grow hair, and group five scalp massage. The single five-group experiment would accomplish as much as a series of two-group experiments, and far more efficiently. (To evaluate the

8. Calculate the degrees of freedom.

   df for SS_t = the total number of scores minus 1; $32 - 1 = 31$

   df for SS_b = the number of groups minus 1; $4 - 1 = 3$

   df for SS_w = the total df minus the between df; $31 - 3 = 28$

9. Divide SS_b and SS_w by their respective df. These give the mean squares for each.

   $ms_b = 173/3 = 57.67$

   $ms_w = 167/28 = 5.96$

10. The F ratio equals ms_b/ms_w.

    $F = 57.67/5.96 = 9.68$

11. Turn to appendix C. For 3 and 28 degrees of freedom, an F value of 2.95 or more is significant at the .05 level.

A significant F value means that at least one group differs significantly from at least one other. To establish which one(s) differ significantly from which other(s), additional computations are necessary. Several procedures are available; the one that follows was developed by Tukey and modified by Snedecor (1956).

12. Divide ms_w by the number of subjects per group. Take the square root of the answer.

    $5.96/8 = 0.745$ and $\sqrt{0.745} = 0.865$.

13. Turn to appendix C, and find the entry under the number of groups in the study and the number of degrees of freedom associated with ms_w. If the number of degrees of freedom isn't listed, use the next lower number.

The table doesn't list 28 df. For four groups and 24 df, the number is 3.90.

14. Multiply the answer from step 12 by the answer from step 13.

    $0.865 \times 3.90 = 3.37$.

15. If the mean of any group differs from the mean of any other group by at least the value found in step 14, the groups differ at the .05 level.

The means of each group are as follows:

$\bar{A} = 44/8 = 5.5$.
$\bar{B} = 80/8 = 10.0$
$\bar{C} = 95/8 = 11.9$
$\bar{D} = 69/8 = 8.6$

B − A = 4.5; C − A = 6.4; D − A = 3.1. So, A differs significantly from B and C. C − B = 1.9; B − D = 1.4. So, B does not differ significantly from C or D. C − D = 3.3, so they don't differ significantly.

⋯⋯⋯⋯⋯⋯⋯⋯⋯⋯⋯⋯⋯⋯⋯⋯⋯⋯⋯⋯⋯⋯⋯⋯⋯⋯⋯⋯⋯⋯⋯⋯⋯⋯⋯⋯

effects of each of five IVs against the effects of each of the other four, 10 two-group experiments would be needed. This would entail 20 groups of subjects and 10 different times and conditions of testing.)

The statistical analysis of multi-group experiments, whether they involve several levels of a single IV or several IVs, is much like that for the two-group case. Box 15.1 gives the procedure.

Factorial Designs

The answers to many questions depend on circumstances: "Which sport, basketball or hockey, is more popular?" The answer depends on whether the question is asked in the United States or Sweden. "Is a daily five mile run good for health?" Maybe, if the subjects are 20; highly unlikely, if they are 90. "Is sexual attractiveness increased when a person changes from faded jeans to an expensive dress?" Maybe, for Daryl Hannah; no, for Mel Gibson. (Even then, the answer depends on the judges.)

When the effects of one variable depend on the value of a second, the two are said to interact. A recurring criticism of social science experiments is that they isolate variables from stimulus complexes in which the variables normally occur; but outside of laboratories, the variables interact in so many different ways and with so many other variables that the laboratory-based laws don't apply. It's easy to find examples:

- No single teaching style is best. The ability of college teachers to convey material, as measured by examination scores of their students, depends on who the students are. When the beliefs of students and their professors about important matters were most similar, the students scored highest (Majasan, 1972).
- McKeachie (1974) reported a great many interactions between teaching styles and students' personalities.
- The effectiveness of persuasive messages cannot be judged in isolation. Effectiveness depends on interactions among characteristics of the persuader, listener, message, and setting (McGuire, 1973).
- Mischel (1973) anticipated that personality research will become the study of higher order interactions: "For example, to predict a subject's voluntary delay of gratification, one may have to know how old he is, his sex, the experimenter's sex, the particular objects for which he is waiting, the consequences of not waiting, the models to whom he was just exposed, his immediately prior experience—the list gets almost endless."
- Aptitudes of salespeople interact with job conditions to such an extent that aptitude tests administered by one firm should never be used as the basis of personnel decisions of another firm without additional validation studies.

Cronbach (1975) suggested that social scientists recognize the importance of interactions and devote more time to their study. In addition, he wrote, social scientists should stop trying to discover general behavioral laws. They should accept the more modest goal of describing in detail the conditions under which their experimental data are collected; and they should analyze their data for local effects arising from uncontrollable conditions.

Table 15.2 *Fictitious Data for a 2 × 2 Factorial Design*

		A	
		Low	High
B	Low	20	30
	High	25	100

With a single study called a factorial design, an experimenter can test not only the separate effects of each of two or more independent variables (factors) but also how they interact. And with fewer computations and subjects than separate experiments would require. Factors can be administered at several levels, for example, low, medium, and high intensity. If one factor has three levels and the other five, the design is called a 3 × 5 factorial. If there are five factors, one with three levels and all the others with two, the design is called a 3 × 2 × 2 × 2 × 2. Experimenters rarely use more than three factors or more than five levels of any one factor. As the number of factors increases, so does the number of potential interactions.

Suppose that two IVs—factor A and factor B—are combined in an experiment. The design is a 2 × 2, meaning that both IVs are administered at two levels (called, for simplicity's sake, low and high). Suppose the DV is amount of weight lifted, and the results are as shown in Table 15.2.

The effects of the A variable can be assessed by combining scores of the two groups that received the low level of A and contrasting them with the scores of the two high-level A groups: 20 + 25 versus 30 + 100. Similarly, the effects of B can be assessed by contrasting 20 + 30 with 25 + 100. Analysis of variance is a statistical method for combining and contrasting that allows precise estimation of the probability that observed differences in dependent variable scores are due to chance. Two × two factorial designs give rise to eight possible combinations of main effects (effects of the IVs) and interactions.

Eight Possible Outcomes of a 2 × 2 Factorial Study

1. There is a significant main effect of factor A. That is, A significantly affects the DV: a low dose produces a different effect than a high dose. B does not affect the DV, i.e., no matter what the level of B, the DV score is the same. There is no interaction. This pattern is diagrammed in Figure 15.2.
2. B significantly affects the DV and A doesn't, as shown in Figure 15.3. (In the previous example, the DV score increased as A went from low to high, but here it decreases as B goes from low to high. This was done to remind you that effects can occur in many ways.)

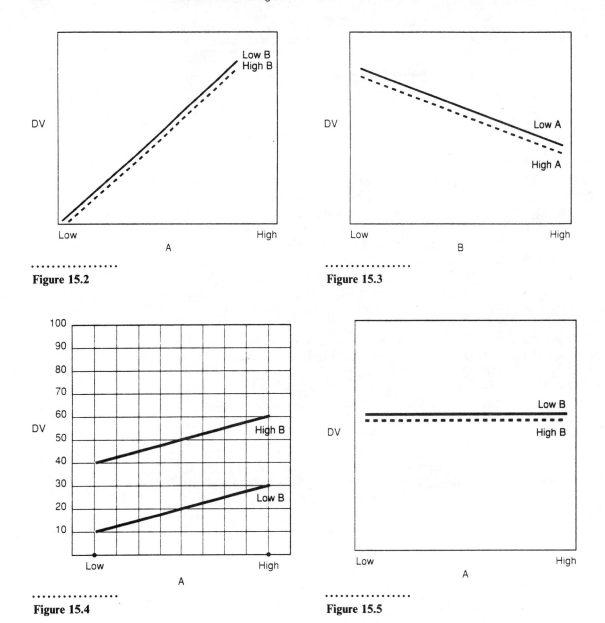

Figure 15.2

Figure 15.3

Figure 15.4

Figure 15.5

3. Both *A* and *B* significantly affect the DV, but the influence of each is independent of the value of the other: there is no interaction. In Figure 15.4, as the *A* level goes from low to high, the DV score increases by 20 units. The 20 unit increase occurs no matter what the level of *B*. Similarly, as *B* goes from low to high, the DV score increases by 30 units—no matter what the level of *A*.
4. Neither variable affects the DV and there is no interaction. See Figure 15.5.

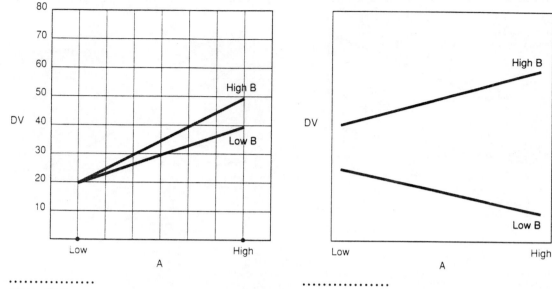

Figure 15.6

Figure 15.7

5. *A* significantly affects the DV, *B* doesn't, and there is an interaction. That is, the effects of *A* depend on the value of *B*. When *B* is at a low level, as *A* changes from low to high the DV score increases by 20 units; but when *B* is at a high level, a change in *A* from low to high increases the DV score by 30 units.

    This pattern, in which an IV (factor *B*) does not act directly but modifies the actions of another IV, may seem improbable but isn't. For example, the drug SKF 525-A, administered alone, does not promote sleeping; but SKF 525-A greatly prolongs the sleep-inducing effects of hexobarbital (Cook, Toner, & Fellows, 1954). See Figure 15.6. Please read the next section for additional comments.

6. *B* significantly affects the DV, *A* doesn't, and there is an interaction. See Figure 15.7.
7. Both *A* and *B* affect the DV and there is an interaction. See Figure 15.8.
8. Neither *A* nor *B* has a significant effect, but there is an interaction. See Figure 15.9.

Interpretation of Interactions

Interactions come in many forms, and for any two IVs the form may depend on the DV being measured. Interactions restrict generalizations, as can be seen by comparing conclusions from the results depicted in Figures 15.2 (no interaction), 15.6, 15.7, and 15.9.

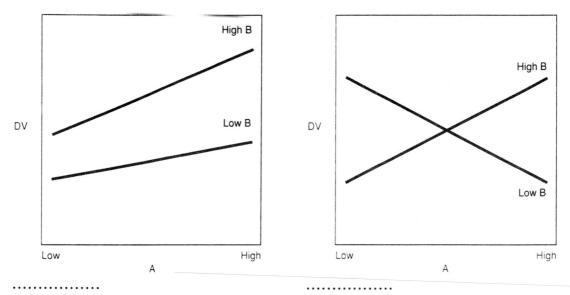

Figure 15.8

Figure 15.9

Conclusion from 15.2: As A goes from low to high, DV scores increase.
Conclusion from 15.6: As A goes from low to high, DV scores increase, and the increase is greatest when the level of B is high.
Conclusion from 15.7: As A goes from low to high, DV scores sometimes increase—when B levels are high, not when they are low.
Conclusion from 15.9: As A goes from low to high, and B is low, DV scores increase; but when B is high, DV scores decrease as A goes from low to high.

Findings of the type shown in Figure 15.9, with an interaction but no significant main effects, should not cause disappointment. The reason the A results are not significant is that A acts in opposite directions at low and high levels of B; so the effects cancel each other out. If A were tested at only one level of B, significance would be reached; and B, likewise, would be significant if tested at only one level of A.

More generally, a significant interaction tells that variables work better under certain conditions. These are readily identified from inspecting a graph, so drawing and analyzing graphs should be done routinely. In addition, a significant interaction implies that the variables would have significant effects if tested under the right conditions.

An Example

Suppose an experimenter wanted to see how teaching method (method X or Y) and length of classes (30 minutes or 60 minutes) affect scores on a math test. He could do two independent studies. For the first, he would randomly

Table 15.3. Results from a Fictitious Study on the Effects of Teaching Method and Class Length on Scores on a Mathematics Test

Group A Method X/30-min	Group B X/60-min	Group C Y/30-min	Group D Y/60-min
8	10	7	10
13	8	7	9
10	12	9	9
12	8	9	12
11	10	13	5
13	7	6	12
9	9	8	10
12	10	10	11
14	7	8	10
12	11	9	13

assign, say, 20 subjects to method X and 20 to method Y. For the second, he would assign 20 subjects to 30-minute classes and 20 to 60-minute classes. The total of 80 subjects would help him draw two conclusions: whether or not teaching method makes a difference, and whether or not class length makes a difference.

Alternatively, he could do a 2 × 2 factorial study. He would randomly assign 40 subjects to four groups: (a) 10 students to learn by method X with 30-minute classes, (b) 10 to learn by X with 60-minute classes, (c) 10 to learn by Y with 30-minute classes, and (d) 10 to learn by Y with 60-minute classes. Scores on the math test in this fictitious example are summarized in Table 15.3. Analysis of variance would enable him to combine scores of the 20 subjects who learned by method X and compare them with the 20 who learned by Y; and the 20 who had 30-minute classes with the 20 who had 60-minute classes. Not only would he economize on subjects in evaluating his two main effects (the effects of the IVs), he would be able to draw a third conclusion: whether or not they interact. What a bargain! Box 15.2 gives the computational procedures.

More on Interactions

A researcher who had run two separate two-group experiments on teaching method and class length would have concluded, probably unhappily, that his results have no educational policy implications: neither factor makes a difference. The factorial design shows that the conclusion would have been a terrible mistake. Teaching method X is much better than Y with 30-minute class periods, Y much better than X with 60-minute periods.

BOX 15.3

Computational Procedures for the 2 × 2 Factorial Design

1. From Table 15.2, get the sum of the scores for each group.

   $\Sigma A = 8 + 13 + 10 + 12 + 11 + 13 + 9 + 12 + 14 + 12 = 114$

   $\Sigma B = 10 + 8 + 12 + 8 + 10 + 7 + 9 + 10 + 7 + 11 = 92$

   $\Sigma C = 7 + 7 + 9 + 9 + 13 + 6 + 8 + 10 + 8 + 9 = 86$

   $\Sigma D = 10 + 9 + 9 + 12 + 14 + 12 + 10 + 11 + 10 + 13 = 110$

2. Get the squares of all the scores and sum them.

   $8^2 + 13^2 + 10^2 + \ldots + 11^2 + 10^2 + 13^2 = 4214$

3. Add all the group sums, square the result, and then divide by the total number of scores. Call this value C. You will use it several times.

   $114 + 92 + 86 + 110 = 402;$
   $402^2 = 161604;$
   $161604/40 = 4040 = C$

4. Subtract C from the result of step 2. The answer is SS_t.

   $SS_t = 4214 - 4040 = 174$

5. Compute the effects of class length by getting two separate sums: for groups A and C (30-minutes) and for groups B and D (60-minutes).

   $A + C = 114 + 86 = 200$
   $B + D = 92 + 110 = 202$

6. Square the two sums from the previous step, divide each by the number of scores in the two groups combined, and add the two answers.

   $200^2 = 40000; 40000/20 = 2000$
   $202^2 = 40804; 40804/20 = 2040.2$
   $2000 + 2040.2 = 4040$

7. Subtract C from the result of step 6. The answer is

   $SS_{class\ length}$

   $SS_{class\ length} = 4040 - 4040 = 0$

8. Compute the effects of teaching method by getting two separate sums: for groups A and B (method X) and for groups C and D (method Y).

   $A + B = 114 + 92 = 206$
   $C + D = 86 + 110 = 196$

9. Square the two sums from the previous step, divide each by the number of scores in the two groups combined, and add the two answers.

   $206^2 = 42436; 42436/20 = 2121.8$
   $196^2 = 38416; 38416/20 = 1920.8$
   $2121.8 + 1920.8 = 4042$

10. Subtract C from the result of step 9. The answer is

    $SS_{method}.$

    $SS_{method} = 4042 - 4040 = 2$

11. Compute the interaction effects of class length and teaching method.

    a. Square the sums of each of the four groups, and divide each product by the number of subjects in the group.

    $114^2 = 12996; 12996/10 = 1299.6$
    $92^2 = 8464; 8464/10 = 846.4$
    $86^2 = 7396; 7396/10 = 739.6$
    $110^2 = 12100; 12100/10 = 1210$

    b. Add the four answers from a.

    $1299.6 + 846.4 + 739.6 + 1210 = 4096$

c. The sum of squares for the interaction between class length and teaching method, written $SS_{\text{class length} \times \text{method}}$, is found by subtracting the sum of $SS_{\text{class length}}$, SS_{method}, and C from the result of 11b.

$$4096 - 0 - 2 - 4040 = 54$$

12. Compute SS_{error}.

$$SS_{\text{error}} = SS_t - SS_{\text{class length}} - SS_{\text{method}} - SS_{\text{class length} \times \text{method}}$$
$$SS_{\text{error}} = 174 - 0 - 2 - 54 = 118$$

13. Calculate the degrees of freedom for all the sums of squares. In a 2 × 2 factorial design, the degrees of freedom for the two independent variables and for the interaction will always be one.

df for SS_t = the total number of scores minus 1; $40 - 1 = 39$

df for $SS_{\text{class length}} = 1$

df for $SS_{\text{method}} = 1$

df for $SS_{\text{class length} \times \text{method}} = 1$

df for SS_{error} = the df for SS_t minus 3; $39 - 3 = 36$

14. Divide each sum of squares except SS_t by its df. This gives the mean squares.

$$ms_{\text{class length}} = 0/1 = 0$$
$$ms_{\text{method}} = 2/1 = 2$$
$$ms_{\text{class length} \times \text{method}} = 54/1 = 54$$
$$ms_{\text{error}} = 118/36 = 3.28$$

15. Compute three F ratios:

a. $ms_{\text{class length}}/ms_{\text{error}} = 1/3.28$ = less than 1, so no need to compute

b. $ms_{\text{method}}/ms_{\text{error}} = 2/3.28$ = less than 1, no need to compute

c. $ms_{\text{class length} \times \text{method}} = 54/3.28 = 16.46$

16. Each of the F ratios is evaluated from Appendix C, with 1 and 36 df. Since there is no listing for 36 df, use 30. The F ratio must be at least 4.17 to be significant at the .05 level. So, neither of the main effects is significant, but the interaction is.

17. The results can be tabled as follows:

Table 15.4 *Summary of Analysis of Variance*

Source	SS	df	ms	F	p
Total	174	39	—	—	—
length	0	1	0	—	n.s.
method	2	1	2	—	n.s.
length × method	54	1	54	16.46	<.05
error	118	36	3.28	—	—

Table 15.5 F *Ratios from Eight Fictitious Studies*

Study	Factor 1	Factor 2	Factor 1 × Factor 2
A	2.57	4.55	1.98
B	8.00	0.46	3.24
C	7.47	5.09	2.44
D	3.12	2.53	1.07
E	2.57	4.55	5.03
F	8.00	0.46	4.18
G	7.47	5.09	7.77
H	3.12	2.53	6.45

Given in Table 15.5 are the F ratios from eight other fictitious 2 × 2 studies, each with 16 subjects per group.

....

1. How many degrees of freedom are there for the error term?

2. How large must the F ratio be to reach statistical significance (0.05 level)?

3. Give a one or two sentence summary of the results of each study.
....

Subject Variables

The effects of an independent variable may change with the subject population. Males and females often respond differently, as do old and young, soldiers and civilians, normal and schizophrenic, and so forth. To learn if an IV affects both sexes similarly, a researcher might use gender as one of her factors in a 2 × 2 design. Her four groups would be (a) males, level one of the IV; (b) males, level two; (c) females, level one; and (d) females, level two. All subjects would be assigned randomly.

....

4. If gender were the first factor of Table 15.5, which pattern(s) would show that gender affects the DV scores?

5. If, in the teaching method/class length study, the means were higher for students exposed to method X and for students taking 30-minute classes, and both F ratios for main effects (ms_{method}/ms_{error} and $ms_{class\ length}/ms_{error}$) were significant, the researcher would have reasonably concluded that both teaching method X and 30-minute class lengths cause improvement in math tests scores. Suppose a second researcher studied psychotherapeutic techniques X and Y with both schizophrenics and manic-depressives. And that schizophrenics and patients exposed to X scored higher on a test of verbal fluency, with both F ratios reaching significance. What reasonable conclusions could be drawn? Don't answer hastily; remember the train example, page 133.

....

Blocking

As subjects become increasingly dissimilar in characteristics relevant to the dependent variable, ms_{error} increases. And the greater the ms_{error} term, the less likely is the F ratio to reach statistical significance. But investigators must often use subjects who differ, such as males and females or students from two distinct school districts, to have a large enough sample. If the trait on which they differ is identified beforehand, and if it correlates at least .30 with the DV, it can be used as one of the factors in a factorial design. So, for example, an investigator interested in the effects of teaching methods X and Y on math scores, even though not interested in gender differences, might nevertheless use a 2 × 2 factorial design with gender as one of the factors. If gender correlated at least .30 with math scores, ms_{error} would be reduced by this approach, called blocking. Subjects within a block would be randomly assigned to methods X or Y.

The Experimenter as a Variable

Rosenthal (1963), McGuigan (1963), and Kintz, Delprato, Mettel, Persons, and Schappe (1965) emphasized that experimenters are neglected variables. Two experimenters following identical protocols may get substantially different results. Below are some examples:

- Two physicians administered placebos; patients of one consistently increased their gastric acid secretions, patients of the other consistently decreased secretions (Wolf, 1962).
- Two experimenters, a young, petite woman and a mature, large man, reinforced subjects for saying hostile words. Significantly more were emitted in the presence of the woman (Binder, McConnell, & Sjoholm, 1957).
- Psychotherapists were divided into two groups on the basis of their perception of movement in Rorschach inkblots. Later, they administered Rorschachs. Those who had perceived lots of movement scored their patients as perceiving significantly more movements (Graham, 1960).
- Clinical psychologists and interns had their personal hostility estimated by an investigator, and then they scored drawings for aggression. The correlation was 0.94 between personal hostility and degree of hostility they saw in the drawings (Hammer & Piotrowski, 1953).

McGuigan (1963) urged researchers to confront the problem by routinely using factorial designs with experimenter as one of the factors. A nonsignificant experimenter effect would increase confidence in the generality of results; a significant one would open up interesting avenues for further research.

Within-Subject Designs

If each subject is exposed to each experimental condition, ms_{error} can be reduced to a minimum. Experiments of this type, in which subjects serve as their own controls, are called within-subject designs.

6. In within-subject designs, each subject serves as her own control. Why does ms_{error} not reduce to zero?

A researcher wishing to compare teaching methods X and Y could randomly assign her subjects to one or the other condition or, alternatively, could expose each subject to both conditions. The savings in subjects of the latter approach is considerable and a major advantage of within-subjects designs. On the debit side, if subjects drop out after receiving only a single treatment, the data collected on them is wasted. In addition, if the two conditions are not equally desirable, subjects receiving the less desirable one first may drop out in greater numbers, making the final results difficult to interpret.

The counterbalanced within-subjects 2×2 design tests the possibility that the order in which subjects receive each of two treatments makes a difference. Subjects are randomly assigned, half to receive method X first and half to receive Y first. Thus, there are two factors—method and order—and the data are analyzed as with any other 2×2 experiment.

Single Case Experimental Designs

In the within-subject design described above, each of several subjects is exposed once to both levels of the IV. Alternatively, a few subjects can be exposed repeatedly to two or more levels of the IV. Garmezy (1982) showed that the development of psychology owes much to single case experimental designs. In the same 1982 book (Kazdin & Tuma, 1982), several distinguished scientists argued that such research has a place within their fields (animal behavior, clinical research and practice, operant conditioning, and psychopharmacology). Single case experimental designs often provide a satisfactory alternative when financial or other considerations rule out properly-conducted, large-scale studies. (Although single case experimenters typically use just a few subjects and sometimes only one, large samples are possible; Kazdin (1982) cited a project involving more than a million people. The key is that each subject is studied intensively.)

Sidman (1960) vigorously advocated such small sample studies. He wrote: "When an organism's behavior can repeatedly be manipulated in a quantitatively consistent fashion, the phenomenon in question is a real one and the experimenter has relevant variables well under control." He added that the control ". . . can be exercised, at will, over the course of time." In experiments with groups of subjects, such control is lacking.

The appropriate type of conclusion following an experiment comparing two groups of subjects is that a treatment either does or does not affect mean scores. If the experiment is successfully replicated with new subjects, and especially if they are of diverse types (young and old, male and female, experienced and naive, etc.), the results are often assumed to have broad generality. But for valid generalizations, samples must be randomly chosen from populations. For psychology experiments, they rarely are. In addition, when data are analyzed at the group rather than individual level, population generality is an illusion.

Review the fictitious data of Table 14.7, page 214, on 27 pairs of identical twins. The sign test showed a statistically significant difference in creativity scores for the group that had received special musical training. Now assume further that the pairs included both sexes and wide ranges in age, education, and socioeconomic status. It would be tempting to conclude from the diversity of the sample that the results apply to a broad population. But doing so would require ignoring pairs 5, 6, 11, 15, 19, and 24. They might have been the only left-handers in the study, or the only women, or the only college graduates. Whatever their special characteristics, they would be swallowed up by grouped data.

Recall Eysenck's contention (p. 78) that psychotherapy is ineffective. He based his position on a statistical analysis that showed little difference in outcomes between clients who had undergone therapy and untreated controls. But Bergin (1966) showed that the average effect of psychotherapy is small partly because many clients get worse; the genuine improvements observed in many others are obscured when outcomes are averaged. Other scientists, focusing on individual cases rather than grouped data, identified client and therapist characteristics associated with positive outcomes and showed that outcomes depend on an interaction between type of therapy and type of disorder (Nietzel & Bernstein, 1987; Strupp & Hadley, 1979).

When a few subjects are studied intensively, experimenters are favorably situated to observe interesting relationships between their treatment and response measures; they can then test other subjects under identical conditions. If a relationship remains unchanged regardless of the age, gender, level of experience, etc., of the new subjects, a strong argument can be made for population generality. And if the new subjects respond differently, the experimenter may be led to testable speculations about the reasons. To quote Sidman again:

Tracking down sources of variability is then a primary technique for establishing generality. Generality and variability are basically antithetical concepts. If there are major undiscovered sources of variability in a given set of data, any attempt to achieve subject or principle generality is likely to fail. Every time we discover and achieve control of a factor that contributes to variability, we increase the likelihood that our data will be reproducible with new subjects and in different situations. Experience has taught us that precision of control leads to more extensive generalization of data.

Arnold (1970) advocated the technique of dimensional sampling for small sample research. The experimenter would select each subject according to some theoretically crucial dimension. For example, if her theory led to the prediction that both level of anxiety and gender were relevant, she might choose four people: an anxious man and woman and a nonanxious man and woman. Differences between them might point the way to rapid identification of the key variables and rapid tests for generality.

Procedures for Conducting Single Case Experimental Designs

The most common methods for single case experimental research begin with evaluation of the subject's performance prior to treatment. Data are collected from this baseline phase until the behaviors of interest stabilize; unfortunately, no formula is available to tell when suitable stabilization has occurred; and investigators taking baseline measurements with people must be sensitive to both practical and ethical considerations. For further discussion, see Sidman (1960) and Kazdin (1978).

The two most popular single case experimental research designs are the ABAB and multiple-baseline designs (Kazdin, 1978). ABAB is shorthand for baseline/treatment/baseline/treatment. A treatment is introduced when baseline behaviors stabilize, and it continues until the behaviors restabilize at (it is hoped) a different level. The treatment is then withdrawn so the experimenter can see if the behavior is reversible; it is reinstated when the behaviors have stabilized again. If the treatment has long-lasting effects, the start of the second B phase may require a lengthy wait. Experimenters must also concern themselves with the ethical implications of withdrawing treatments that have been effective in reducing harmful behaviors.

In the multiple-baseline design, a researcher obtains baseline data on at least two behaviors, and the intervention is applied at different times to each. For example, Kazdin (1978) described a treatment program for a young boy with several behavioral problems. Training was first focused on his poor eye contact and inappropriate body position when in the company of others. Those two behaviors improved while his poor speech and bland affect remained unchanged. When treatment was extended to them, they improved as well. Thus, control was demonstrated over all four problem behaviors.

Data Analysis

The results of single case experimental research are often evaluated by visual inspection, but formal statistical analysis is possible with even one subject if the treatment schedule is specified in advance. Edgington (1972) showed how several commonly used statistics could be adapted to the single-subject experiment. Suppose, for example, that a single subject ran a 50-yard dash, once in the morning (A) and once in the evening (P) every day for 14 days. The data could be presented and analyzed exactly as given in Box 14.3 (p. 215), example 2.

Experimenter as Subject

Two experiments that helped establish the legitimacy of single subject studies (those of Ebbinghaus (1885) on learning and Stratton (1897) on perception) were noteworthy in another way as well—the experimenter was also the subject. Experimenter as subject (E/S) studies offer several advantages over other research designs.

Experimenter/subjects, aware of the research hypothesis, have no need to guess about it. So demand characteristics (see p. 50), while still operating, differ from those faced by other subjects and can be estimated from performance differences between the experimenter and others. The experimenter need not worry about misinterpretation of instructions nor about failure of subjects to do their best because of lack of interest. He receives immediate feedback so he can adjust the values of important variables until they are optimal. He can use powerful yet inexpensive motivators that would be of dubious value with unknown subjects. (Some people like pistachio nuts, some like peanuts.) In addition, the experimenter who serves as subject has a particularly valuable source of feedback about the effectiveness of the IV manipulation.

In many experiments, only the acute effects of a manipulation are measured, but there seems to be an implicit assumption that the chronic effects would be equivalent. Thus, research on the effects of a single dose of LSD on creativity might receive the title "The effects of LSD on creativity." The assumption of equivalence is often unwarranted. Moreover, the effects of chronic treatment, although rarely studied, may have greater theoretical and practical implications. For example, people who try heroin often respond with euphoria to their first dose but feel little more than lethargy after frequent use. The frequent users are of greatest concern to society. Chronic preparations are relatively easy with small samples, especially with E/S studies. Carlson (1971) complained that psychologists neglect many important problems, especially those in which subjects are studied over long time periods. The reason is that longitudinal studies require careful tracking of subjects, sometimes transporting them long distances, and paying and keeping them motivated throughout the course of the project. With E/S research, the problems don't arise.

Certain important questions can be answered only through intensive study of individuals, as in biorhythm research. The study of biorhythms requires that subjects be monitored continuously over long time periods. It seems unlikely that many people would volunteer for a project that required daily visits to a laboratory for several months; but the inconvenience would be minimal, the commitment maximal, for experimenters who normally work everyday in their laboratories or whose laboratories travel with them.

E/S studies minimize the likelihood that interesting and unexpected effects will go unnoticed. Once stable baseline responding is established, disruption will become apparent and can be followed up inexpensively. Thus, hunches can be indulged and many variables tested at a time. Unlike the group situation, where variability is regarded as a nuisance factor, variability in single

Box 15.3

Two Experimenter/Subject Studies

Linton (1982) was interested in the course of forgetting of normal, everyday events. Over a 6-year period, she recorded on individual cards at least two personal events each day such as where she ate or who she saw, and she rated the importance, affective quality, or novelty of each one. From her stock, which eventually reached 5,500 cards, she randomly picked 150 each month. She read the description of each event and then tried to recall when it had occurred and how she had rated it. By the end of each year, she had forgotten about 1 percent of that year's events and by the end of the study, she had forgotten 31 percent of the events.

Garfield (1973, 1976) recorded her dreams, several thousand of them, over a 25-year period starting when she was 14. Although many personality theorists and therapists rely heavily on dream material, they have paid little attention to the methods of data collection. Garfield described her method in detail. She provided unique and valuable data on fluctuations in amount and contents of dream recall during her menstrual cycle and in relation to fatigue, illness, and major life changes.

INTENSIVELY STUDYING THE SYNERGISTIC EFFECTS OF ULTRAVIOLET RADIATION AND SEVEN PERCENT ALCOHOL SOLUTION ON COMPREHENSION OF COMPLEX PICTORIAL INFORMATION

subject data should be welcomed; it increases the possibility that previously unrecognized determinants of behavior will be discovered. All small N studies are valuable in this respect, E/S studies especially so since the variables of interest can be isolated more rapidly from the enormous number of potentially relevant ones. In addition, natural experiments (the experimenter is sick, dieting, euphoric, despondent, sleepy, etc.) can be exploited as they occur. Interactions between variables can be studied easily.

Box 15.3 gives two examples of E/S research that made significant contributions to their fields.

Summary

Three variations on the two-group design were described. First, an experimenter can test more than two values of an IV or more than one IV in a single study. Second, an experimenter can use a factorial design to test two or more different values of two or more IVs. Third, a within-subjects design can be used to economize on subjects.

There are two advantages of using more than two levels of an IV. First, the experimenter increases her chances of picking an effective level. Second, she learns how the DV changes with changes in the IV.

Factorial designs also have two major advantages. A single 2 × 2 factorial requires fewer subjects than two two-group experiments and it indicates whether or not an interaction has occurred.

Within-subject designs reduce error variance, which means that fewer subjects are needed for a given level of power. This can be a major consideration if subjects are expensive or must undergo painful or time-consuming procedures. Single case experimental designs, in which individual subjects are studied intensively, are a special class of within-subject design that help researchers gain control over sources of variability. Research in which the experimenter is also the subject is perhaps an underused approach.

Key Terms

ABAB Design A subject's baseline behavior (A) is recorded until it becomes stable. Then a treatment (B) is introduced, and behavior is recorded until it again stabilizes. Then the treatment is withdrawn (second A phase) and reintroduced (second B).

Blocking If subjects come from two distinct populations, and they are likely to differ on the DV as a result, population can be used as a factor in a 2 × 2 factorial design. The two populations are called blocks. Blocking is used to reduce ms_{error}, which in turn reduces the number of subjects needed.

Counterbalanced Within-Subject 2 × 2 Design Each subject is exposed to two treatments, with half receiving treatment *A* first and half receiving *B* first. Thus, there are two factors (treatment and order) and subjects serve as their own controls.

Dimensional Sampling Subjects are not selected randomly, but because they have characteristics relevant to theory.

Factorial Design An experiment in which two or more factors (IVs) are studied simultaneously.

Interaction An interaction occurs when the effects of one IV depend on the value of a second.

Main Effect The effect of an IV independent of any interactions.

2 × 2 Factorial Design An experiment in which two IVs are tested, each at two levels. Thus, four groups are formed, and each subject is randomly assigned to one of the four.

Multiple Baseline Design Baseline data are collected on at least two behaviors, and treatments are applied to each separately. Thus, control can be demonstrated over specific behaviors.

Pilot Study A small-scale study to try out the research procedures and design.

Sign Test A test of whether differences between means of two related samples are statistically significant.

Single Case Study Research in which data are analyzed at the individual rather than group level.

Within-Subjects Design An experiment in which each subject is tested at each level of the IV. This procedure reduces ms_{error}.

....
Answers
....

1. 60
2. 4.00
3. A. Factor 2 causes a statistically significant change in the DV (but see answer 5 below on subject variables; if factor 2 is a subject variable, the proper conclusion is not that it causes changes in, but only that it correlates with, the DV; in the remaining examples, the same caution applies). Factor 1 does not cause significant changes and there is no interaction.
   B. Factor 1 causes a significant change and there is no interaction.
   C. Both factors cause significant changes and there is no interaction.
   D. Neither factor causes a significant change and there is no interaction.
   E. Factor 2 causes a significant change and there is an interaction.
   F. Factor 1 causes a significant change and there is an interaction.
   G. Both factors cause significant changes and there is an interaction.
   H. Neither factor causes a significant change but there is an interaction.
4. In studies B, C, F, and G, the F ratio for factor 1 is significant, showing that gender affects DV scores. In studies E, F, G, and H, the significant interaction also shows an effect of gender. Only patterns A and D do not.

5. It would be reasonable to conclude that psychotherapeutic technique affects scores of verbal fluency. But the study would not justify the conclusion that type of psychopathology affects fluency. Subjects were not randomly assigned to be manic depressive or schizophrenic; that aspect of the study was correlational, and the same restrictions on causal inferences apply as in other correlational studies.

6. The ms_{error} term applies to all unintended causes of differences in scores between subjects: slight changes in conditions of testing, measurement errors, and pre-treatment differences between subjects. Within-subject designs eliminate only the last named source of error.

CHAPTER 16

Experimenting—Sequential Sampling

By the time you finish reading this chapter, you should be able to answer the following questions:

What is the advantage of sequential sampling over more traditional statistical analyses?

What type of data are not suitable for sequential sampling?

What is the difference between open and closed designs?

Can sequential sampling be used for both interval and ordinal data?

The research designs described so far all require a predetermined sample size. That is, the experimenter decides on the number of subjects before he collects data. By contrast, in sequential experiments the experimenter analyzes data as he goes along, deciding after each pair of subjects whether more information is needed; this generally results in a considerable savings of time, money, and subjects compared with nonsequential designs. Consider a fictitious example: a researcher tests a potentially valuable drug that may, however, produce a serious side effect. Following conventional procedures, she injects the drug into 100 monkeys and placebo into 100 others, all within a short period of time. She decides that it will merit further development only if the experimental group shows less than 10 percent excess sickness. Suppose 75 of the experimentals and none of the controls get sick. Had she tested one pair at a time, she could have saved a lot of monkeys.

Wald (1947) introduced sequential analysis in 1947. His treatment involves complex mathematics, well beyond my competence and that of most social scientists. It did not catch on. In 1960, Armitage (1960) published a short book in which he gave procedures for graphically analyzing results from sequential designs. That too didn't catch on and to my knowledge has not been discussed before in a social science research methods book. But, as efficiency is one characteristic of good scientists, and sequential designs are considerably more efficient than conventional ones, they are too important to ignore. The approach may seem strange and difficult to researchers trained to do analyses of variance. But the mathematics is rigorous and, with a bit of practice, the procedure is simple.

For sequential experiments to be most useful, subjects' responses should be available shortly after treatments are administered. The methods would not be practical, for example, for studying the effects of diet on longevity.

The four sequential designs described below are for comparing two treatments; they require that observations be made in pairs. For within-subject comparisons (each subject is her own control), successive observations on the same subject can be paired, with the order of treatments assigned randomly. For between-subject comparisons, successive subjects entering the experiment can be paired, also with random assignment to experimental or control treatments.

In two of the cases, the data are on an ordinal scale. That is, for each pair of subjects the data indicate only who did better or if there was a tie. The sign test described in chapter 12 is one of several conventional statistics for analyzing ordinal data. In the other two cases, measurement is on an interval scale; the data show the amount by which one subject outperformed the other. The F-test described in chapter 14 is a conventional alternative to sequential methods for interval data.

Both ordinal and interval sequential measurements can be analyzed with open or closed designs. With open designs, the researcher does not set an upper limit on the number of trials. If the treatments don't differ, the experiment is usually brought to a quick close. With closed designs, the experimenter decides beforehand on the maximum number of trials. Because researchers who use open designs risk the possibility of a long series of trials, Armitage prefers closed designs. With both open and closed designs, reductions in sample size are greatest when differences between treatments are largest.

Overview of Procedures

In all of what follows, the probability of both Type I and Type II errors has been set at .05, two-tailed. For other values, consult Armitage's book. All analyses require graph paper.

From a set of equations and numbers found in a table, straight lines called boundary lines are plotted on graph paper. Data, as collected, are represented graphically on the same sheet. The data path is extended until either it crosses one of the boundary lines or a predetermined maximum number of trials is reached. Crossing certain boundaries indicates that statistical significance has been reached. Crossing other boundaries indicates that the experiment must be terminated without statistical significance.

Steps in Sequential Analysis

1. Decide on the minimum effect size you want to detect. For ordinal data, effect size is expressed as the proportion favoring one treatment over the other. For interval data, effect size is defined as the difference between the means divided by their standard deviation.
2. Tables 16.1 and 16.4 give numbers that must be substituted into simple equations. Each of the four sequential designs has a different set of equations. Use the appropriate table and equations.
3. From the equations, draw lines on the graph paper. These represent boundaries. As data are collected, plot the results on the paper. Start at the origin and move one unit diagonally upward (northeast) each time treatment *A* has a greater effect than treatment *B,* one unit diagonally downward (southeast) each time treatment *B* produces the greater effect. For ordinal data, exclude ties.

For Ordinal Data

Open Design   With $n =$ the number of trials that have been run at any given time, excluding ties, plot four lines to satisfy the following equations:

1. $y = a_1 + bn$
2. $y = -a_1 - bn$
3. $y = -a_2 + bn$
4. $y = a_2 - bn$

Table 16.1 Coefficients for Ordinal Data

Minimum Acceptable Proportion Favoring One Treatment or the Other	Coefficients in Equations for Boundaries		
	a_1	a_2	b
.55	36.25	29.61	.0501
.60	17.94	14.65	.1007
.65	11.75	9.60	.1524
.70	8.59	7.01	.2058
.75	6.62	5.41	.2619
.80	5.25	4.29	.3219
.85	4.19	3.42	.3882
.90	3.31	2.70	.4650
.95	2.47	2.02	.5640

From P. Armitage, *Sequential Medical Trials*, 1960. Courtesy of Charles C. Thomas, Publisher, Springfield, Illinois.

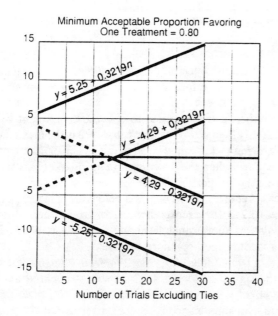

Figure 16.1

Table 16.1 gives coefficients for a_1, a_2, and b for different effect sizes. For example, if you decide beforehand to accept results as significant only if treatment A produces a greater effect than treatment B in at least 80 percent of the pairs, then $a_1 = 5.25$, $a_2 = 4.29$, and $b = 0.3219$. The four equations for a minimum effect size of 80 percent are plotted in Figure 16.1.

Equations 1 and 2 provide the outer boundaries. If the path drawn from the data crosses either of these, the experiment stops with a significant difference established. If the path crosses either of the inner boundaries representing equations 3 and 4, the experiment stops with no significant differences. The dotted lines extend the inner boundaries; if the path crosses both of them, the experiment stops with no significant differences. To draw the data path,

Table 16.2 Comparison of Trials Required for Different Effect Sizes for Sequential and Nonsequential Procedures

Minimum Acceptable Proportion	Approximate Mean Number of Trials Required When Actual Proportion is .5 (A), Same as Minimum Acceptable Effect Size (B), and Midway Between .5 and Effect Size Chosen (C)			Number of Trials Required in Equivalent Nonsequential Procedure
	A	B	C	
.55	870	660	1,080	1,294
.60	215	160	270	319
.65	95	70	115	138
.70	51	40	63	75
.75	31	25	38	46
.80	21	17	25	30
.85	15	12	17	20
.90	10	9	11	14
.95	7	7	7	9

From P. Armitage, *Sequential Medical Trials*, 1960. Courtesy of Charles C. Thomas, Publisher, Springfield, Illinois.

start at the origin and move one unit diagonally upward for each pair in which treatment A has a greater effect, one unit diagonally downward each time treatment B has a greater one.

The mean number of trials needed to stop the experiment will vary as a function of the actual effect size. Given in Table 16.2 is the mean number when the proportion favoring treatment A is .5 (that is, when experimental and control groups are equal); when the proportion favoring A equals the effect size chosen; and when the proportion is midway between .5 and the effect size chosen, which case requires the most trials before stopping. Given in the last column of Table 16.2 is the average number of trials when a nonsequential procedure with similar significance level and power is used. In all cases, the nonsequential procedure requires more subjects.

Example: Data from Table 14.7, page 214, from 27 pairs of subjects, were analyzed with the sign test. Figure 16.2 shows a sequential analysis of the same data. The effect size chosen is a proportion of .70 favoring one of the treatments. Statistical significance is reached after testing 22 pairs.

Closed Design

1. Decide on the maximum number of trials (n) excluding ties that you will run. Make a dot by this number on the abscissa.
2. Draw the outer boundary lines to satisfy the equations

   $y = a + bn$ and $y = -a - bn$; a equals the corresponding a_1 value from Table 16.1, and b equals b in Table 16.1. Thus, for an effect size of 0.70, $a = 8.59$ and $b = 0.2058$.

3. To get the inner boundaries, draw a vertical line from n to intersect both outer boundary lines. Call the upper intersection point u and the lower point L.

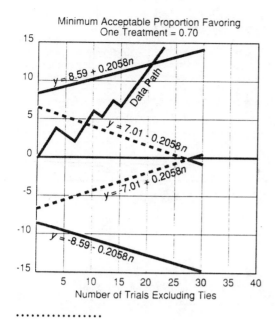

Figure 16.2

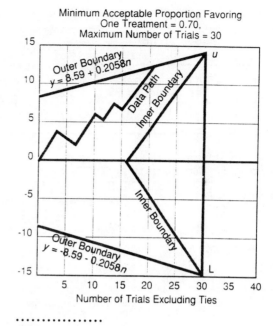

Figure 16.3

4. Start from the nearest integer value of y just below u and draw a line at an angle of 45 degrees to the horizontal extending to the abscissa. Draw a second line starting from the nearest integer y-value above L, also at an angle of 45 degrees to the horizontal and extending to the abscissa.

Figure 16.3 shows how a closed design with a minimum acceptable effect size of .70 and a maximum n of 30 would handle the data of Table 14.7 (p. 214). Statistical significance is reached after only 21 pairs of subjects.

For Interval Data
Open Design

1. Find the difference between each pair of responses (d), the cumulative sum of the differences (Σd), and the quantity $(\Sigma d)^2/\Sigma d^2$ (z). Table 16.3 shows calculations of d, Σd, and z for data on four pairs of subjects. The graphs have number of trials (ties not excluded) on the abscissa and z on the ordinate.
2. Two boundary lines are used. Before you can draw them, you must first decide on the minimum acceptable effect size. Effect sizes are defined as the difference between the means divided by their combined standard deviation. Cohen (1969), using the same definition, assigned labels of small, medium, and large to effect sizes of 0.20, 0.50, and 0.80 respectively. Table 16.4 gives upper

Table 16.3 *Fictitious Data on Four Pairs of Subjects for Open Design*

Pair	Treatment A	Treatment B	d	d^2	Σd	$(\Sigma d)^2$	Σd^2	z
1	6.3	3.7	2.6	6.76	2.6	6.76	6.76	1.00
2	4.2	4.4	−0.2	0.04	2.4	5.76	6.80	0.85
3	8.4	7.1	1.3	1.69	3.7	13.69	8.49	1.61
4	3.7	2.0	1.7	2.89	5.4	29.16	11.38	2.56

Table 16.4 *Coefficients for Interval Data*

Minimum Acceptable Effect Size (ES) ES	Number of Observations n										Closed Design with Fixed N (N) N
	20	40	60	80	100	120	140	160	180	200	
.2	18.94	11.80	9.45	8.35	7.76	7.44	7.26	7.17	7.15	7.17	448
	——	——	——	——	——	——	——	.09	.22	.37	
.3	10.51	7.78	7.12	7.00	7.10	7.32	7.59	7.91	8.26	8.62	201
	——	——	——	.22	.55	.91	1.29	1.68	2.08	2.49	
.4	7.75	6.81	6.98	7.45	8.03	8.68	9.36	10.06	10.79	11.52	116
	——	.09	.66	1.32	2.02	2.74	3.47	4.22	4.97	5.72	
.5	6.68	6.81	7.62	8.60	9.69	10.78	11.91				74
	——	.74	1.77	2.87	4.00	5.15	6.31				
.6	6.29	7.26	8.67	10.18	11.76						53
	.22	1.62	3.18	4.80	6.43						
	Number of Observations										
	5	10	15	20	25	30	35	40	45	50	
.7	——	6.03	6.01	6.25	6.61	7.03	7.49	7.96	8.46	8.96	40
	——	——	.25	.68	1.16	1.66	2.17	2.69	3.21	3.74	
.8	——	5.65	5.94	6.41	6.97	7.57	8.19	8.83	9.48	10.14	32
	——	.09	.64	1.25	1.89	2.54	3.21	3.88	4.56	5.24	
.9	——	5.46	6.01	6.69	7.44	8.21	9.01	9.81	10.63	11.44	26
	——	.37	1.10	1.88	2.69	3.51	4.33	5.16	6.00	6.83	
1.0	4.98	5.39	6.16	7.05	7.97	8.92	9.88	10.85			22
	——	.69	1.61	2.57	3.54	4.52	5.51	6.50			
1.2	4.48	5.45	6.63	7.88	9.16	10.45					17
	.22	1.42	2.70	4.00	5.30	6.62					
1.4	4.23	5.64	7.20	8.79	10.39						
	.61	2.19	3.80	5.42	7.05						

From P. Armitage, *Sequential Medical Trials*, 1960. Courtesy of Charles C. Thomas, Publisher, Springfield, Illinois.

and lower boundaries for different effect sizes. (Ignore the last column of the table for now.) The upper number of each pair refers to the upper boundary and the lower number to the lower boundary. Both boundaries are nearly straight lines, although the upper boundary dips slightly at first.

Table 16.5 is a single row from Table 16.4, when effect size equals 0.7. The upper boundary does not start until 10 pairs have been observed, the lower boundary at the 15th pair. To draw the upper boundary, connect successive points given by the (z, n) coordinates 6.03, 10; 6.01, 15; 6.25, 20; etc. To draw the lower boundary, connect 0.25, 15; 0.68, 20; 1.16, 25; etc.

Table 16.5 A Single Row from Table 16.4

ES	Number of Observations										N
	5	10	15	20	25	30	35	40	45	50	
.7	—	6.03	6.01	6.25	6.61	7.03	7.49	7.96	8.46	8.96	40
	—	—	.25	.68	1.16	1.66	2.17	2.69	3.21	3.74	

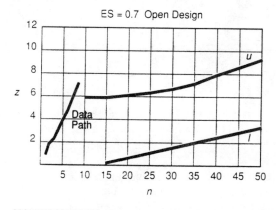

Figure 16.4

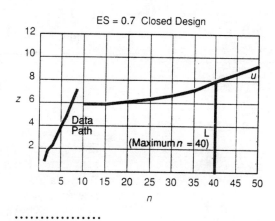

Figure 16.5

3. After each trial, calculate $z = (\Sigma d)^2 / \Sigma d^2$ and plot the result on the graph paper. If z crosses upper boundary U, the results are statistically significant. (The sign of d will indicate whether treatment A or B is superior.) If z crosses lower boundary M, the experiment is stopped—significant differences have not been found.
4. Continue plotting successive values of z until a boundary is crossed.

Closed Design Table 16.4 is only a small part of an extensive set of tables published by the National Bureau of Standards. Because it is limited, the table can be used for only a few fixed values of n. The last column of the table gives the value for n needed to detect an effect as small as that in the first column, with probability of Type I and Type II errors set at 0.05.

1. Use Table 16.4 to get the upper boundary.
2. Draw a line parallel to the z-axis at n. This is the lower boundary.

Data from Table 14.4, experiment 1, page 210, were analyzed previously. Figures 16.4 and 16.5 show open and closed (with maximum $n = 40$) sequential analyses of the same data. Table 16.6 shows calculations of the necessary quantities. Although no pairing was done originally, adjacent subjects have been paired for the sequential analysis. With an effect size of 0.70, the minimum number of trials before a boundary can be crossed is 10. Only eight pairs were used originally, so the sequential data are not significant. Had the pattern held for two more pairs, the results would have been significant.

Table 16.6 *Sequential Analysis of Data from Table 14.3*

Pair	Treatment A	B	d	d²	Σd	(Σd)²	Σd²	z
1	10	7	3	9	3	9	9	1.00
2	11	6	5	25	8	64	34	1.89
3	9	8	1	1	9	81	35	2.31
4	12	8	4	16	13	169	51	3.31
5	9	6	3	9	16	256	60	4.27
6	9	7	2	4	18	324	64	5.06
7	10	7	3	9	21	441	73	6.04
8	10	7	3	9	24	576	82	7.02

Notes:

1. Researchers who do a series of experiments with the same number of subjects and minimum acceptable effect size in each can save time by making multiple copies of a sheet with the appropriate boundaries. Refer to appendix H.
2. When data are collected bit by bit, as in sequential analyses, researchers may be affected by the early returns. They may change in subtle ways throughout the remainder of the experiment. These potentially serious experimenter effects can be minimized by double blind procedures (see p. 47).

Summary

Sequential sampling is a data collection and analytic technique that generally requires fewer subjects than do more conventional approaches. Researchers randomly assign pairs of subjects to experimental and control groups, then draw and evaluate the results on graph paper before testing the next pair. Four sequential sampling designs are considered:

1. Open design, ordinal data. The researcher does not specify an upper limit on the number of subjects. For each subject pair, the only information recorded is which one did better.
2. Closed design, ordinal data. Before collecting any data, the researcher decides on the maximum number of subjects to be run. For each subject pair, the only information recorded is which one did better.
3. Open design, interval data. The researcher does not specify an upper limit on the number of subjects. The size of the difference between the two subjects in each pair is recorded.
4. Closed design, interval data. Before collecting any data, the researcher decides on the maximum number of subjects to be run. The size of the difference between the two subjects in each pair is recorded.

CHAPTER 17

Analyzing Existing Data—Secondary Analysis

By the time you finish reading this chapter, you should be able to answer the following questions:

What are the advantages and weaknesses of secondary analysis?

What are the different types of secondary analysis?

How can material be located?

What types of units can be used for content analysis?

How is material sampled for content analysis?

Suppose a researcher wishes to learn about changes over a 30-year period in attitudes of Americans on the size of the ideal family. His concern is not just with overall changes, but with variations throughout different regions of the country and between men and women of both reproductive and nonreproductive ages. The question seems best answered with a survey, so he might dutifully plan his life's work: He would make up a questionnaire, administer it to a pilot group, revise it, distribute it across the country for 30 years, and collect and analyze the data. In the process, he might spend hundreds of thousands of dollars and build a small empire of students and other staff. Alternatively, he might do what Judith Blake (1966, 1967a, 1967b) did.

Blake searched through survey archives until she found 13 national surveys conducted by American agencies, spanning a period of 30 years, that had asked an appropriate question about ideal family size. Then she combined the data and published a series of papers. With a relatively small budget, and despite collecting no original data, she did creative and important research. Her approach, using raw data for other than their original purposes, is called secondary analysis.

Reasons for Using Existing Data Sources

Several scientists (cf. Bryant & Wortman, 1978; Holsti, 1969; Hyman, 1972) have pointed out advantages of secondary analysis over other forms of research. The weaknesses of secondary analysis are discussed later in the chapter.

1. Compared with the collection of original data, solving problems by analyzing existing materials is economical of money, time (both total duration of a study and speed of getting results), and personnel. Scholars without established reputations, whose limited resources preclude them from conducting large-scale primary studies, can do secondary analysis.
2. Collecting data on sensitive issues may heighten community tensions, either directly or by providing ammunition for political partisans. By contrast, secondary analysis uses no new subjects; it is nonreactive.
3. Certain questions can be answered best by analysis of old documents. For example, White (1966) developed a theory about factors that promote war. He could have waited for a new war to break out, but he was able to support his argument by citing already available materials. Other scientists may similarly test their insights about causes of present-day events by using available data that had not previously been connected to the events.
    Scientists who study long-term trends in attitudes cannot rely exclusively on their own data. But information about trends may be interesting and, by means of secondary analysis, easily collected. Similarly, few scientists compare attitudes of people in different

countries; collecting cross-national survey data is difficult. But the information, which would interest many personality theorists, economists, political scientists, and others, can be gathered by secondary analysis.

Even though the size of a research sample is adequate, the sizes of subgroups may be too small to support generalizations. With secondary analysis, investigators can enlarge the samples. Also, they may be able to locate the original subjects of a research project and persuade them to participate in a longitudinal study.

4. People who collect original data generally measure their concepts with narrow focuses. Secondary analysts must make do with what is available, so they examine a wide array of possible measures. This forces them to think of the concepts in odd ways and through the minds of people often far removed in time and place. As a result, the concepts may be elevated to higher and more theoretically useful levels of abstraction.

As noted in chapter 9, playing with data facilitates scientific discovery; secondary sources provide a rich lode of data with which to play. And even investigators who intend to conduct primary research can get ideas from preliminary secondary analyses about how to define their concepts and design their studies efficiently.

Locating the Appropriate Materials

On pages 162–163 is a short listing of survey archives. Hyman (1972) gives a much more extensive list. He noted that searches should not be restricted to archives, as every survey agency and investigator is a potential source of useful data. Bryant and Wortman (1978) urged that archives be established to house the data from psychological research, but their recommendation has not been followed. Government documents are an excellent source for demographers and other researchers in need of vital statistics. Books by Hauser (1975) and Stewart (1984) tell where to find statistical information on births, deaths, marriage, divorce, education, labor, delinquency, crime, and many additional topics.

Secondary analysts are an eclectic lot. They study books, newspapers, magazines, suicide notes, television shows and commercials, tape recordings, diaries, bathroom graffiti, speeches, photographs, letters, and any other written or oral materials that bear on their research questions. If dogs conducted the analyses, they'd study odor trails. In short, all data sources are fair game.

Types of Secondary Analysis

At least five types of secondary analysis can be distinguished: direct use of statistical material, documents used to develop and support a position, experiments and surveys, meta-analysis, and content analysis.

Direct Use of Statistical Material

Data from public or private documents are searched for possible relationships. The published statistics can be used as IVs, DVs, or both.

Phillips (1972) speculated that people sometimes postpone their deaths until some special event, such as their birthday, has occurred. From various sources of biographical information, he found both birth and death dates of more than 1,200 people. The evidence supported his view: fewer deaths occurred in the month prior to the birth month than would be expected under the hypothesis that month of death is independent of month of birth. (He classified dates by month rather than day to simplify the analysis.) He also found fewer deaths than would be expected in New York City, which has a large Jewish population, before the Jewish Day of Atonement, and in the entire United States before Presidential elections.

Since 1977, Bill James (1986) has analyzed baseball box scores in unique ways for his annual baseball abstracts. Box 17.1 presents excerpts from his 1986 edition. James' work illustrates a compelling feature of secondary analysis—from the comfort of their living rooms, scientists can read material of special interest to them and both develop and find ways to test surprising generalizations.

Documents Used to Develop and Support a Position

Material extracted from letters, diaries, newspaper accounts, and so forth can be used to support arguments about causes and sequences of events and to develop personality profiles of historic characters. Phillips used a document to strengthen his claim that people often postpone their deaths until an important event has occurred. He noted that both Jefferson and Adams died on July 4th, 50 years after the Declaration of Independence was signed. He cited Jefferson's last words, quoted by his physician:

About seven o'clock of the evening of that day, he [Jefferson] awoke, and seeing me staying at his bedside exclaimed, "Oh Doctor, are you still there?" in a voice however, that was husky and indistinct. He then asked, "Is it the Fourth?" to which I replied, "It soon will be." These were the last words I heard him utter.

White (1966) speculated that wars are caused in large part by misperceptions that each side has about the other. He identified six forms of misperception and used documentary evidence to show their importance as causative agents in World Wars I and II. The six forms are (a) an image of the enemy as diabolical, (b) a virile self-image, (c) a moral self-image, (d) selective inattention (positive qualities of the enemy are glossed over and attention is focused only on the negative), (e) absence of empathy, and (f) military overconfidence. White cited *Mein Kampf* and several of Hitler's speeches to demonstrate misperceptions by both Americans and Germans prior to World War II. Many Americans believe that Hitler publicly glorified war,

BOX 17.1

Excerpts From the Bill James Baseball Abstract, 1986

Baseball people often claim that good teams win close games. Actually, the smaller the margin of victory, the more likely that the better team will lose. Teams that win 60 percent of their games win only about 54 percent of their one-run games; teams that win 40 percent of their games win about 46 percent of one-run games.

James developed a formula for estimating the number of wins for teams in a given season. He then studied all American League teams that had changed managers between 1961 and 1985, to see if teams that fire their manager before the start of a season tend to do better or worse than the projections. Gains and losses balanced pretty evenly. But teams that changed managers were much more likely to have won-loss records that differed sharply from the projections.

James reported the results of a study of a single team, the 1985 Toronto Blue Jays, to show how an offense changes at home as opposed to the road. In their home games, the Blue Jays outscored opponents 37–32 in the first inning and 40–20 in the second. In away games, they outscored opponents 36–35 in the first inning but lost the second inning 35–27. A team's better hitters bat in the first, their weaker hitters in the second, so James speculated that weak hitters are more affected than good hitters by being on the road.

Further analysis of the 1985 Blue Jays showed how stolen base attempts affect the hitter at the plate. After successful steal attempts, Toronto hitters batted .198 (21/106) and their opponents .127 (9/71). After unsuccessful attempts, they batted .100 (4/40) and .154 (4/26) respectively. Each average was considerably lower than when steals hadn't been attempted.

James analyzed the effects of artificial turf on career length in two ways. I'll describe only the simpler, as both methods point to the same conclusion. He counted the number of players excluding pitchers born in 1933 or earlier and on a major league roster in 1970 (the year in which widespread use of artificial turf began); and the number born in 1948 or earlier and on a roster in 1985. There were only eight in the first group, 38 in the second. His surprising conclusion: Artificial turf probably prolongs careers.

An even more unexpected conclusion followed his analysis of ballpark effects: Teams tend to develop those characteristics least favored by the parks in which they play. The reason, according to James, is that teams playing in parks that favor pitchers tend to have good pitching statistics even if the pitchers aren't very good; and teams that play in hitters' parks tend to have good hitting statistics even if their hitters are mediocre. Because the team executives don't take sufficient account of ballpark effects, they identify their needs incorrectly; they concentrate on acquiring hitters in the first case, pitchers in the second. As evidence for his argument, James offered the following: The Chicago Cubs, Atlanta Braves, and Boston Red Sox play in hitters' parks. Over the most recent 10-year-period, each team has been among baseball's leaders in runs scored in its home park, which created the illusion that their offenses were excellent. But performances on the road yield a much truer measure of a team's abilities relative to those of other teams; on the road, over the same period, the Cubs and Braves were at the bottom and the Red Sox near the bottom in runs scored. On the other hand, the three teams gave up lots of runs at home, leading most baseball people to conclude that they desperately needed pitching help. But on the road, their pitchers did far better than the league averages. James also presented statistical evidence from teams playing in pitchers' parks such as the A's, showing that those teams have much better hitters and weaker pitchers than is generally realized.

but his propaganda after coming to power in 1933 was characterized by extreme adherence to conventional moral standards (moral self-image). For example:

> I wish to point out first, that I have not conducted any war; second, that for years past I have expressed my abhorrence of war and, it is true, also my abhorrence of warmongers; and third, that I am not aware for what purpose I should wage a war at all.

White acknowledged that conclusions from this type of research are much less certain than those from double blind methodology, but that is no reason to ignore them; readers can decide for themselves how convincing the evidence is. Formulated during the American war in Vietnam, White's arguments were provocative on a topic of great urgency.

Barber (1985) studied available documents—biographies, diaries, speeches, and so forth—that bore on the lives of twentieth century American presidents. From the materials, he characterized each president as active or passive (a measure of the energy they devoted to their jobs) and having positive or negative affect (a measure of their feelings about their jobs). He concluded that Roosevelt, Truman, Kennedy, Ford, and Carter were active and positive; Taft, Harding, and Reagan were passive and positive; Wilson, Hoover, Johnson, and Nixon were active and negative; and Coolidge and Eisenhower were passive and negative. He further concluded that presidents within each group dealt in similar ways with problems of the day.

As a final example of the use of documentary evidence, I turn to fiction. Several reviewers called the book *Daughter of Time,* by Josephine Tey (Dell Publishing Co., 1951), one of the greatest mystery stories ever written. The hero, a police inspector hospitalized with a broken leg, becomes convinced that history books are grossly inaccurate in their accounts of the life of Richard III. The inspector, from his bed, supervises a search of archival evidence and ultimately makes a persuasive case for the correctness of his own interpretation. Tey's book clearly demonstrates that documents can be used to test hypotheses.

Experiments and Surveys

The reanalysis of original data allows investigators to assess the credibility of research findings. Wolins (1962) wrote to 37 authors of published studies, requesting their raw data; he received seven sets and found serious arithmetic and conceptual errors in three of them.

Old data, evaluated from new perspectives and with advanced analytic techniques, may reveal relationships that had been hidden from the original investigators. Read Cronbach and Webb (1975) for an interesting example.

As noted above, data archives do not exist for experiments, so secondary analysis is difficult or impossible. Surveyors are more fortunate and can get information from archives.

Meta-Analysis

The results of published studies on a given topic (without the original data) are combined quantitatively and an overall conclusion derived. This approach, meta-analysis, is discussed in chapter 24.

Content Analysis

Content analysis is a group of techniques for making inferences from messages. Qualitative data are systematically analyzed to yield quantitative findings. For example, a task force of the National Organization for Women systematically analyzed 2,760 stories from 134 textbooks for young children (Women on Words and Images, 1972). The task force counted each occurrence of male and female characters, what the characters did, and whether the stories were male-centered or female-centered. They found that males outnumbered females and were more frequently portrayed as smart, brave, strong, independent, and ambitious. The study, evidence that our schools foster stereotyped gender roles, led to discussions about the harm caused by such stereotypes and to increased sensitivity on the part of writers, publishers, parents, and teachers.

Sampling of Sources   The task force had to sample from an enormous population of textbooks. Content analysts must frequently select small samples from large populations. The solution is to follow the lead of surveyors and develop strategies for simple random, stratified, systematic, or cluster sampling.

In surveys of voters, all respondents are considered equal in that they have one vote (except perhaps in Chicago). But the sources for a secondary analysis, such as newspapers and television shows, vary greatly in the size, geographical location, and political orientation of their audiences. So, stratified sampling may be essential. The sources themselves are often sampled rather than studied in their entirety. Thus, an investigator interested in newspaper coverage of a political campaign would first choose an appropriate cross-section of papers, then decide the editions and pages to study from each sampled paper. Decisions about sampling, including what to do if some of the randomly selected sources are unavailable, must be made before collecting data and then adhered to strictly.

For many research questions, the population is small enough that it can be sampled in its entirety. In fact, despite the many written and oral forms available for analysis, the population may be too small. Materials bearing on a particular question may not exist. So, content analysis is often infeasible.

Scoring   Consider two of the units used by the NOW task force: frequency of occurrence of male and female characters, which was obtained simply by counting; and portrayals of males and females as smart, brave, and so forth, which required complex judgments. Counters rarely need make difficult judgments, but they may encounter problems if units are referred to in more than one way. For example, Krippendorff (1980) noted that Richard Nixon has been called, among other things, "Tricky Dick," "the 37th President of the United States," "the first president to visit China," "the occupant of the White House between 1969 and 1974," and "he." Therefore, carefully specified guidelines for counting should be drawn up in advance. And guidelines for judgment decisions should be made in great detail.

Units of any type are acceptable as long as they can be scored reliably. For example, Speed (1893) used entire newspaper columns; he reported an increase in sports and gossip articles in New York daily newspapers between 1881 and 1893.

The *Federalist Papers* were published anonymously in 1787–88 by Alexander Hamilton, John Jay, and James Madison to persuade the citizens of the State of New York to ratify the Constitution. Prior to 1964, scholars generally agreed on the authorship of 70 of the papers, but disagreed whether Madison or Hamilton had written the remaining 15. Then Mosteller and Wallace (1964) did a convincing content analysis. From the undisputed papers, they searched for items that discriminated between Madison and Hamilton. These became their units of analysis. Sentence length turned out not to be a good unit: Hamilton and Madison averaged 34.5 and 34.6 words respectively per sentence. But several individual words were effective discriminators: In the papers known to have been written by Madison, the word *whilst* occurs in eight, the word *while* in none. *While* occurs in 15 of Hamilton's essays, *whilst* not at all. So, the presence of *whilst* in five of the disputed papers is evidence in favor of Madison's authorship. Mosteller and Wallace counted the frequencies with which all other words in each undisputed paper occurred per thousand words of text. Several had been used much more by one author or the other, so frequency counts of those words in the disputed papers became further basis for assignment of authorship.

Units used by Scott and Franklin (1972) were entire issues of *Life, Look, McCall's, Newsweek, Reader's Digest, Saturday Evening Post,* and *Time* for 1950, 1960, and 1970. They reported a slight increase over the 20-year time span in sentences about sex.

Two other simple and generally highly reliable units involve space (as in column inches within a newspaper) and time (as in minutes of radio programming). Box 17.2 describes research requiring a scoring system on the other end of the complexity spectrum.

Reliability Content analysis is a method of measurement and, like other measuring devices, must be reliable to be useful. When material is analyzed on two different occasions by either a single rater or different raters, the scores should be essentially the same. Reliabilities of judgments are usually lower than those of counts; still, if units are carefully defined and raters trained, a sufficiently high reliability can generally be attained. Tetlock reported correlations between raters of .84, .87, and .91 in the three studies in Box 17.2. Reliability is often high enough so that content analysis scales, developed by

> **BOX 17.2**
>
> ## Integrative Complexity
>
> Tetlock and his associates, in a series of content analyses (Tetlock, 1984, 1985; Tetlock et al., 1984), measured a construct called "integrative complexity." People at the low end of the integrative complexity continuum make evaluations based on only a few factors. For example, they lump all legislative proposals into one of two groups: good (patriotic) and bad (defeatist). People higher on the continuum recognize that national policies can have effects not easily categorized on a single dimension—effects on immediate and long-term balance of power, strategies of both allies and opponents, sectors of the domestic economy, and political constituencies. People at the highest end of the continuum not only recognize several factors, but also that the factors interact with each other in complex ways.
>
> Tetlock, Hannum, and Micheletti (1984) scored integrative complexity of conservative and liberal United States senators in five Congresses. The units were their speeches while in office. Conservatives showed relatively low, stable levels across Congresses; liberals generally had higher levels, but exhibited a pronounced downward shift during the two Congresses in which conservatives had the balance of power.
>
> In a related study, Tetlock (1984) scored transcripts of interviews that had been conducted with members of the British House of Commons in 1967.
>
> Moderate socialists were the most integratively complex; moderate conservatives were next; extreme socialists and conservatives, the two groups lowest in integrative complexity, were not significantly different from each other. Tetlock's interpretation is that people who value both freedom and equality highly are under greater pressure to think about these issues in integratively complex terms than are people who place much greater weight on one value or the other: extreme socialists value equality much more than freedom, extreme conservatives freedom much more than equality.
>
> Tetlock (1985) reviewed research showing that different negotiating strategies are used by people of low and high integrative complexity: the first group tries to gain advantages by standing firm and using pressure tactics; the second tries to achieve agreements in which both sides are reasonably satisfied. This conclusion was supported by his analysis of American and Soviet foreign policy statements in each of the 154 quarter-year periods between 1954 and 1983. In general, upward shifts in integrative complexity were associated with cooperative American/Soviet behaviors, downward shifts with political or military interventions abroad. Complexity shifts tended to precede policy initiatives for the Soviets and to coincide with them for Americans.

a procedure described by Viney (1983), have been used to assess people's psychological states. Briefly, after giving the precise characteristics of a state such as anxiety or sociability, a scale-developer specifies detailed cues from which the state can be inferred. She then analyzes verbal communications of any type for those cues and assigns scores. Data are then compared from different samples of people and situations.

Data-Analytic Techniques No unique statistics are needed for evaluating content analysis data. When data are expressed numerically (counts, column inches of newspaper, minutes of television time), each independently sampled item can be treated as one subject. For example, to test the hypothesis that newspaper coverage of ballet has increased over the past decade, a researcher might count articles about ballet on a single day in 50 newspapers in each of two years. (The 1980 and 1990 batches should be from the same edition, month, and day of the week.) The data would consist of 50 matched pairs and could be analyzed with a sign-test, analysis of variance, or similar test.

A different analysis would be required if the goal were to compare favorable and unfavorable portrayals of males and females in a sampling of children's readers. Suppose the figures were as follows: 264 favorable and 103 unfavorable characterizations of males; 93 favorable, 20 unfavorable characterizations of females. To test whether favorable and unfavorable characterizations depend partly on gender, the results could be put in a 2 × 2 table and analyzed with a phi coefficient (see pp. 189–190). (If the objective were to see only if the number of male and female characters per reader differ significantly, the sign test would be sufficient.) Krippendorff (1980) mentioned several other techniques, and Tetlock et al. (1985) used a time-series analysis, all beyond the scope of this book.

Weaknesses of Secondary Analysis

Many research questions can't be answered by secondary analysis, as the appropriate data are unavailable. And, as is true of all non-experimental forms of research, the methods (except for secondary analyses of experiments) do not answer causal questions unambiguously. But they often provide support for the position that one event has caused another. For example, Phillips (1982, 1983) reported an increase in suicides following television episodes depicting suicides, and in homicides after extensive media coverage given to championship heavyweight boxing matches.

Because secondary analysts don't participate in the process of data collection, they are poorly equipped to judge how adequate it was. Because they use materials housed by others, they can't control for the possibility of selective retention. *Daughter of Time,* referred to previously, offers a realistic account of both the methods of documentary analysis and the possible distorting effects of selective retention. Some future historian, after randomly sampling editorials housed in the archives of one of our presidents, might conclude that the President had been universally loved. But since favorable editorials are more likely to be saved than unfavorable ones, the interpretation might not be accurate. Some secondary analysts are probably discomfited by a situation in which quality control standards have been implemented by unknown predecessors.

Secondary analysis of surveys poses two problems that usually have satisfactory remedies. First, each surveyor measures key concepts idiosyncratically, so different surveys are rarely worded identically. The remedy is to modify concepts and lump questions in meaningful ways. Second, older surveys are likely to be unsophisticated by current standards, and even modern ones may have methodological weaknesses; so in many cases, some are excluded from the analysis. The problem is that selective exclusions may be done in a biased

way. To increase the likelihood that decisions about comparability of questions and exclusion of studies are not influenced by the research hypothesis, they should be assigned to a scientist unaware of the hypothesis.

Although steps such as the above are important, the possibility of bias is everpresent and a major problem for all forms of secondary analysis. Formal experimentation limits and double blind experiments almost entirely eliminate unintentional bias. But secondary analysts work in relative privacy with an enormous quantity of data, and have so much discretion in choosing and interpreting it that bias may easily intrude. Overt cheating, though probably rare, is also easy. The reason is that secondary analysts can sample and resample, combining data in infinite ways and discarding anything that doesn't support the research hypothesis, before submitting their results to public scrutiny. As a result, secondary analysis may be more effective as a tool for convincing the researcher herself rather than for convincing others.

Summary

Secondary analysis is an economical and nonreactive research method that can be used by people with limited resources. It allows scientists to assess the credibility of published findings and to reevaluate old data from new perspectives. It facilitates longitudinal studies.

There are several types of secondary analysis: direct use of statistical material, use of documents to develop and support hypotheses, reanalysis of surveys and experiments, meta-analysis, and content analysis.

Secondary analyses are not always possible, as data may not be available for answering a particular research question. Secondary analysts have no control over the original collection of data, the possibility that they were subsequently edited, and whether they were selectively retained. The specter of bias, both unintentional and deliberate, limits secondary analysis as a research tool.

Key Terms

Content Analysis A research technique for quantitatively analyzing verbal materials.

Secondary Analysis A group of techniques for analyzing existing records and public and private documents.

CHAPTER 18

Designing, Implementing, and Evaluating Social and Medical Programs

By the time you finish reading this chapter, you should be able to answer the following questions:

How does evaluation research differ from other forms of scientific research?

What are the three broad types of evaluations?

How should program developers collect information on the problems they propose to correct?

What reasons are there for evaluating ongoing programs?

What sources of data are available for evaluating ongoing programs?

What principles should dictate the choice of criteria for evaluating the outcomes of studies?

Why is it often very difficult to choose good outcome criteria?

How are programs evaluated by the cost-effectiveness approach?

Is it essential that program participants be randomly assigned to treatment and control groups?

What are the consequences of nonrandom assignment?

Both the development and evaluation of social and medical programs requires research skills. But skilled researchers who work with the programs must often feel frustrated from several features that distinguish program development and evaluation from other forms of scientific research. First, political overtones are inevitable: Decisions about funding programs and policies are influenced only partly by the programs' potential for helping clients. A more important factor is the willingness of taxpayers to provide such help. Still more important are the perceptions of elected and appointed officials about the public's attitudes.

A second distinguishing feature is that the audience for both grant proposals and evaluation reports is comprised primarily of political decision makers rather than scientists. Third, whereas the designers, implementers, and analyzers of experiments are generally one and the same, program developers do not usually evaluate their products. Fourth, because unfavorable reports may result in severe cutbacks and even the demise of programs, administrators often impede evaluation efforts. Finally, and of major importance, programs are often poorly designed for tests of their effectiveness.

Designing, implementing, and evaluating outcomes of programs are distinct tasks and are considered separately, but they must be integrated. Too often in the past, decisions about expensive programs with far-reaching consequences have been made from uninterpretable data—data that could have been adequate had the programs been designed properly. What follows is presented from the perspective of evaluation. Program developers who adopt that perspective enhance their chances of receiving favorable evaluations.

EVALUATION OF SOCIAL PROGRAMS

YOU CALL THIS EVALUATION FAVORABLE! WE'D HAVE TO RAISE TAXES TO TRAIN A BUNCH OF DELINQUENTS AND OLD PEOPLE WHO DON'T EVEN VOTE. I SAY LET'S FORGET IT AND GO EAT!

Evaluating Program Design

Is a Problem Documented?

People develop programs to deal with problems, so their first task should be to document the problem. They should collect information on the segment of the population affected, the seriousness of the effects, the availability of treatments through existing programs, and the probable demand for the proposed services. They should consider the perspectives of various community members, as Rossi and Freeman (1985) did in listing five groups whose inputs are important for assessing mental health problems: agencies and individuals that treat people with mental health problems, agencies that make referrals to mental health services, people who are using or have used mental health services, community groups organized around a common goal, and area residents.

To evaluate the documentation of a problem, determine how many of the following sources have been used effectively:

- Knowledgeable community leaders and experts. Have they been identified and questioned?
- Attendees at open forums sponsored by respected individuals and organizations. Have they had a chance to express their views?
- Statistics on the number of people in treatment for the problem in a similar community.
- Both direct and indirect indicators. (A direct indicator of incidence of child abuse is the number of child abuse arrests. An indirect indicator is the number of hospital admissions of children for treatment of serious bruises. Indirect indicators can be economic or social conditions with which child abuse correlates, such as periods of high unemployment.)
- Data from surveys and censuses.

Ask if the collected data have been converted to statements of problems, arranged in the order of priority the program will give to them. Each statement should deal with a single issue. For example:

1. Too many 3- to 6-year-old children in our community suffer beatings at home.
2. Too many of the children are fed diets inadequate for maintaining health.
3. Too many of the children miss an unacceptable number of school days.

Have the Program Goals and Objectives for Accomplishing Them Been Stated Clearly?

Problem statements can be reworded to establish the broad goals of the program. For child abuse, two broad goals might be, "Reduce the number of beatings of children in their homes" and "Identify victims of abuse rapidly, and give them immediate emotional support." Objectives should be listed for accomplishing each goal. They should be specific, realistic, measurable, and time-limited. Some objectives for the first goal might be:

1. The number of arrests for child abuse during 1990 will be reduced by 25 percent compared with arrests during 1989.
2. The number of children admitted to University Hospital for treatment of severe bruises will be reduced by 40 percent during the same period.
3. The number of absences from school will be reduced by 25 percent.
4. The "Hansel and Gretel Fear of Adults Test" will be administered to all elementary school children in May, 1989, and again in May, 1990. There will be a 25 percent decrease in the number of children scoring in the range that indicates excessive fear.

Some objectives for the second goal might be:

1. Among children brought to physicians for treatment, there will be a 20 percent reduction in the number of bruises judged more than three days old.
2. As judged by admitting physicians, there will be an average decline of 25 percent in the severity of injuries.
3. Attendance at our monthly outings, which is voluntary for children and their parents, will exceed 80 percent.
4. The "Hansel and Gretel Fear of Adults Test" will be administered to all elementary school children in May, 1989, and again in May, 1990. There will be a 25 percent decrease in the number of children scoring in the range that indicates excessive fear.

Both goals and sets of objectives are admirable, but they are different. Evaluators should demand explicit goals that cannot be misunderstood.

Have the Activities for Meeting the Objectives Been Clearly Stated?

Ask how the activities of the program will lead to accomplishment of its objectives. Require three testable statements of relationships:

1. Problem situation X is caused (at least in part) by condition Y. (Continuing with the child abuse example: "One reason that many single parents abuse children is that they have nobody to help them when they feel stressed.")
2. The program will change condition Y. ("Our 24-hour crisis hotline, drop-in counselling center, and classes in stress reduction will teach parents how to cope with stress.")
3. The program will change situation X. ("Implementation of the activities mentioned above will reduce child abuse.") Note that the beneficiaries of a program (children at risk of abuse) need not be the direct recipients of services (their parents).

Even if hypotheses one and two are supported by data, the third does not follow inevitably. Rossi and Freeman cited a study by Festinger (1964) to make this point. Festinger hypothesized that (a) discriminatory hiring practices of white employment managers toward black applicants are caused in part by the managers' lack of understanding and knowledge of blacks; (b) educational experiences provided by Festinger would increase the managers' understanding and knowledge; and (c) after finishing the program, the managers would discriminate less. Although there is strong evidence in favor of hypothesis one, and data collected during the study support hypothesis two (the program did increase understanding and knowledge), the managers who were most affected discriminated more than ever. Festinger suggested that the increased knowledge made it impossible for them to continue behaving as before, but did not ensure direction of change. Whatever the reason, hypothesis three proved false.

Once a clear rationale for a program has been developed, activities must be planned in great detail. Who will do what, and where and when? If there is doubt about the feasibility of certain activities, pilot test their feasibility.

Have Criteria Been Established for Deciding if the Objectives Have Been Met?

Worden (1979) criticized directors of drug abuse prevention programs for accepting testimonials by participants as proof of effectiveness. Unless a program's activities are rigorously evaluated with measurable criteria, the activities

> . . . may or may not be effective in preventing alcohol or drug abuse. We do not know. They may or may not have much to do with the etiology of alcohol and drug problems. We do not know. They may or may not produce better mental health. *We do not know.*

Worden's criticism applies to any research endeavor that doesn't provide criteria for distinguishing between favorable and unfavorable outcomes. Yet program directors often fail to specify or test outcome criteria. Tims (1982) mailed questionnaires in 1979 to a random sample of programs funded by the National Institute on Drug Abuse. He asked if, as required of programs funded by NIDA, they had performed evaluations in the preceding year. Only 54 percent of the questionnaires were returned, and only 44 percent of the respondents answered affirmatively. Only 19 percent of these followed up on clients once they had left the program. Surprisingly, Tims' conclusion was optimistic: "Thus, it is evident that a significant number of programs conducted both in treatment and follow-up evaluation."

The requirements of good outcome criteria are discussed below, under the heading "Evaluating Outcomes."

Evaluating How Well a Program Has Been Implemented

Administrators, although leery of evaluators, can profit from analysis of the day-to-day functioning of their programs. This should be pointed out to secure their cooperation. Analyses (a) give feedback about problem areas; (b) document successes and failures—information that is required on most applications for additional funding; (c) guide continuous fine-tuning of programs; (d) show which program components are being under- and over-used, which facilitates efficient allocation of resources; and (e) provide bases for reappraisal of objectives. Analyses also help developers of similar programs reproduce crucial features.

The evaluator should seek answers to questions covering all phases of program operation. The questions below are only a sample:

Staff: How many are there? What are their qualifications? Do they seem to get along with each other? What is the hierarchy of command? Do people working on different aspects of the program communicate with each other?

Participants: What is the average age and educational background of program participants? How are they recruited to the program? What percent of the target population comes to the program? If, as is generally the case, a program can't serve the entire target population, are nonparticipants within the community resentful? Have measures been taken to gain their support? What is the dropout rate?

Other questions, as relevant, might include items on drug history, job experience, prison record, command of English, and so forth.

> Activities: What are the daily activities and responsibilities of each staff member? How are the different aspects of the program interrelated?
>
> Facilities and Equipment: Where is the program located? How many square feet of working space are available? What is the equipment inventory?
>
> Community: Might any special demographic or socioeconomic characteristics of the community affect the program?
>
> Standards: What are the minimum resources required for program success? (Measures should be taken, as they apply to specific programs, of the following types of things: staff/patient ratio, qualifications of staff, number of rooms that permit small discussion groups, number of personal computers, number of entryways that offer wheelchair access.) What are the minimum standards for treatment? (Measures should be taken of such things as hours of weekly contact between each patient and a senior staff member, time between a patient's entry into the program and a tentative diagnosis, time between a job-seeker's submission of a resumè and his first job interview.)

Records: What sort of records are being kept? What measures have been taken to preserve the privacy of data on participants?

Laws: Are all laws being observed?

Trends: Are some aspects of the program being over or underused? Is the participant population changing? Have any costs been surprisingly large or small? Have any activities been particularly popular, unpopular, effective, or ineffective? Are the activities that were scheduled to be carried out being carried out? Has the philosophy of the program changed?

Rossi and Freeman listed five sources of data for evaluating program implementation: direct observations, records kept by program staff of services they delivered, records kept by management (management information systems), materials prepared specially by staff to answer evaluators' questions, and interviews and questionnaires given to recipients of services.

Evaluating Outcomes

Evaluators must decide if the lives of the program's target population are significantly different from what they would have been without the services. The principles of inference are the same as those governing the interpretation of laboratory experiments. Conclusions are most secure about programs in which (a) subjects have been randomly assigned to receive or not receive services and treated identically in all other respects, and (b) a valid measure of the desired effect has been used. Consider the second point first.

Outcome Criteria

Evaluators use the term "outcome criteria" to refer to their dependent variables. Whatever the name, the measures should be reliable, sensitive, valid, and, when possible, unobtrusive. Multiple measurement, always desirable, should be routine in outcome evaluations. For one thing, the chain of inference from a manipulation to its presumed effects is generally longer and weaker for evaluations than laboratory experiments, so changes in a single measure may be incorrectly interpreted. Second, if only a single measure is used, staff members may focus on it to the detriment of the program's overall goals. For example, employment interviewers rated exclusively by the number of interviews they conduct may lose sight of their primary goal, placing clients in jobs. Third, Jaffe (1984) cited evidence that for drug abuse treatment programs, several dimensions of outcome—drug use, alcoholism, general health, work, crime, and social and psychological well-being—are relatively independent; beneficial programs that measure only a single outcome may be discontinued if that one is negative. Fourth, impacts of programs often extend beyond the direct recipients of services; these secondary effects must also be documented.

Problems In Choosing Good Criteria   Many programs have been established to deal with social problems such as mental disorder, suicides, and child abuse. But changes in incidence rates are affected by many factors and may not accurately reflect a program's performance. Windle (1986) argued that "number of persons served" and "impact of service on those contacted" are often more appropriate measures.

Measurable criteria are hard to find for goals such as attitude change and the production of small though long-lasting effects in large populations. Moreover, lengthy measurement procedures may be resisted by staff; untrained in research methodology, they may believe their time is best spent seeing clients. They may not measure objectively, since the continued existence of the program may depend on favorable evaluations.

Cowen (1978) raised the issue of invasion of privacy as a stumbling block to the use of certain criteria. For programs seeking to improve impaired sexual functioning or disrupted parent/child relationships, questions about clients' sexual adjustment or relationships with children are valid but unacceptable. Cowen also noted that evaluators of studies that compare different groups or settings (such as school districts) must be sure that the criterion measures are equally appropriate for each of the groups or settings.

Sjoberg (1983) charged that criteria are imposed upon experimental programs by the dominant bureaucratic structures, and the criteria are often inadequate. One of his many examples concerned the Head Start program designed to provide compensatory education for disadvantaged children. Evaluators focused on the children's cognitive changes and concluded that the program was unsuccessful. Sjoberg wrote that ". . . it would be surprising, on theoretical grounds alone, to have found major cognitive improvement in children who had spent a year or less in the Head Start program. The utilization of traditional categories for measuring educational achievement, plus the diversity of the social order and the factors that impinge upon a child's socialization and learning patterns, would lead one to hypothesize that such a program as Head Start would have little, if any, significant impact upon educational improvement."

Sjoberg pointed out that parents of enrollees increased their participation in the activities of educational centers, which may have had a powerful long-term impact upon the educational careers of their children. Had increased parental participation been used as a major criterion, conclusions about the success of Head Start would have been different.

Criteria are often selected by vote among a group of evaluators who differ widely in their views of what constitutes a favorable outcome. Bennett and Lumsdaine (1975) raised the concern that the group criterion may become a sort of lowest common denominator, like the political candidate nominated because he is acceptable to everybody without being the first choice of anybody. Criteria of this type may be positive for the client population as a whole but have a substantial negative impact on many of them.

Principles for Choosing Good Criteria Windle (1986), Cowen (1978), and Jaffe (1984) all offered advice for choosing good outcome criteria. Some of their principles, for example 5 and 6 below, conflict; Windle urged balance among them.

1. Focus on program goals while considering all stakeholder groups. Measure effectiveness for managers, efficiency for taxpayers, quality for staff, and responsiveness for clients.
2. Recognize that adequate measurement of certain goals may be impossible, and the use of simple measures to approximate the goals may cause program distortions. At the least, some people may confuse the measure with the goal; even worse, the measure may become the basis for funding decisions and so displace the original goal.
3. Spend as much time as necessary to convince all staff that careful, honest measurement is crucial to the success of the program.
4. Build reliability and validity checks into the data collection and analysis procedures.
5. Build useful and accepted measures gradually. A new measurement system should have only a few simple criteria. Experiment with their use.
6. Use multiple measures.
7. Both the immediate and long-range effects of programs should be measured. This requires follow-up studies. People who seem improved immediately after a program ends may quickly relapse without its continued support. On the other hand, some program benefits may not be detectable until many months later. Cowen gave the hypothetical example of a preventive program for children experiencing current life crises who have not yet shown major signs of maladjustment. Jaffe wrote: "Each followup study of the large multicentered evaluations...has enlarged our appreciation of the importance of monitoring behavior over time (not months, but years) and of the need to measure outcome along several dimensions."
8. If criteria are in conflict, seek solutions acceptable with respect to several of them rather than optimal for a single one.
9. Programs may have impacts, both positive and negative, beyond their listed goals. These should also be measured. For example, many people who successfully withdraw from opiates become alcoholic (Kolb, 1962) and many who successfully withdraw from alcohol deteriorate in other areas of life (Gerard, Saenger, & Wile, 1952). The Head Start program, mentioned above, had few immediate measurable effects on children's cognitive performance, but changed the behavior of parents.

Cost Effectiveness Analysis

Most evaluators base their decisions primarily on one of two grounds: whether (a) the programs met their stated objectives, or (b) the target population scored significantly higher than controls on outcome criteria. But neither ground may accurately reflect a program's worth. Imagine a program designed to feed poverty-stricken families that lists a single outcome criterion: a 50 percent reduction in diagnoses of malnutrition at the county free clinic over a one-year period. Suppose that by the year's end the number of diagnosed cases has dropped 50 percent and is significantly less for the food recipients than the controls. Whether judged by *a* or *b* above, the program would be considered successful.

But what if the program involved daily deliveries by chauffered limousines of caviar, pheasant, and lobster to each recipient? Although undisputedly benefitting a few needy people, the program's resources would have been badly used. Conversely, some programs may yield small effects but great returns on investments. Weiss, Jurs, LeSage, and Iverson (1984) spent a total of 12 hours teaching smokers techniques for stopping and for maintaining abstinence. About 16 percent of the participants stopped entirely, and an additional 63 percent reduced their smoking. The results compare favorably with those from more expensive programs—if costs are considered.

All competent evaluators take program costs into account when assessing worth, but few do so formally. Levin (1987) stressed the need for formal analyses while at the same time documenting their rarity. He noted that for every three billion dollars spent by the health sector in 1984, one cost-effectiveness study was conducted; and only two of 500 presenters at the 1986 Annual Meeting of the American Evaluation Association focused on cost-effectiveness.

When costs are considered, the question "Does this program work?" is replaced by the successively more useful questions, "What is the average cost per effective unit?", "What is the cost for each additional unit of effectiveness?", and "For a given expenditure of resources, which program yields the best results?"

Cost-effectiveness analyses require three major steps: listing costs, listing benefits, and converting all of them to the same (usually monetary) units. Costs and benefits differ for each stakeholder. For example, a program for treating abused children has important consequences for the children, their parents, their playmates, taxpayers, schools, and the juvenile correction and justice systems. Separate cost-effectiveness analyses can be done for each perspective, but perspectives should not be mixed within an analysis.

When both direct and indirect costs and benefits can be listed without dispute, the method is powerful. But many and major assumptions are often required—in which cases the quantitative precision of cost-effectiveness analysis is only an illusion. For the smoking cessation study mentioned above, costs included participants' fees and lost wages—which varied among them—while they were in the program. The benefits, all anticipated for the future, included lowered insurance rates (fire, health, life, and workman's compensation), reduced absenteeism, and increased productivity. These varied according to participants' success rates in stopping smoking and their age, gender, and projected years of future employment. Monetary values were placed on each benefit, taking account when it would be earned. (A given amount of money is worth more today than if it will be received at a future time.)

Despite its limitations, cost-effectiveness analysis offers an important perspective for evaluating programs. Programs must always compete for limited resources, so some system must be used to compare them. The cost-effectiveness of programs should be an important consideration. Interested readers might wish to consult Thompson (1980) or Levin (1983).

Randomization

Clients are often assigned nonrandomly to treatment and control groups. Many methodologists have argued that outcome analyses of such programs are of little value, and others disagree. Some of the points of contention are discussed in the following sections.

Random Assignment Is Not Always Possible Program administrators, confident that their treatments work, are appalled by suggestions that some needy clients be given only placebos. They impose their views, one of several reasons that evaluators must often consider data from subjects assigned nonrandomly to groups.

- Many funding agencies don't insist on random assignment, and program administrators may be unaware of its importance. They may not have the time, skills, or resources to do the required bookkeeping and staff coordination.
- The salaries and prestige of administrators are often related more to the size than the effectiveness of their programs. So they fight for program growth rather than evaluation.
- Staff may intentionally or unintentionally sabotage assignments. Blumenthal and Ross (1973) designed a study in which judges would randomly assign people convicted of drunk driving to receive one of three penalties. The judges agreed to cooperate if they were allowed to make exceptions for special cases. Then, despite the pleas of the researchers, they made repeated exceptions. In a second study, Blumenthal and Ross arranged that motorists stopped for moving violations would be randomly assigned to receive one of four penalties; but many patrolmen issued tickets according to their feelings about what the motorists deserved.

- Laws and policies vary from place to place. Evaluators can compare school districts, cities, counties, and countries that have different policies on teaching methods, gun control, gambling, prostitution, drugs, and automobile speed limits. But they can't assign randomly.
- Randomly assigned clients may drop out in unequal numbers from different groups.

People assigned randomly may be treated unequally in ways other than administration of the independent variable. Cowen raised the possibility that teachers (or therapists or counselors) asked to use an innnovative method may allow their ratings of the recipients of services to be influenced by their attitudes toward the method. But teachers of controls, believing that ratings reflect their own teaching abilities, may tend to give high ratings. They may feel threatened by the new technology, so work harder than usual. Supervisors may respond more favorably to requests from control schools, as compensation for the special attention lavished on the experimentals.

Random Assignment Facilitates Effective Decision-Making The importance of random assignment was discussed briefly on page 202. Its costs are justified if they lead to better policy decisions. Gilbert, Light, and Mosteller (1975) wrote:

How much should society be willing to pay for an evaluation that does *not* lead to firm inferences about how well a program is working? Not much. Yet for years we have been paying enormous amounts for evaluations of this second kind. Had we earmarked a small portion of the budget expended on these efforts for a few randomized field studies, we might have realized three benefits: first, we might have learned with greater reliability how well or how poorly a particular program was working; second, we might have spent less on overall evaluation expenses; and third, we might have been able to use the results of the field trials to improve the operation of the program, so that it could better serve its intended beneficiaries.

Gilbert et al. noted that the Federal Government, between 1963 and 1971, spent 6.8 billion dollars on manpower training programs: 6.1 million people received training and 180 million dollars were spent on evaluations. Yet reports to the United States Congress (Goldstein, 1972) and by the National Academy of Sciences (1974) state that little is known about their effectiveness. The evaluations involved nonrandom samples and other methodological inadequacies, so were inconclusive.

Many innovations produce small but important and long-lasting changes. These can't be accurately measured except with powerful research designs that include random assignment. Only if the changes are documented and their causal agents identified can they be built upon and improved. As an example, Gilbert et al. cited the small, cumulative changes in treatment of childhood leukemia. The median survival time following diagnosis was about three months until 1953, 13 months from 1955 to 1964, and 30 months from 1964 to 1966. By the time of their article (1975), writers were claiming that childhood leukemia was no longer incurable.

Decision makers should be concerned not just with the question "Does treatment X work?" but with questions such as "Does X work well at some sites and poorly at others?" and "Does X work better for some types of people than for others?" Such interaction effects (see pp. 232–237) are common. For example, McLellan, Luborsky, Woody, O'Brien, and Druley (1983) found that the personality characteristics of patients starting treatment programs for drug abuse were strong predictors of the eventual outcomes of treatment. Without randomization, inferences about interactions are weak.

Foulds (1963) reported that four times as many studies without than with adequate controls get positive results. Chalmers, Block, & Lee (1972) observed the same pattern and emphasized the harm it does: 57 published papers on a possible anti-cancer drug failed to convince scientists of the drug's worth, even though many more patients were used in these inconclusive studies than would have been needed for a proper randomized experiment. Boruch (1987) gave several examples in which the results from nonrandomized and randomized experiments differed greatly. Gilbert et al. noted that poorly controlled experiments inevitably leave room for plausible alternative explanations, so policy makers are reluctant to use them as the basis for important decisions. Gilbert et al. concluded:

Nonrandomized trials may not only generate conclusions with doubtful validity—they may actually delay the implementation of better evaluations. If a series of nonrandomized studies indicates that a certain treatment is highly effective, some investigators will point out the ethical difficulties of withholding this apparently valuable treatment from some persons in a randomized trial. Yet . . . nonrandomized studies may often artificially inflate our estimate of a treatment's value. Whenever this happens, and a randomized trial is postponed as a consequence, an ineffective treatment or program may be administered for years.

Humanitarian reasons have motivated many program administrators to assign the neediest people, or the people from the most disadvantaged group, to receive treatment. But, as Campbell and Boruch (1975) have shown, this type of nonrandom assignment is likely to lead to an underestimation of treatment effects—especially when children are involved.

1. If two tests are highly but not perfectly correlated with each other, people who score extremely well or poorly on one tend to score closer to the mean on the other. The reason for this regression to the mean is explained in introductory statistics books. When disadvantaged people are given a pre-treatment test, others in the general population who score equally poorly are often picked as controls. But since each subgroup regresses to the mean of its own group, the controls can be expected to regress more; in the absence of treatment effects, they will score higher on a posttest.
2. Differences in children's scores on pretests are caused in part by differences in their growth rates in previous years. If differential rates continue at least for a time, children from disadvantaged groups will show the smallest pretest/posttest growth changes; this will obscure effects due to treatments.

3. Differences between groups are demonstrated best with reliable tests, and test reliability increases as children get older. So in the absence of treatment effects, post-treatment tests will show greater differences between advantaged and disadvantaged groups.
4. Test reliabilities are generally lower for disadvantaged groups, so their gains are harder to detect and seem smaller.
5. The running speeds of 1- and 12-month-old babies are the same—zero. But a year later, children in the first group outrun those in the second. This information does not lead to the conclusion that older babies outgain younger ones in motor ability over the one-year span. The reason is that 1- and 12-month-olds differ in motor ability, but a test of running speed is insensitive to the difference. Similarly, insensitive pre-treatment tests may not detect differences between the advantaged and disadvantaged. Posttest differences may then be falsely attributed to treatment failure.
6. Behaviors have multiple causes. Even if a treatment produces positive effects, other environmental features continue to act and may increase the disparity between advantaged and disadvantaged people.

Random Assignment Is Usually Possible   Gilbert et al. reviewed nine social, eight socio-medical, and 12 medical innovations, both large- and small-scale, in which randomization was used. Conner (1977), Boruch, McSweeney, and Soderstrum (1978), and Boruch (1987) cited others conducted in diverse fields. The conclusion is inescapable: randomized experiments can be done. Campbell and Stanley (1966) pointed out that the number of eligible candidates for a treatment often exceeds available resources. In such cases, random assignment doesn't even pose an ethical problem. Below are some suggestions, extracted from the articles by Gilbert et al., Conner, and Boruch, for designing social programs that will yield interpretable results: for designing randomized experiments.

1. An individual in charge of research should control the randomization process. Program administrators and staff often fail to devote proper attention to what seems to them a minor detail.
2. Only those exceptions specified before treatments begin should be made to the random assignments. If necessary, make sure that the neediest people receive prompt treatment, but then exclude them from the study.
3. Guarantee if possible, perhaps with monetary incentives, that everybody will end up at least as well off as they would have been had they not participated.
4. Random assignment does not require that people be assigned in equal numbers to control and treatment groups. Unequal assignment, so that most people receive treatment, does however complicate the statistical analysis and may complicate interpretation as well.

5. Randomized trials can be embedded in large-scale, nonrandomized programs. Thus, there need be no delay in implementing the programs.

Alternatives to Nonrandom Assignment Despite the unquestioned superiority of randomized over nonrandomized experiments, circumstances (mentioned previously) sometimes dictate use of the latter. Campbell and Stanley called them quasi-experiments and, in a classic work (1966), described several research designs for maximizing the information they yield. Scientists, though they strive to conduct flawless experiments, should not neglect data collected under imperfect conditions. So quasi-experimental analyses have become popular and are the subject of a 1979 book by Cook and Campbell (1979).

Before briefly describing three of the many quasi-experimental designs mentioned by Campbell and Stanley and Cook and Campbell (see their works for directions on statistical procedures), I'll re-emphasize their limitations with a quote from Campbell and Boruch (1975):

It may be that Campbell and Stanley (1966) should feel guilty for having contributed to giving quasi-experimental designs a good name. There are program evaluations in which the authors say proudly, "We used a *quasi*-experimental design." If responsible, Campbell and Stanley should do penance, because in most social settings there are many equally or more plausible rival hypotheses than the hypothesis that the puny treatment indeed produced an effect. In fact, however, their presentation of quasi-experimental designs could as well be read as laborious arguments in favor of doing randomized assignment to treatment whenever possible. Admittedly, there are also encouragements to do the second best if the second best is all that the situation allows.

Interrupted Time-Series Design

In the interrupted time-series design, a reform is put into effect abruptly and comparisons are made between pre- and post-reform records. A widely cited example is Connecticut Governor Ribicoff's 1956 crackdown on speeding, following a record 324 traffic fatalities in that state in 1955. The following year there were 284 fatalities, a drop of 12.3 percent. Ribicoff attributed the decline to the crackdown. Campbell agreed, but pointed out plausible alternatives.

....
1. How many plausible alternatives can you think of?
(Several types were discussed in Box 14.1, p. 204).
....

Control Series Design

The interrupted time-series design has no control group, as everybody within the relevant population receives the experimental treatment. All people who drove in Connecticut were subjected to the new laws against speeders. But a control group, although not assigned at random, can often be found. Then, the procedure is called a control series design. (But don't try to match experimental and control groups on pretest scores. Campbell made the point that this common practice leads to regression artifacts.)

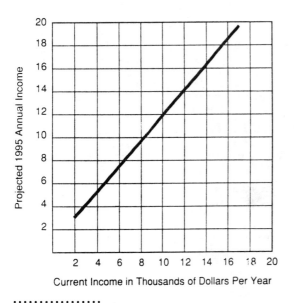

Figure 18.1

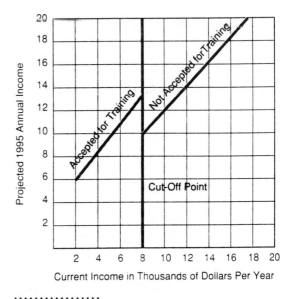

Figure 18.2

Campbell used four neighboring states of Connecticut as controls. They did not show a similar sharp drop in fatalities in 1956, and their fatalities remained higher than those in Connecticut in 1957, 1958, and 1959.

Regression Discontinuity Design

Scarce resources are often awarded according to ratings of applicants on a dimension that judges deem relevant. For graduate fellowships, the dimension might be scores on the Graduate Record Exam; for Job Corps training, it might be family income. Outcome criteria might be graduate school grade point average in the first case, annual income five years later in the second. If a significant correlation exists between the prescore and outcome criterion, the two can be fitted to a regression line (see pp. 183–185) as in Figure 18.1.

Suppose that a job training program opens for people whose annual income is less than $8,000. If the training has no effect on future earnings, then applicants whose pretraining income is slightly less than $8,000 should eventually earn about as much as rejected applicants whose initial incomes slightly exceed $8,000. The regression line showing the relationship between initial and future income should be uninterrupted. But if the program improves earning capacity, the regression line will be discontinuous at the cut-off point. See Figure 18.2. (Figure 18.2 shows that, even though the program has a positive effect, the average 1995 salary of controls exceeds that of trainees; unsophisticated analysts might conclude that the training was harmful.)

2. What if a supposedly relevant dimension turns out unrelated to the outcome criterion?

Using Evaluations

Generalizing

Policymakers use outcome evaluations to predict the future effectiveness of programs. As long as a program's client population, personnel, and treatment philosophy are stable, its performance is unlikely to change much over a short time period. But generalizations from evaluations are often farther ranging: they may provide evidence for decisions about whether to discontinue or expand pilot projects, and they may be entered into the basic science literature. When evaluations are intended for such use, they should include information on unique features of a project's staff, activities, facilities, or clients.

For Policymakers

Ogborne and Smart (1982) told a discouraging story about the lack of impact of an expensive evaluation of alcohol detoxification centers in Ontario, Canada. The evaluation committee, finding that the centers had drifted from their original objectives, made several specific recommendations. But many of the staff at the centers were former alcoholics and members of Alcoholics Anonymous who placed more emphasis on the AA philosophy than on research findings; they had poor job prospects outside the detoxification field and found the evaluation threatening. Serious conflicts arose between the evaluators and staff, and the recommendations were ignored.

Ogborne and Smart's advice to prevent similar occurrences is directed at policymakers: Involve evaluators in programs from the beginning, and seek their help in designing small-scale experiments to explore alternative ways of coping with social problems.

Rules for Evaluators

Lipton and Appel (1984) advised evaluators to follow several rules.

1. The paramount purpose of evaluations is to help policymakers. So talk with them. Ask what questions they want answered and the best way to present the answers.
2. Move swiftly. Long-term evaluation projects are tolerable only when short-term products are also available. Partial answers are useful.
3. Accept imperfect evidence; the search for perfection in evaluation is fruitless.
4. Be willing to make crude estimates initially. These can be refined as work progresses.

5. Produce reports in four parts:
   a. Put the summary and recommendations, no longer than two pages, first. Answer the important questions in jargon-free language.
   b. Present the body of the report like a table of contents with summarizing headlines. Provide legislators with conclusions, not with recommendations unless asked for.
   c. Write conclusions in short paragraphs. Use simple statements that can be quoted in legislative debates.
   d. Put all technical material, including computations and methodological details, in an appendix.
6. Organize results by political jurisdiction. Policymakers can't make decisions about areas that have no political reality.

Summary

Evaluation research is distinguished in several ways from other forms of scientific research: (a) it is done in a political context, (b) the audience is primarily policymakers, (c) program designers and evaluators are often different people, (d) program staff may be uncooperative, and (e) the evaluator may have to make do with an inadequate research design.

Evaluation activities change during the three stages of a program's life. For the initial (design) phase, the evaluator should make sure that the existence of a problem has been documented; data have been collected on the incidence, prevalence, and effects of the problem on various groups of people; evidence has been presented that the program's activities will reduce or eliminate the problem; and methods have been given for measuring outcomes.

Evaluators are sometimes called on to monitor all phases of ongoing program operations.

Evaluators must decide on proper outcome criteria. These should be reliable, sensitive, valid, and unobtrusive. Multiple measures are always desirable.

Cost-effectiveness analysts try to show not just whether a program works but also (a) how programs compare in their costs per unit of effectiveness and (b) how much each additional unit of effectiveness would cost.

Random assignment of a program's participants to treatment and control groups:
   a. is not always possible.
   b. is always desirable.
   c. is generally possible.

Key Terms

Cost-Effectiveness Analysis A technique for specifying and evaluating in terms of cost the relative effectiveness of alternative programs and policies.

Outcome Criteria The dependent variables of evaluation research. Outcome criteria tell how a program's results will be measured.

Quasi-experiments Studies in which subjects are not randomly assigned to groups.

Regression to the Mean If two measures are partly but imperfectly correlated, people who receive extreme scores on the first tend to score closer to the mean on the second.

Answers

1. History: Weather conditions might have been particularly hazardous in 1955. Perhaps there was a big increase in liquor taxes in 1956, which resulted in fewer people driving while under the influence. (Campbell emphasized that awareness of other possibilities should not lead researchers to refuse to use the evidence; instead, they should investigate further to find out if any of the plausible hypotheses is supported by additional data.)
Maturation: Perhaps death rates were steadily going down year after year, and 1956 simply continued the trend. (Campbell analyzed data from several years prior to and after the crackdown, and he found no support for this hypothesis.) Data collected at several points prior to and after institution of a reform provide a convincing test of the maturation hypothesis.
Regression: Policy makers are most likely to urge reforms and support new programs following strong demonstration of a need. And needs appear strongest after extreme scores. Traffic fatalities were extremely high in Connecticut in 1955 and, even had there been no policy change, would probably have declined in 1956. The reason is regression to the mean, which was mentioned briefly on page 283.
Testing: If the data on traffic fatalities had been collected amid great publicity, the publicity itself might have changed people's driving habits even without a speeding crackdown.
Instrumentation: Recording practices often change when a new policy is instituted. Suppose that the 1955 practice was to report as a *traffic* fatality any death that occurred within seven days of a car crash; and that in 1956, deaths were counted as traffic fatalities only if they occurred within 24 hours of a crash.
2. That would mean that assignment to treatment and control groups was essentially random, and the study could be treated like a true experiment.

CHAPTER 19

Studying the Development of Behavior

By the time you finish reading this chapter, you should be able to answer the following questions:

How are cross-sectional designs used for studying the development of behavior?

How are longitudinal designs used for studying the development of behavior?

What are the limitations of cross-sectional designs?

What are the limitations of longitudinal designs?

How else, besides cross-sectional and longitudinal designs, can development be studied?

Developmental psychologists study how people change with age. Suppose an investigator wanted to find out whether 20- or 70-year-olds have larger vocabularies. He might select 100 20-year-olds and randomly assign them to two groups, with age as the independent variable. People in the first group would be allowed to age normally for 50 years, those in the second treated exactly the same except that they'd be commanded to stay at age 20. Then everybody's vocabulary would be tested and . . . Hmm, he might run into technical difficulties.

Cross-Sectional Designs

Undaunted, he might try a second approach. He'd randomly select 50 people from each age group, randomly pair them, and test them as follows: Each pair would sit at a table and listen while he'd slowly read a set of 100 vocabulary words. For each word, the first subject to call out an accurate synonym would earn one point; the winner would be the one with the most points out of 100. If 20-year-olds won significantly more of the contests, the investigator would conclude that vocabulary declines with age.

But strong alternative explanations about developmental changes in vocabulary would exist. They always do for cross-sectional studies, that is, studies in which two or more age groups are compared at the same time. Twenty-year-olds might hear the words better as they are read. Or, even if the 70-year-old member of a pair thinks first of a synonym, the 20-year-old might respond quicker. And, 20- and 70-year-olds might differ in motivation to perform well or in cautiousness in calling out uncertain answers. These possibilities indicate a general problem with cross-sectional studies: a given stimulus may not mean the same thing to people of different ages.

A second frequently encountered problem relates to sampling procedures. The investigator's weren't mentioned. Suppose he went to a college campus for the 20-year-olds, to a nursing home for the 70-year-olds. Neither setting would yield a representative sample. If he looked hard enough on college campuses, he could find enough 70-year-olds to fill his quota. But they would certainly not be representative. In fact, virtually no method of selection would eliminate the potential of a biased sample. He might surmount the problem by limiting his objective to a comparison of 20- and 70-year-olds in a small town. He could purchase a list of all residents, write down the ages of each one, and then randomly sample until reaching his quotas for the groups. Although that would produce data of limited generality, he could combine it with other information, for example, samples from several different communities and grandparent/grandchild pairs. If all gave comparable results, the possibility that observed age differences were due to nonrepresentative samples would become less plausible.

If the problems of sampling and possible nonequivalence of stimuli are handled properly, cross-sectional studies can lead to justifiable conclusions of the type "The average 20-year-old woman of today is two inches taller than today's average 70-year-old woman." But the studies should not be interpreted further. Conclusions such as "The average woman shrinks two inches between the ages of 20 and 70" would be unwarranted, because there are always at least two competing alternatives. Sometimes the alternatives are highly plausible, sometimes less so, but they can never be entirely discounted.

1. Not all the people who were 20-years-old half a century ago are 70 today; many are dead, and the survivors are a biased sample. If all 20- and 70-year-olds alive today were given a test to measure masculine versus feminine tendencies, I would confidently bet that the 70-year-olds would on average receive higher femininity scores. But the conclusion that individuals become more feminine with age would have to confront the fact that a greater proportion of males than females die at every age, so the ratio of females to males continually increases. Similarly, if any other variable correlates with longevity, conclusions from a cross-sectional study about developmental changes in that variable would be invalid.
2. People of different age groups have invariably had different experiences. School environments have changed greatly in half a century. So have styles of parenting. There have been important medical developments, wars, riots, assassinations, changes in the quality of the air, trends in art, music, motion pictures, diet, size of families, and in virtually every other facet of life. Twenty- and 70-year-olds may differ because of any of these and not because individuals normally change in certain ways with age.

Longitudinal Designs

Convinced that cross-sectional designs are unsatisfactory, the investigator might try a new approach and persuade the 20-year-olds in his psychology class to be his subjects for the rest of their lives. He would give them a battery of tests and have them return once a year to fill out questionnaires and retake the tests. The procedure, in which the same people are studied over many years, is called longitudinal research. Suppose that subjects scored higher on average on a vocabulary test at 40 than they did at 20. What would it mean?

Longitudinal studies preserve subjects' identities, so allow for research on both the stability of individual differences and later correlates of earlier behaviors. Kagan (1986) found that temperaments were fairly stable. Many shy and timid children became less so as they grew older, but they rarely surpassed in either sociability or boldness others who had been outgoing and bold. Eron, Huesman, Lefkowitz, and Walder (1972) found that children's third grade preferences for watching violent television programs accurately predicted their twelfth grade aggressiveness.

Data such as, "John Jones scored five points higher on a vocabulary test at age 40 than at age 20" can be extracted from longitudinal studies. But even if all subjects showed the same improvement, the conclusion that "People's vocabularies grow as they age from 20 to 40" would be unwarranted. There would always be plausible alternatives.

1. Subjects drop out of the study. They lose interest, move, die. The ones remaining constitute a biased sample. (The extent of differences between those who drop out and remain can be estimated by checking their initial scores.) What if no subjects drop out? Is the problem eliminated?
2. Repeated testing of the same subjects may yield improvements because of practice, or decrements because of boredom, nervousness, or resentment at having to fulfill an annoying obligation. The pattern may not be known in a given case, but some effect of repeated testing is probable.
3. People who were 20 at the study's start constitute a unique group, as explained in the last paragraph of the previous section. Their unique experiences may be responsible rather than changes due to development.

Longitudinal studies require a major commitment on the part of investigators as well as subjects. The hypothetical cross-sectional vocabulary study could be completed, once subjects were assembled, in a day or two. The corresponding longitudinal study would take 50 years. As Wohlwill (1970) noted, there are other practical problems. Special provision must be made for data storage and record-keeping. Personnel move, investigators develop other interests, and advances in knowledge and techniques that take place over the course of a project can rarely be incorporated once the project is launched.

Further Comments on Cross-Sectional and Longitudinal Designs

Both cross-sectional and longitudinal researchers have at their command all the data collection methods available to other social scientists. But in light of the foregoing criticisms, it should come as no surprise that their conclusions often conflict. For example, most cross-sectional studies show that performance on IQ tests peaks at about age 30 and then steadily declines. But longitudinal studies (cf. Schaie & Labouvie-Vier, 1974) show little or no change in IQ scores between ages 30 and 60.

Even when results agree, they may be wrong. Goodman (1977) reported that more than three dozen research studies support the belief that IQs of the institutionally retarded decline with age. But, she noted, the brighter members of institutionalized populations often earn eventual release. So, cross-sectional comparisons of young, newly admitted people with older, permanent residents are biased in favor of the younger. Longitudinal studies have the same bias, because brighter people are lost from the sample when released.

I CAN'T WAIT TO SEE WHAT THESE GUYS WILL BE LIKE WHEN THEY'RE SEVENTY!

Goodman located 284 retardates who had been given a routine IQ test upon institutionalization and again on at least one later date. The initial scores of the 144 who were being returned to the community were significantly higher than the initial scores of the other 140. That supports Goodman's position on selective drop out. In addition, she compared initial and last tests of each retardate. For both groups, scores on the later tests were higher. So the conventional wisdom was false, an artifact of method of testing.

Conclusions about the typical course of development are more plausible from longitudinal than cross-sectional studies, because only in the first case are they based on actual developmental changes within specific people. If a cross-sectional comparison shows a two inch height differential in favor of 20-year-olds over 70-year-olds, it doesn't necessarily follow that people normally shrink as they get older. Not even that the ones in the sample did. Because of improved diets, today's 20-year-olds may average three inches taller than today's 70-year-olds when they were 20; this would mean that the 70-year-olds actually grew an inch during the past 50 years.

By contrast, data from logitudinal studies are of the form "Jane Jones shrank two inches between ages 20 and 70." The generality of Jane Jones' development can be doubted, but the specific form that hers took is not in dispute. A reasonable hypothesis is that others develop similarly.

Advanced Approaches

Cross-sectional and longitudinal approaches can be combined in a single design. For example, 70- and 20-year-olds can be simultaneously tested (cross-sectional). Then, every 10 years until they reach 70, the 20-year-olds can be retested (longitudinal, giving a developmental function from ages 20 through 70). In addition, a new group of 20-year-olds can be added to the study every

10 years and tested until they reach 70. The statistical analyses of such designs are beyond the scope of this book (see Baltes, Reese & Nesselroade, 1977; Nesselroade & Reese, 1973).

Miscellaneous Techniques

At least two sources of evidence other than conventional designs are relevant to hypotheses about development. One is animal research. Laboratories can be standardized so that animals are reared under identical conditions, then tested at different ages. Sampling is straightforward. Loss of subjects is minimized and due only to death; analyses of causes of death provide clues about bias among the survivors. Below are three among thousands of findings from animal research that bear on hypotheses about human development.

- Periodic fasting accompanied by dietary supplements of vitamins C and E slows down aging in mice (Leibovitz & Siegel, 1980).
- Birds have critical periods for learning songs, which has led to the speculation that humans may have critical periods for learning languages (Marler, 1970).
- Monkeys raised in isolation from their peers do not develop normally either sexually or socially (Harlow & Harlow, 1962).

The second approach is historical. The possibility that today's 20- and 70-year-olds differ not because of age but from their unique developmental experiences can be tested by reading descriptions of 20- and 70-year-olds of other times.

Summary

Two research designs for studying how subjects change with age are the cross-sectional and the longitudinal. The cross-sectional design involves testing two or more age groups at the same time, whereas longitudinal researchers study the same subjects repeatedly over long periods of time. Differences in behaviors between the groups or time periods have often been attributed to the effects of aging, but plausible alternatives always exist.

Several serious problems complicate the interpretation of cross-sectional designs.

1. A given stimulus may affect subjects differently at different ages.
2. Research subjects are typically drawn from different subsets of their age-group populations.
3. Older subjects, having already lived past an earlier age, are a biased subset of all subjects of that age.
4. Subjects differ not only because of age but also because they grew up during different times.

Longitudinal studies require more work and a far greater commitment on the part of the investigator than do cross-sectional. Problems with longitudinal studies include the following:

1. Subjects may drop out before completion of a study and render useless the data already collected on them. Those who remain constitute a biased sample.
2. Repeated testing may affect performance.
3. Experiences that affect current behavior may have been particularly weak or strong for subjects who grew up during a certain time period.

Combined cross-sectional and longitudinal designs enable researchers to draw much stronger conclusions than they can from either one alone.

Key Terms

Cross-Sectional Study Studying age-related changes by comparing different age groups at a single time.

Longitudinal Study Studying age-related changes by testing the same subjects repeatedly over time.

Answer

The problem of generalizability of findings remains even if nobody drops out from a longitudinal study. Suppose, for example, the study began when subjects were 20 and ended when they were 70. If mortality was zero, the subjects could hardly be considered representative.

PART V

Controversies

CHAPTER 20

Using Nonhuman Animals in Research

By the time you finish reading this chapter, you should be able to answer the following questions:

What advantages accrue to researchers who work with nonhuman animals?

How do scientists evaluate animal models of human behaviors?

What value is there in studying animals that have unique properties?

What are the advantages to researchers of concentrating on a single species such as the Norway rat?

What are the advantages of comparing behaviors across several species?

What is the difference between ultimate and proximate questions about behavior?

What are the advantages of doing research from an evolutionary perspective?

How do scientists study the importance of environments in shaping behaviors?

You will also be able to liven up dull conversations with tales of armadillo quadruplets, blood sucking moths, and rotating eyeballs of frogs.

The principles and methods of research discussed in previous chapters apply with equal force whether the subjects are human or nonhuman animals (except that some nonhumans are notoriously uncooperative when it comes to filling out questionnaires about their personal lives). The use of nonhuman animals (called 'animals' throughout the chapter) confers many advantages. The most important is control. Researchers can control the administration and measurement of independent, dependent, and extraneous variables much more fully than in work with humans. And they can test the variables over a greater range. In addition, animals such as Norway rats fit conveniently into small cages yet are large enough to permit most kinds of operations. They breed readily in captivity and have short generation spans, which is valuable for developmental and behavioral genetic research. They are resistant to infections and easily handled.

Because of the advantages, scientists experiment on a staggering number of animals. In the United States alone, an estimated 70 to 90 million are used per year (Gendin, 1986). The number would probably be even larger except for two factors. First, many people are skeptical about the relevance of animal research to humans. Second, others take exception to animal research on ethical grounds (see discussion in chapter 8).

Why Use Animals in Research

Several overlapping purposes for studying animals are listed below and expanded upon in the sections following. Each of the first four purposes is often achieved efficiently even when a single species is studied. Achievement of the last purpose requires at least two and often several species.

- To estimate how an independent variable affects or would affect humans
- To create, for observational purposes and testing of IVs, an animal model of a particular condition
- To study naturally occurring systems in animals in order to learn about comparable systems in humans
- Because a given species has an interesting property
- To gain a better understanding of how our evolutionary past and current environments shape our behaviors

Using Animals to Test How Independent Variables Affect or are Likely to Affect Humans

Animals are used to test the beneficial and hazardous effects of new products: drugs, cosmetics, surgical techniques, safety helmets, and so forth. And to test environmental conditions with potentially important health consequences, such as pollution, crowding, noise, and nutritional deficiencies. The assumption is that results will improve predictability about human responses and thus lead to more effective treatments and wiser decisions. But the assumption must constantly be tested anew.

Recall the research of Zucker and of Falk, discussed in chapter 6. Zucker measured biorhythms in rats. He kept his animals in temperature-controlled, light-controlled, largely soundproof environments and monitored their eating and wheel-running behaviors 24 hours per day. Falk deprived rats of food and then fed them at one minute intervals for three hours per day. He measured their food and water intakes both during test sessions and when the rats were in their home cages. Both lines of research have important implications for humans; but they required continuous, precise control and so could not have been conducted on human subjects.

If your goal is to use animals to predict how an IV will affect humans, choose the species carefully. For drug research, generalizations to humans are better from cats than from other animals. At three times the oral dose (adjusted for body weight), cats respond remarkably similar to humans. But their reactions are not similar toward morphine and a few other drugs.

Bustad (1966) encouraged the use of miniature (150 to 200 lbs) pigs for anatomical and physiological studies. Pigs and humans are strikingly similar in heart, circulatory system, diet, digestive system, and teeth.

For many years, the Norway rat seemed an essential part of psychology experiments. Richter (1954) wrote that Norway rats are very similar to humans in dietary needs; physiology of nerves, muscles, and glands; geographic distribution; world population; and colony formation.

Even bacterial responses can be relevant to humans. Certain bacteria change their genetic information, i.e., they mutate, when exposed to substances that cause human cancers. The Ames test measures bacterial response and provides an inexpensive method for screening potential human carcinogens.

When the goal of an animal study is to improve predictions about how an independent variable will affect humans, focus on that goal. Why bacteria mutate in response to certain chemicals is irrelevant when screening for carcinogens, although perhaps posing a stimulating problem for other scientists.

Creating Animal Models of Human Conditions

Many of the greatest medical success stories, including those about polio, scarlet fever, strep throat, and malaria, begin with the development of an animal model. Animals were infected with the diseases and then exposed to various test drugs until cures were found. But behavioral models, though they have a long history dating back to Pavlov's 1921 research (1928), are rarely so simple.

Pavlov required hungry dogs to distinguish between a circle and an ellipse. When successful, they were fed. Then he made the figures more and more similar to each other until the dogs could no longer discriminate. Previously normal dogs soon began to constantly struggle and howl, and Pavlov claimed to have created an animal model of neurosis. Reines (1982) believes that such claims are made too easily, and that animal models of conditions like schizophrenia, depression, neurosis, and ulcers have different causes, symptoms, neurobiological mechanisms, and responses to treatments from their supposed human analogues. But a model's usefulness does not necessarily depend on how closely it mimics a human syndrome. For example, Zucker (1988) acknowledged that models of seasonal affective disorder (SAD, a depression that occurs in wintertime) differ in many ways from the expression of SAD in people. Still, he argued, they have considerable practical value. They have suggested ways to measure, classify, and manipulate the disorder in humans and to get at its physiological basis. Zucker reviewed several principles that emerged from research on biological rhythms of animals and are relevant to research on SAD; and he noted that specific treatments have been developed as a result.

The development of a model is a beginning, not an end. Scientists must assess their models for relevance to humans and try to improve them.

Learned Helplessness Seligman (1975) gave dogs a series of randomly-occurring, inescapable electric shocks. The next day, he sounded a tone prior to each shock; the dogs could have escaped by jumping over a small hurdle, but most just crouched in the apparatus and passively received each shock. Seligman asserted that exposure to inescapable shock had taught the dogs that nothing they did made a difference. They became inactive and unaggressive even outside the test apparatus and performed poorly when required to make responses for food reward. Seligman called their response learned helplessness and said that it resembles human depression. It appeared to be an excellent model. But several criticisms, summarized in an excellent review by Willner (1985), show how tricky modeling can be.

Willner noted that passivity and inactivity characterize only a certain kind of human depression, called bipolar endogenous depression. Human victims do not respond readily to treatment, but the dogs do. If they are repeatedly dragged across the hurdle at the onset of shock, they eventually start crossing on their own.

People exposed to a series of insoluble problems subsequently do poorly on problems they otherwise would have been able to solve. The finding has been assumed to bolster the learned helplessness/depression connection. But the procedure also causes anxiety, which is not a regular feature of endogenous depression; and anxious depressives are not characterized by passiveness.

The validity of the helplessness model rests on three assumptions. First, that animals exposed to inescapable shock learn that their responses don't matter. But they may learn only to be inactive, which would interfere with learning of new responses. Another alternative for which some evidence exists is that inescapable shock depletes the brain of neurochemicals necessary to initiate responding—much more so than does escapable shock. The second assumption is that a similar state is induced in people confronted with uncontrollable, unpleasant stimuli. But it is not clear that people given a series of insoluble problems perceive themselves as being helpless. The third assumption is that learned helplessness is the central symptom of depression. But, though nondepressed volunteers subjected to helplessness-induction procedures show performance deficits similar to those of severely depressed patients, they are no more likely to become depressed than to become hostile or anxious.

Alcoholism   Some scientists have claimed that animals other than humans don't drink alcohol spontaneously. Evans (1979) pointed out that they said "animals" but meant domesticated rats and mice. Several species, including rhesus monkeys, miniature pigs, and chimpanzees, voluntarily drink. (Some mice, given a choice between a 10 percent ethyl alcohol solution and water, drink much more than others. There appears to be a strong association between their alcohol preference and activity of an enzyme within their liver for metabolizing alcohol (McClearn, 1965). The finding may have important implications for understanding human alcoholism.)

Whether alcoholism is caused by spontaneous drinking or daily injections of high doses of ethanol is in some cases irrelevant. A scientist may, for example, wish to test the possibility that a new wonder drug reverses liver damage in alcoholics; how they got that way would make no difference. The important consideration is that the model reflects the purposes of the scientist. According to Falk (1977, 1985), if the purpose of an animal model is to promote understanding of human alcoholism, it should incorporate several features. They include the following:

- Overdrinking should be voluntary, not forced.
- Drinking should be sensitive to environmental conditions. For example, human alcoholics given a choice between continued alcohol drinking and money abstained voluntarily for seven days when the payment was high enough (Cohen, Liebson, Faillace, & Speers, 1971). Their intakes were affected by the work required to get the drinks. Analogous sensitivity should be demonstrable for animals.
- Animals should not drink just to avoid withdrawal symptoms, because that is not what maintains drinking in humans (cf. Mello & Mendelson, 1972).
- Drinking should not be induced by nor serve as a means of relieving stress, because that is not how it functions in humans (Mello, 1972).
- Drinking should be chronic, lasting for months or years; and cyclic, with amounts drunk fluctuating greatly from day to day (Mello & Mendelson, 1972).

....

1. Can you figure out how Falk adapted his method for generating excessive water-drinking in rats to produce a rat model of alcoholism that has all the features listed above?

....

Using Animals to Learn About Human Systems

Our understanding of many human behavioral and physiological systems has been increased by studies of corresponding systems in animals. The strategy is to pick an appropriate species, study the system of interest thoroughly, and see how the results might apply to humans. Dolphins, master communicators and highly social, have yielded important insights into human communication and social behavior. Observations of lions, hyenas, wolves, and African hunting dogs, all cooperative hunters, may help scientists understand how primitive human hunters lived. Pigeons and other birds are ideal subjects for research on color vision. Many satisfactory generalizations have been made from fruitfly genetics to human genetics.

Chimpanzees, our closest relatives, are used in studies on learning, motivation, and mother/infant attachments. But chimpanzees are expensive and hard to maintain in laboratories. Norway rats are far more convenient and, as a result, more experiments have been done with rats than chimpanzees—or any other animal.

The simplicity of the rat brain may make it a more rather than less appropriate species than the chimpanzee for certain kinds of research. Simple preparations are often better than complex ones for revealing how things work. That's why many researchers interested in the mysteries of human aging study aging in single-celled animals. Similarly—maybe—rat experiments can illuminate human psychological processes. But domesticated laboratory rats are in many respects an unusual species. Richter (1954) extolled their virtues but also detailed a host of large anatomical, physiological, and behavioral differences between laboratory and wild rats. A pattern emerges: laboratory rats have larger gonads and a greater reproductive capacity than wild ones, but smaller adrenal glands and a less efficient response to stressful situations. Richter raised the intriguing possibility that changes in humans with the emergence of civilization parallel those in rats from domestication. He speculated that many modern diseases of humans, such as hypertension, rheumatism, and arthritis, are caused by reduced adrenal gland size; and that increases in certain types of tumors are caused by larger gonads.

Lockard (1968) cited Richter's article and concluded that domesticated rats, having evolved under less challenging conditions than wild ones, are inferior and should not be used by behavioral scientists. But laboratory rats may not be inferior (cf. Price, 1972; Pavlov, 1928). In any event, the laws of physics and evolution apply to all species. Whether or not behavioral principles will also generalize is an empirical question.

Henshel (1980) urged sociologists to study phenomena even if they occur only inside laboratories. He wrote: "Sociologists have consistently overlooked a fundamental feature of the experiment in natural science, something we shall call *unnatural experimentation. Unnatural experimentation is the deliberate exposure of phenomena to conditions or circumstances not found anywhere else."* (his italics). Henshel noted that scientists learned a great deal about normal brain functioning by studying brain-damaged people and people exposed to radiation. They learned about the nervous system by studying the effects of sensory deprivation and the aftereffects of limb amputation and resulting "phantom sensations." Henshel's argument, which he supported with many additional examples from both the natural and the biological sciences, is relevant to discussions about the "artificiality" of experiments with domesticated rats.

If the purpose of an animal research project is to make discoveries that apply to human systems, applicability to humans should be the criterion of success. According to Lowe (1983), the most extensive studies in animal behavior have been on schedules of reinforcement. Early proponents (Morse, 1966; Skinner, 1969) claimed enormous generality for their results, but later reviews have been less enthusiastic. Lowe, summarizing recent research, wrote: ". . . the behavior of humans with respect to scheduled reinforcement differs in fundamental respects from that of other animal species. The differences appear so extensive, applying as they do to all the basic schedules of reinforcement, that they throw into doubt much of the theorizing about human learning which is based upon animal experimentation. And if the animal model does not hold good for human operant behavior under controlled experimental conditions why should it do so in the hospital, school, or stock exchange?"

Lowe's pessimism notwithstanding, behavior modifiers have successfully applied principles learned from rat experiments to treatments of human behavior problems.

Failed generalizations don't necessarily mean failure. On the contrary, they may generate research into why organisms that are similar in some ways differ in others; this in turn may lead to more profound generalizations. For example, early behaviorists believed that animals of all species respond similarly to all reinforcing events. When this proved false, the deeper generalization emerged that animals' responses to stimuli are shaped by the environments in which they evolve. Exceptions to the generalization that animal species respond essentially the same to drugs stimulated research into species differences in rates of absorption, metabolism, and excretion. This led to the generalization that drug effects are predicted better from knowledge of how a drug concentrates at its site of action than by administered dose, which has important consequences for human pharmacology: it helps explain differences in responsiveness to drugs due to gender, race, age, and many other factors.

Studying a Species Because It Has an Interesting Property

The animal kingdom is breathtaking in its diversity, and many animals have properties that seem bizarre from the human perspective. Scientists welcome the bizarre and organize research projects around it. One reason is that an unusual property may provide a window to an important but otherwise inaccessible phenomenon. Secondly, unusual properties pose puzzles—which scientists love. As discussed in chapters 1 and 6, scientists whose sole motivation is to solve puzzles may change our perceptions of the world and enrich our culture.

All species have evolved intricate and intermeshing adaptations for surviving in their particular ecological niche. The specific examples below should not obscure that fact. So all animals are potential puzzles to be solved. (I've listed only a single reference for most of the examples, but none are based on a single study. Solving puzzles about behavior, like solving jigsaw and crossword puzzles, requires fitting many pieces together.)

Studying a Species Because It Offers Special Access to a Phenomenon

Because of squids, we know a lot about nerve cells. Each nerve cell has an axon that carries the nerve impulse away from the cell body to adjacent cells. Most axons are about 1 millionth of a meter in diameter, but squids have several giant nerve cells with axons about one thousand times as large. The axons can be dissected out of the squid and kept alive for many hours in a special water bath. Hodgkin and Keynes (1956) and Hodgkin and Huxley (1952) took advantage of these properties to study the sequence of events leading to the firing of nerve cells. Their work, which generalizes to all nerve cells, earned a Nobel Prize. (The squids left the awards ceremony in disgust, feeling they should have been at least co-winners.)

Marsupials (pouched animals like kangaroos and opossums) are born prematurely compared with other mammals. The red kangaroo reaches more than six feet in height, but it is about three-quarters of an inch long at birth. So, marsupials are valuable subjects for embryological and developmental research. See Sharman (1973).

Storrs and Williams (1968) took advantage of a unique property of nine-banded armadillos: they typically give birth to genetically identical quadruplets. At the births of several sets of quadruplets, Storrs and Williams measured weights of various anatomical organs and activity of several neurotransmitters. They found many substantial (up to 140-fold) differences between the highest and lowest scoring newborns of each set. The results indicate that the intrauterine environment plays an important role in development. Behavior geneticists might profit from armadillo research.

Frogs and toads regenerate body parts, so research on them may provide clues to the healing process. Sperry (1943) removed the eyes of mature frogs and rotated them 180 degrees. The nerve endings regenerated and soon the frogs could see again—but their visual worlds were displaced 180 degrees. When a fly was offered above and to the left of them, their tongues flicked down and to the right. Sperry's work is of interest to scientists concerned with behavioral plasticity and nerve growth.

Some species have an artificially unique property: they have been studied a great deal. As Beach (1950) noted, scientists who work with such species can compare, combine, and correlate their findings, which helps ensure accuracy and accelerates the acquisition of data. The popularity of rats stems in part from the accumulation of information and techniques for using them. Similarly, the early focus of behavior genetics researchers on mice promoted the breeding of inbred mouse strains with special properties. Suppliers like the Roscoe B. Jackson Memorial Laboratory list more than 200 inbred strains. Do you want a mouse strain susceptible to auditory convulsions? With a low level of spontaneous activity? Likely to experience retinal degeneration? Go through their Sears-like catalog and place your order. The availability of so many strains ensures continued use of mice.

Studying a Species Because It Poses a Puzzle Some behavioral and anatomical properties of animals were so puzzling that they stimulated the launching of research projects to find answers. These sometimes led to important practical consequences: new techniques, later used in applied research; inventions, patterned after sophisticated feats of biological engineering; increased understanding of how all animals adapt to their environments; and generation of ideas for additional research and theory, often with relevance to humans. The references below are to popular, highly readable accounts of the various projects. They show excellent scientists at work, observing, correlating, and experimenting according to the problems at hand. The scientists continually considered alternative explanations to their own and designed studies to test the alternatives. Serious students of methodology should read the original accounts.

Bats fly in the dark, avoiding obstacles and capturing prey with marvelous efficiency. To learn how they do it, Griffin and his colleagues (1950) observed them hunting under natural conditions, then brought them into the laboratory. The scientists measured the rates at which bats capture insects and the minimum distances at which the prey are detected. In one experiment, the bats had to fly around a big screen, then through a series of narrowly-spaced bars, to enter an area where they successfully caught small, moving insects. In another, placed in a room with a swarm of mosquitos, they caught an average of 14 per minute. Selective masking experiments with low and high frequency noise conclusively demonstrated that bats intercept at least some of their prey by echolocation. That is, bats emit sounds and then orient to the sounds as they are reflected off objects.

In the early 1900s, scientists had concluded that reptiles and insects, including honeybees, are colorblind. Karl von Frisch (1950) thought that inconceivable. "Why," he asked, "would flowers have evolved so many colors except to attract bees to pollinate them?" The puzzle started von Frisch on a career studying honeybees, which culminated in 1973 with a Nobel Prize. His small book, highlights of which are given in Box 20.1, is important in three respects. First, he solved the puzzle. Second, the book communicates the joy of research. Third, it is a wonderful model of sound methodology.

Studying Animals from an Evolutionary Perspective

The studies described so far can be carried out with a single species such as rats, though investigators often use two or more species to (a) improve the accuracy of generalizations to humans, (b) create better models, and (c) make sense of unusual properties of closely related species. Researchers in such cases rarely concern themselves with the ecological niches of the species or their biological relatedness. By contrast, those factors are preeminent to biologists and psychologists interested in the evolutionary histories and functions of behaviors. Such scientists ask questions about ultimate rather than proximate causes, about evolutionary origin and survival value rather than current physiological or psychological controlling mechanisms. (A question about ultimate

BOX 20.1

Bees

Von Frisch put honey out on a colored piece of cardboard on a table and waited, sometimes for several hours, for bees to find it. Once a bee had found the rich food source, it sucked up some honey and returned to the hive. Soon the table swarmed with bees. Von Frisch let them feed for awhile and then placed out two clean, new pieces of cardboard, one of the original color and one a different color. Incoming bees alighted on the card of the original color, indicating that they can both distinguish and remember colors.

In his first experiments, he fed bees on blue cards. He considered the possibility that they always have a preference for blue, so he tried many different colors; they invariably returned to the color on which they had discovered food.

Colorblind people can distinguish between red and blue, because red appears much darker. So von Frisch fed bees on a blue card surrounded by gray cards of all shades from white to black. When fed on blue, they came again to clean blue cards.

He considered the possibility that blue has a distinctive odor, so he placed the cards under a glass plate. Preferences were unaffected.

Von Frisch cited a study by Kuhn, in which bees were fed in a spot irradiated with ultraviolet light. The bees then went to every spot so irradiated, proving that they can detect UV light although it's invisible to humans.

Flowers come in a wide assortment of shapes, so he drew different patterns on pieces of cardboard and pasted them on the entrance to a box containing sugar water. To other boxes, without food inside, he pasted papers of the same color but different patterns. Bees learned quickly in some cases but not others; they didn't, for example, learn to distinguish between a triangle and a square or circle. Later studies showed that the crucial characteristic is the degree of "brokenness" of a pattern. Bees can distinguish all the figures in the top

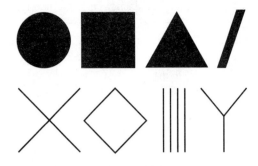

Figure 20.1

row of Figure 20.1 from all the figures in the bottom row, but they can't distinguish figures within a row from each other.

Von Frisch wondered why other bees arrived so quickly after one or two had discovered a food source. He realized that they must have a communication system. He couldn't observe inside a normal bee hive, so he constructed an artificial one with a side made of glass. He painted the bees with small spots of different colors on different parts of their bodies, which enabled him to distinguish 599 individuals at a time.

He put a rich source of food near the hive and waited for a bee to find it, at which time she was marked with a spot and followed back to the hive. Returning bees began what von Frisch called a "round dance." They turned once to the right, once to the left, repeating again and again, often for 30 or more seconds; then they moved to another spot on the honeycomb and did the same thing. The vigor of the dance intensified as the food was made sweeter or increased in volume. The other bees followed the dancer and became very excited; soon one, then more and more, left the hive and flew to the feeding place. When they returned, they also danced.

Von Frisch placed dishes of two fragrant flowers, cyclamen and phlox, near the hive. He put sugar solutions on one of the flowers. When a bee discovered the sugar and soon afterwards danced in the hive, her recruits flew to the flower that had had the food. Similarly, when pieces of cardboard were impregnated with distinctive smells such as peppermint oil, the bees returned to the smell that had been associated with food. He observed that bees within the hive hold their antennae to the dancer's body; the upper surface of the bee's body has the ability to hold scents for long periods.

The scent from a flower is carried on the bee's body. Von Frisch considered the possibility that sugar water placed directly on the flowers can take up their scent, which would then be carried internally within the bee's stomach. So, when bees landed on cyclamen, he fed them with sugar water saturated with the odor of phlox blossoms. Thus, the scent adhering to the bee's body was cyclamen, the scent from the stomach was phlox. When the feeding place was close to the hive, recruits visited both types of flowers; when the feeding place was a half-mile away or more, they visited the flowers whose scent was carried in the stomach. He concluded that scented material adhering to the body is lost during long flights.

Bees produce scents themselves, from special glands. Von Frisch put food out on two pieces of unscented cardboard and put shellac over the scent glands of bees that visited one of them. The shellac prevented them from depositing scents. Ten times as many recruits returned to the other cardboard.

He put different amounts of food out at 10 and 300 meters from the hive. Most recruits came to the richer source, showing that the dance must communicate distance. He observed that all the dancers from a nearby source perform a round dance, but dancers from a distant source do a "wagging dance." They run a short distance in a straight line while wagging their abdomens rapidly from side to side. Then they make a complete 360 degree turn to the left, run straight ahead once more, turn to the right, and repeat it over and over again. He kept moving the nearer feeding place to greater and greater distances, and somewhere between 50 and 100 meters the round dance changed to a wagging dance; and he brought the far feeding place closer until, between 100 and 50 meters, the wagging dance changed to a round dance.

The wagging dance announces a rich source of food at a definite distance, communicated by the number of turns per second. Von Frisch observed 3,885 dances. At 100 meters, dancers made 9 to 10 complete cycles in 15 sec; at 200 meters, 7; at one kilometer, 4½; at 6 kilometers, 2; so, with a stopwatch, he could tell how far the dancing bee had flown. Bees have dialects, with different rates for different colonies.

Bees aroused by round dances search the vicinity of the hive in all directions. But wagging dances communicate direction as well as distance. To show this, von Frisch placed food out in a particular direction from the hive and observed that recruits flew only in that direction. He found that the direction of the straight part of the wagging dance is related to the direction of the food source. Dancers change direction throughout the day, to indicate where the food source is with reference to the sun. If a dancer heads directly upward on the honeycomb during the straight part of her dance, this means the feeding place is in the same direction as the sun. Straight downward means "Fly away from the sun," 60 degrees to the vertical means "Fly 60 degrees away from the sun."

BOX 20.2

Rodent Copulatory Behavior

Dewsbury captured animals of 31 rodent species and observed their copulatory behaviors in a standard laboratory test situation. He cited data indicating that their behaviors in the wild are quite similar. He developed a classification system with four key elements (he noted that behaviors are defined in terms of male behaviors, but females exert considerable control over the occurrence, timing, and topography of copulations):

1. Some rodents lock while copulating, that is, have a mechanical tie between penis and vagina. Most rodents don't lock.
2. Laboratory rats have a single thrust during each penile insertion; some rodent species show repetitive intravaginal thrusting during a single mount.
3. Animals of some species ejaculate the first time the penis is inserted into the vagina, but many require multiple intromissions preceding ejaculation.
4. Laboratory rats ejaculate more than once before reaching satiety, but many species ejaculate only a single time.

The four attributes, each with two alternatives, give rise to $2^4 = 16$ possible patterns. Dewsbury observed seven of the patterns among his 31 species. There was little variability within a species, even when he tested animals from different natural populations of the same species. He wrote, "If a vole of unknown species escaped its cage, I would be more confident of correct identification on the basis of copulatory pattern than of morphology."

Dewsbury related copulatory patterns to anatomical and ecological features. For example, in species which lock, the glans penis is thicker than in species that neither lock nor thrust. Species that thrust are intermediate. He studied 12 new species and correctly predicted in each case whether they would lock, thrust, or both.

He found no simple correlations of copulatory patterns with feeding habits, habitat, or type of social organization. Two tentative conclusions were (a) a safe nest site is necessary, though not sufficient, for the evolution of a locking pattern; and (b) patterns characterized by ejaculation on a single brief insertion represent adaptations to life with sparse cover. These hypotheses can be tested. They require both laboratory studies and field data.

---

cause might be, "Why are monkeys more likely to reproduce if they eat ripe rather than unripe fruits?" A question about proximate cause might be, "What area of the brain controls a monkey's preference for ripe over unripe fruit?") Scientists who focus on ultimate questions sometimes conduct experiments, but their primary methods are observational and correlational.

Making Comparisons to Order Information and Generate Hypotheses Why do males of only some primate species provide extensive parental care? That interesting question could not be answered by observations of individual species. But when data became available on many species, Clutton-Brock and Harvey (1977) showed that male parental care is associated with monogamous mating systems.

Dewsbury (1975) also ordered information from many species in the hopes of finding correlations between behavior patterns and anatomical and ecological features. His work is summarized in Box 20.2.

Studying the Evolutionary History of a Behavior   Many scientists interested in behaviors of ancient animals study fossils. From analysis of the tracks left by dinosaurs in soft mud come inferences about how fast they moved and their social structure (clusters of tracks pointing in the same direction indicate that they traveled in groups); and the fossil remains of several young dinosaurs in an apparent nest suggest parental behavior. Similarly, study of fossil footprints of our Australopithecine ancestors indicates that they walked upright much as we do today (Alcock, 1984).

A second approach to tracing the historical pathways of present-day behaviors depends on three assumptions: (a) differences between individuals in the behavior under study are accounted for at least partly by genetic differences between the individuals, (b) closely related species that have a similar behavior pattern probably had a common ancestor with the pattern, and (c) behaviors displayed by only one or a few species within a large group of closely related ones probably evolved later than the ancestral pattern. To use the method, a researcher must identify the related species and observe and classify their behaviors. Alcock (1984) gave several examples, leading to plausible accounts of the evolution of blood sucking in a species of moth, courtship displays in balloon flies and bowerbirds, and dancing in honeybees.

Relating Behaviors to Ecology   Behavior patterns evolve in response to ecological pressures, so behavior ecologists collect data on related species living in diverse environments and unrelated species in similar ones. Two books by Krebs and Davies (1981, 1984) provide extensive discussions of the aims and methods of behavior ecologists. (The methodological problems are considerable. Ecological variables cannot always be defined with operational clarity. The choice of species is somewhat arbitrary, and the number of animals observed within each species may vary greatly. So may the proportions of males and females and adults and juveniles, and the conditions of observation. Data must often be collated from publications of investigators with different observational skills and recording criteria. These and other concerns are discussed by Clutton-Brock and Harvey in the first chapter of *Behavioral Ecology: An Evolutionary Approach*.)

Comparing and Contrasting Behaviors in Related Species
Kittiwake gulls build their nests on narrow ledges of high cliffs, whereas most species of gulls nest on the ground. As a result, nesting kittiwakes are virtually immune to predation and other gulls are not. Cullen (1957) identified a great many behavioral differences between kittiwakes and ground-nesting black-headed gulls and related them to their contrasting nesting sites.

Glickman and Sroges (1966) placed novel objects in the cages of more than 100 species of animals in Chicago's Lincoln Park Zoo, then recorded how many times the animals oriented toward and manipulated the objects. See Box 20.3 for their procedures and results.

BOX 20.3

Curiosity in Zoo Animals

Glickman and Sroges studied more than 200 zoo animals representing more than 100 different species. They tested each animal individually, during its active period, in its home cage or a familiar adjacent cage. They presented five types of novel objects, scaled to size for each animal, in a standard order during a six-minute test session; and they measured how frequently the animals oriented toward and made contact with the objects. They found a trend toward increased responsiveness toward all five objects as they moved from reptiles to primitive mammals and rodents to carnivores and primates. They tested 53 mated pairs of animals and found no systematic sex differences. In 12 of 17 cases in which they compared a subadult with an adult of the same species and sex, the subadult responded more.

Glickman and Sroges hypothesized that reactivity to novel objects is related to activities the animals perform in their daily lives. Two critical activities are feeding and evasion of predators. Feeding patterns that require extensive manipulation of the environment favor the development of curiosity, ready availability of food does not. Any factor that reduces predator danger—effective defensive weapons, organized social structures, low incidence of predators, or an environment offering good protective cover—favors a high level of reactivity. Their hypothesis was supported by data on different species of monkeys. Species with a varied diet are much more reactive than those that live exclusively in trees and are strictly herbivorous.

Comparing and Contrasting Ecologies

If kittiwake and black-headed gulls behave differently because of the ecology of their nesting sites, then other species with protected sites should show many of the adaptations of kittiwakes, and vulnerable species should be more like black-headed gulls. Alcock reviewed evidence in support of this position. More generally, explanations of why closely related species differ behaviorally can be tested on other species with both similar and different ecologies.

Some scientists use ecology as a starting point. Once an appropriate dimension has been identified, closely related species that vary along the dimension can be tested for behavioral differences. Crook (1964) contrasted forest, savannah, and grassland ecologies. His findings are presented in Box 20.4.

2. Why must the species be closely related?
3. Identify five ecological dimensions.

BOX 20.4

Weaver Birds

Crook observed about 90 species of weaver birds in both the forest and the savannah. Food in the forest is dispersed and difficult to find, whereas in the savannah seeds are clumped in rich patches. In the forest, nests can be carefully hidden but in the savannah they must be exposed. Crook hypothesized that the two ecological variables—availability of food and exposure to predation—account for many behavioral differences between forest and savannah dwellers.

Forest dwellers generally feed on hard-to-find insects. Because the supply is limited, a single parent would be unable to feed the young. So both parents contribute, which favors a monogamous breeding system. Because the male frequently visits the nest, he is dull-colored to avoid attracting predators. Another way to avoid predators is to space nests far apart, which promotes territoriality.

Females on the savannah can feed their young alone, so males don't show parental care and instead constantly compete for more females. They are brightly-colored and polygamous. Clumps of seeds, once located, can feed an entire flock; so savannah dwellers, unlike forest birds, search for food in groups. Because they can't hide their nests, they build them in protected sites such as spiny acacia trees. These are relatively rare, so many birds build nests in a single tree. The males with the best nest sites attract the most females.

Jarman (1974) explained behavioral differences among African ungulates in terms of diet and exposure to predators. He thus both tested and extended Crook's approach.

Comparing and Contrasting Behaviors

Box 20.4 suggests how behavioral differences can be a starting point for comparative research. The investigator asks, "How can behavioral differences between closely related species be explained by differences in ecology?"

Comparative Methods Applied to Humans Almost every known society has had the custom of marriage. Because marriage is universal, studies restricted to humans cannot explain why. So Ember and Ember (1979) read life histories of 135 free-ranging mammal and bird species, coding for presence or absence of male-female bonding and several other variables. Their conclusion, that bonding occurs most frequently when mothers' feeding requirements interfere with their baby tending, applies to humans.

Many anthropologists use the methods of behavioral ecology, but within a single species: humans. They correlate behavioral and physiological traits to the ecologies of cultures. For example, Edgerton (1965) gave a battery of tests to herdsmen and farmers of four tribes in East Africa and found several statistically significant differences: farmers were more likely than herders to make group decisions; they placed greater value on hard work; and they were more suspicious and hostile toward others. Edgerton related the traits to the methods of earning a living. Netting (1977) gave several other examples.

Summary

Researchers use animals in their projects to gain maximum control over the administration of independent variables, measurement of dependent variables, and constancy of all other variables. In addition, they can pick species that are inexpensive to buy and maintain, easily handled, and have short generation spans or unique properties. Treatments can be imposed on them that would be unacceptable with humans. This last point is not universally accepted and has caused concern about the rights of animals and the ethics of doing experiments on them.

For several types of animal research, a single species suffices. Be sure to pick one appropriate to your purposes.

- Measure the effects of an IV and generalize to humans.
- Create an animal model of a human condition. (Even if a model has many superficial similarities to the human condition, its value must be tested. Researchers should seek ways to improve their models.)
- Study a system in animals to learn how a corresponding system operates in humans. Even if generalizations from one to the other are not substantiated, they may lead to deeper generalizations.
- Pick a species because it has an interesting property: the property may provide a special advantage for studying an important phenomenon or may pose an intriguing puzzle.

Research conducted from an evolutionary perspective is directed toward answers of ultimate questions. Studying fossil evidence and analyzing behaviors of related species enable scientists to plausibly reconstruct the evolutionary history of the behaviors. Correlating behaviors with ecologies enables them to understand how environments act as selective forces to shape behaviors.

Answers

1. Falk placed hungry rats in a chamber in which one tiny food pellet was dispensed every two minutes. The rats were placed in the chamber for three hours per day with a 5 percent alcohol solution available for drinking. They drank much more than animals in various control groups and became physically dependent.
2. Unrelated species differ in many ways and for many reasons. A particular difference between them may have little to do with ecology. The more closely related the species, the greater the likelihood that behavioral and ecological factors are interdependent.
3. The number of potential factors is infinite. Here are a few: degree of crowding, mean temperature, altitude, annual rainfall, type of diet, distribution of food, primary predators, degree of isolation from related species, and extent to which the seasons change.

CHAPTER 21

Using Private Data

By the time you finish reading this chapter, you should be able to answer the following questions:

What is introspection?

Why did the early behaviorists reject introspective data?

What are some ways of evaluating the accuracy of self-report data?

What are some methods for collecting self-report data?

Please answer the following questions: "London is to England as Paris is to _____?" and "Fingers are to hand as toes are to _____?" Did any images form in your mind as you answered? In the early 1900s, psychologists called structuralists hotly debated the possibility of imageless thought. With a goal of learning how mental processes are interrelated, the structuralists introspected—they observed their own conscious processes as they solved problems, recalled experiences, and summoned up emotions. Another group of psychologists, the functionalists, had as their goal the study of the constant stream of consciousness. They also supported introspection. Functionalist William James wrote in 1892:

It is in short obvious that our knowledge of our mental states infinitely exceeds our knowledge of their concomitant cerebral conditions. Without introspective analysis of the mental elements of speech, the doctrine of Aphasia, for instance, which is the most brilliant jewel in Physiology, would have been utterly impossible.

But introspection requires people to report on their own mental experiences, so excludes inarticulate observers—the mentally retarded and disturbed, children, and animals—and material inaccessible to consciousness. A second problem is that people can't simultaneously experience emotions and observe them with detachment and objectivity. Third, because all the data come from within, introspection appears unverifiable. A person who claims to have imageless thought can maintain his position regardless of anybody else's observations. As it turned out, a German group reported imageless thought, an American group denied its existence, and the matter was unresolvable. The groups disagreed on other issues fundamental to introspectionists.

PRIVATE DATA IS PRIVATE

The first two problems limit the method, the third seemed to doom it. In 1913, John Watson (1913) exposed the weaknesses of reports of inner experiences and urged psychologists to adopt a new strategy: to restrict themselves to objective and publicly verifiable data, i.e., to descriptions of overt behavioral acts. He said they should exorcise mentalistic concepts like consciousness, imagery, and mind. Watson realized that self-reports can be treated like other forms of behavior, with no assumption on the part of scientists that they directly mirror inner mental events. But they weren't part of his program. Other psychologists, swept away by the great initial success of the behaviorist revolution, became reluctant to use them. Lieberman (1979) gave several examples of their reluctance:

- Systematic desensitization is a technique used to treat phobias. Patients are taught to relax, then to maintain relaxation while imagining frightening scenes involving their phobias. The technique is highly effective and has been studied intensively; yet there have been almost no attempts to examine the nature of the images formed by different patients or the extent to which their content and vividness correlate with therapeutic outcome.
- For many years, psychologists who studied hypothesis formation assigned problems to subjects and inferred their hypotheses from their solutions. Not until 1971 (Karpf & Levine) did anybody ask the subjects directly what their hypotheses were.
- Mischel (1974) asked children to try different ways to imagine rewards, for example, to imagine marshmallows as either fluffy white clouds or soft and chewy. The instructions affected the children's ability to tolerate delays in receiving the rewards, but Mischel never indicated in his publications that he interviewed the children to find out what images they had actually used.
- An additional example comes from Barker (1971–1972). He told about a colleague who applied for a research grant on problem solving. The hopeful investigator proposed as a first step to question subjects about their preferred strategies for handling certain kinds of learning problems. The proposal was turned down; he later learned that the primary reason was his intention of asking about the mental life of his subjects.

The situation is changing, and today even some behaviorists study aspects of consciousness. Hilgard (1980), summarizing current attitudes, wrote:

The opening up of psychology, without sacrifice of the gains that have been made in tight theorizing and precise experimentation, is all to the good. The exploitation of the new freedom by those who have a distaste for the discipline of science will have to be guarded against, but this risk must not discourage those who have retained a curiosity about all aspects of the mind and human behavior, and at the same time are determined to retain and advance psychology's stature as a scientific enterprise.

Despite Hilgard's optimism, mentalistic concepts have not yet infiltrated introductory research methodology textbooks. Of 18 texts published since 1980 on my bookshelves, the index of only one lists either introspection or phenomenology; the one exception contains a short paragraph criticizing the early introspectionists. The omissions are unwarranted, as analysis of self-report data is a legitimate and valuable scientific research method. But many psychologists probably distrust introspection because of its historical associations. Also, debate continues over whether self-reports provide unique access to the working of the mind. Their use as a research tool, however, need not await resolution of the controversy. Many traditional researchers probably avoid introspection because its proponents have often contented themselves with scientifically inadequate studies.

Some Introspective Reports Are Verifiable

We can never be sure if a subject's report of her thoughts, perceptions, or feelings are honest or accurate. Nonetheless, some aspects of such reports are verifiable.

The Test of Self-Reports Is Objective and Verifiable

Both the existence of self-reports about experiences and their precise wordings are verifiable. If people who take a hallucinogenic drug claim to be having inner conversations with God, competent scientists will agree that the claims were made. They can be skeptical about the truth of the claims, and refuse to prostrate themselves and offer sacrifices. But by collecting objective, verifiable data on the frequency and nature of the claims, they can learn a great deal about the drug experience.

The Correspondence Between Descriptions of Inner Experiences and Other Data Can Often Be Evaluated

Whether a self-report faithfully describes an inner experience can never be known for sure. We can't have another's experiences, so can only speculate about his or her true feelings. But relevant evidence—objective and verifiable—can often be found. A person claiming to have inner conversations with God might be persuaded to divulge some of Her revelations about the immediate future. If these were highly specific and reliably came to pass, we might begin to accept the claim as true.

Lieberman (1979) noted that the accuracy of a report, whether it concerns phenomena of internal or external origin, is judged by its correspondence with other events. To verify a person's claim to have witnessed an external event—say, a meteor passing overhead in a secluded forest—investigators might trace radar records at nearby airports, look for impact craters, check seismographic reports, and so forth. Lieberman described an experiment by Cohen (1960) to show that a similar approach can be used to check reports of internal events. Cohen's subjects, after lengthy exposure to a uniform visual field, sometimes reported being unable to see. Other, observable, events occurred at the same time: Their brain wave activity changed and saccadic eye movements disappeared; and they didn't detect anything (or so they claimed) when various visual stimuli were presented. Cohen might have conducted further tests such as exposing them repeatedly to a light paired with a severe electric shock; after such training, people eventually show strong galvanic skin response changes to the light alone. Had the light then been shown during the period of "not seeing," their galvanic skin responses would have provided evidence about their truthfulness.

Richardson (1984) offered three strategies for testing the accuracy of self-reports.

Test for Criterion Veracity Reports of experience can be compared with independently verifiable events, as in the examples given above. The independent events can be physiological, such as changes in brain wave patterns, or behavioral, such as absence of fear responses to a light previously paired with a shock. Facts can often be checked, as when hypnotized adults regressed to earlier ages give detailed answers to questions about schools, friends, and so forth. (The subjects can't recall the information in their normal waking states.) Sources of information for validating the hypnotically-induced memories include school records, newspaper clippings, and the testimony of parents and older siblings.

Test for Content Veracity Subjects can be asked to make additional comments about an experience some time after they make an initial report. The two sets can be compared. The second inquiry can be conducted much later in time, even several years later. (Note: Richardson gave as an exemplar of this method Kinsey's studies of human sexual experience. Please reread pages 171–172 for criticisms of Kinsey's work.)

Test for Construct Veracity Experiential reports can be compared with predictions based on theory. One theory of development leads to the prediction that people who form vivid mental images can recall more early childhood events than can people with poor imagery. A positive correlation between reports of vivid imagery and reports of childhood memories would provide evidence for the accuracy of the reports. (Richardson acknowledged that other reasons for the correlation would be plausible; for example, people highly motivated to remember (or make up) early memories might also be highly motivated to induce (or pretend to have) vivid images.

Hypotheses Generated from Introspective Data Can Be Falsified

Falsifiability is the criterion used in this book for distinguishing science from nonscience. Althugh introspective data are often unverifiable and never fully verified, they may suggest falsifiable hypotheses. For example:

- The suicide rate is high among people who claim to experience powerful depression.
- Ninety-seven percent of people exposed to a uniform visual field for more than five minutes will claim to be unable to see.
- There is a positive correlation between claims of having inner conversations with God and voting Republican.

Some Uses of Self-Report Methods

1. Self-Report Methods May Provide Unique Insights into Mental Phenomena

   Hayes (1982) gave subjects several problems and asked them to think aloud as they sought solutions. He learned enough to be able to program a computer to solve problems the way people do. Whimbey (1976, 1980) also asked people to give him feedback as they worked through problems, and from their answers he developed a powerful method for improving problem-solving skills. For both research programs, self-report data were crucial.

   Of course, as is true of all types of analysis, self-analysis is limited. Hefferline, Keenan, and Harford (1959) attached electrodes for recording tiny muscle twitches to various parts of the bodies of volunteer subjects. The subjects were asked to listen through headphones to music accompanied by an unpleasant loud noise. They were told that the purpose of the experiment was to see if the noise affected their bodily tension. But the subjects were deceived. The real purpose was to see if they could be conditioned to make tiny thumb twitches without being aware of it. Each twitch (so small as to be unnoticed by the subjects) turned off or delayed the reappearance of the noise for 15 seconds. During the course of the experiment, all subjects increased their twitching without realizing it. In the postexperimental interview, all were amazed when told that their responses had controlled the noise.

   Self-analysis is fallible, like all other research methods. Results should be checked against other data, as Rogers and Layton (1979) did when they asked heroin addicts in treatment to judge their likelihood of using drugs in the near future. Rogers and Layton also collected information about the occurrence of certain events, such as being around drug users. The events predicted more accurately than the addicts' judgments.

2. **Self-Reports May Be the Only Possible DVs**
Singer (1975) described a variety of methods that he and his colleagues developed for their extensive investigations of daydreaming. Most of them require some form of self-report, as does much current research on the meanings of other types of experiences.

Werth and Flaherty (1986) explored the phenomenon of deception. They asked women to describe incidents in which they had been deceived or had deceived others. The investigators extracted five dominant themes, which they called self-deception and collusion, ramification (the deception extended beyond one person), pervasiveness, feelings, and motivations (explanations of why people remained in relationships in which they were being deceived); and they documented each with extensive quotes. Samples from people who had been deceived are reproduced below.

Self-deception: "I fell for [the deceit] because I wanted to. I just didn't want the thing to happen that was going to happen. I desperately wanted to believe her and so I chose to believe her."

Ramification: "She was pretending there was nothing going on also.... She would call him. She even called me a couple of times... to find out things about him but never said that was her intention."

Pervasiveness: "... It was as though it was a twenty-four-hour-a-day thing. It totally affected my whole life. It was in control of my whole life."

Feelings: "The feelings that it evokes are really thunderous. They are not just little quivers every once in a while. They are deep. Like I said, they took over my whole being, they took over how I felt constantly."

Motivations: "I think there was a million reasons why it lasted so long. Part of it was I kept believing he'd change.... Maybe he'll change. Maybe it's a stage he's going through."

When the women's words are read collectively, they give a vivid portrayal of how it feels to be deceived. But so does Shakespeare's *King Lear*. Scientific descriptions, although not necessarily better than literary ones, are held to different standards. From a scientific standpoint (which may not have been the authors' perspective), the research on deception is unacceptable. However, the shortcomings indicated below are not an inevitable part of the introspective method.

- Subjects were not chosen randomly.
- No indication was given of how something qualified as a dominant theme. The authors had no formal scoring system to guide them, so their biases almost certainly played a role. They themselves wrote that different investigators might have emphasized other aspects of the materials.

- No numbers were used. Readers don't know how many words were collected from which the dominant themes were abstracted.
- No precautions were taken to insulate the subjects from the researchers' views on deception. Perhaps the subjects mentioned deception only because they were asked about it; perhaps they spoke about deception in a certain way because the interviewers were especially attentive when they did.
- There was no control group. People who have never been painfully deceived might give similar accounts (in which case the supposed benefit of introspection—providing access to direct experience—does not apply). And the subjects might have used the same themes in talking about experiences unrelated to deception.
- The subjects reported from memory, and memories are often inaccurate. See pages 170–171.

3. Self-Reports May Predict Behavior More Accurately Than Do Other Dependent Variables

The study on drug addicts cited above shows that introspective reports are not invariably reliable. But they often are. Perhaps researchers will some day be able to accurately predict people's voting patterns from their brain waves. At present, however, it's far better to ask them. Carefully planned surveys turn self-reports into accurate predictions about voting behavior.

4. Self-Reports May Have Heuristic Value

The deception study suggested certain relationships that can be tested experimentally. Future researchers might study how reactions to deceptions, in terms of the five themes, are influenced by who deceives, who is deceived, and the nature of the deception.

Barker described in some detail the heuristic value of self-report information. He wrote, "It seems to me that in many areas, the only intelligent way to begin is to spend some time in observing life situations and in interviewing people about the topic under consideration *from their own internal frame of reference*." (his italics)

Barker told of a student, Gunkel, who spent weeks having informal conversations and tape recording interviews on the topic of loneliness: asking people to describe and compare the experiences of being lonely, being alone, being with a group, being with a close friend, etc. Then she collected essays on lonely and not-lonely experiences. She constructed adjective check lists from the interviews and administered them to new subjects. Barker and Gunkel spent several months, and developed and informally tested several hypotheses, before beginning formal procedures.

5. Self-Reports Provide a Distinct Kind of Measurement

Operational definitions are essential for measuring scientific concepts, but no single, concrete, operational definition can measure an abstract concept in all its richness. Sometimes, as in the hypothetical amphetamine/bar pressing example of chapter 4, single measures are misleading. That's why methodologists urge scientists to measure concepts in more than one way. For the same reason, the measures should be distinctive. Reports of inner states tap a different dimension from the behavioral and physiological measures used by traditional researchers; this increases their value. More than that, many researchers collect data so they can make inferences about processes that are not directly observable, such as anxiety, self-esteem, love, and hostility. Self-reports about these inner processes can yield stronger inferences.

6. Self-Reports May Reveal the Key Features of a Situation
Lieberman (1979) and Richardson (1984) gave several examples in which subjects' reports of their thoughts and strategies during an experiment helped researchers identify important features. At the end of a study, subjects' reports can help the experimenter evaluate whether manipulation of the independent variable induced the intended state.

Methods of Collecting Self-Report Data

Each method given below is described in detail in Richardson's book.

The Focused Interview Method

People who have had unusual experiences are asked to describe them. The interviewers are non-directive, listening in a friendly manner without expressing opinions or arguing. They talk if necessary to relieve the subject's anxiety, to give approval when an experience has been described successfully (even if the description does not touch on the interviewer's interests), and to direct the subject's attention to possibly overlooked aspects of the experience.

Interviewers may videotape subjects as they have an experience, and also record various physiological measures; played back before the subjects are asked to introspect, the tapes may improve the specificity of memories. Less elaborate reconstructions of the original experiences may also help.

The Association Methods

People's spontaneous thoughts can be recorded and analyzed for whatever themes interest the researcher. Richardson cited a study on sadness by Cicchini (1976), in which thought samples were scored for negative view of self, of the future, and of external situations and experience. Richardson quoted the instructions, which make the subjects' task clear:

Shortly, I'm going to ask you to do a bit of talking. Your task will be to tell me something about your thoughts—I'll explain what I mean. Soon I'm going to ask you to think out loud. Although this task may sound hard, it's not—all *you* have to do is say whatever comes into your head. So when I say "begin," I would like you to think out loud and say whatever comes into your head and I will tell you to stop in about five minutes. Now, if something comes into your head that you'd rather not report, you're quite free to withhold it. Just let me know by saying there was something in your head that you don't wish to report. Of course anything you say will be kept confidential and will not be identified with you, personally. Okay. So what you are to do is to think out loud, or report whatever comes into your head, and I will tell you when to stop. Do you have any questions now, before we begin, because I'd rather not answer any questions during your talking period? (Pause) Okay. Just relax, close your eyes, if you want, and say whatever comes into your head. Begin.

Another association method requires subjects to focus their attention on a type of experience and speak or write down all the ideas that occur to them as they free associate to it. The subjects then analyze reports from others who have done the same thing, trying to extract common themes. Then they free associate again, followed by another period of analysis. The cycle may be repeated several times.

The Phenomenological Method

The goal of this method is to describe the essence of an experience in as few words as possible. Fischer (1975), trying to describe the essence of "being in privacy," made detailed notes of each instance she could remember when she was in that state. From the notes, she identified core features. Then she asked other people to provide instances of their own experiences of privacy, and she discussed all the submissions with colleagues. Her definition follows:

The watching self and world fade away, along with geometric space, clock time, and other contingencies, leaving an intensified relationship with the subject of consciousness lived in a flowing Now. The relationship is toned by a sense of at-homeness or familiarity, and its style is one of relative openness to or wonder at the object's variable nature.

Science needs descriptive studies, and Fischer described how her subjects defined the experience of "being in private." But whatever feelings her summary statement may evoke, it is unreliable: other investigators, even if they used identical procedures, would probably find other core features (dominant themes) to being in private. In their present form, Fischer's words are not operational, can't be tested, and do not seem to lead to any falsifiable hypotheses. So they are not part of science.

Imagine a subject who reported that "being in private" means having a sharp pain in the stomach. Fischer probably would have done one of two things: (a) tried to persuade readers that having a sharp pain in the stomach really means the same as "an intensified relationship with the subject of consciousness lived in a flowing Now;" or, more likely, (b) concluded that the subject didn't understand what being in private means. But *b* suggests that her criterion for deciding whether a subject understands is that the subject's report corresponds to her own. Her view is thus protected from falsification.

The difference between science and nonscience is at issue here, so imagine a bit further. Suppose that a reader of the American Heritage Dictionary, second college edition, reported that being in private means being "secluded from the sight, presence, or intrusion of others." Fischer would probably have responded with one of the two strategies of the previous paragraph. But the first implies a criterion for deciding if the subject's words can be fairly translated to mean the same as Fischer's; none is available. And use of the second strategy might result in arguments about who "really" understands what "being in private" means. In neither case would there be an agreed-upon method, *based on empirical evidence,* for resolving the controversy.

The Critical Incident Method

The researcher describes an experience, such as a particular emotional state, in enough detail so that subjects have no doubt what it is. Then he asks subjects questions about times when they felt the same emotion. Richardson described a study in which McKellar (1949), after recording and analyzing instances over a 47 day period in which he became angry, formulated tentative hypotheses about anger behavior. He then asked 120 subjects the following questions:

1. Describe in your own words the most annoying thing that has happened to you in the last day (or two days)
2. How did you feel when this occurred?
3. Did you blame anybody? If so, whom?
4. What did you actually do as a result of the situation?
5. Describe in your own words one other annoying thing that has happened to you recently.
6. State whether this was more, less, or equally annoying in comparison with the previous situation described.

Questions 2, 3, and 4 were repeated for the second situation. McKellar's hypotheses, such as that about half of all incidents involving the experience of anger would be unaccompanied by any expression of verbal or physical aggression, were confirmed.

Summary

In the early 1900s, many psychologists asked subjects to contemplate their own thoughts, feelings, and emotions. The subjects' reports were assumed to provide direct access to the workings of their minds.

John Watson, the founder of behaviorism, criticized introspection as being unobjective and unverifiable. The behaviorist movement became the dominant force in psychology, and studies using self-reports became rare. They remain underused today and are not discussed in traditional research methods textbooks. But self-reports can be studied scientifically; and under appropriate circumstances, they generate falsifiable hypotheses.

Self-report methods have many uses:

- They may provide unique insights into phenomena.
- They are the only source of data for topics such as content of dreams.
- They often improve predictions about behavior.
- They have heuristic value.
- They provide an additional, distinctive way to measure concepts.
- They may indicate key features of a situation.

Several different self-report methods are available.

Key Terms

Association Method Subjects are asked to verbalize their spontaneous thoughts about an experience.

Construct Veracity A method for estimating the accuracy of an introspective report—the report is compared with predictions based on theory.

Content Veracity A method for estimating the accuracy of an introspective report—versions of the report given at different times are compared.

Criterion Veracity A method for estimating the accuracy of an introspective report—the report is compared with independently verifiable events.

Critical Incident Method Subjects answer questions about times during which they had a certain experience.

Falsifiability The philosophical position that a system is a science only if its hypotheses can clash, should the facts fall that way, with evidence.

Focused Interview An interviewer encourages subjects to relax and, in nonthreatening circumstances, describe their experiences.

Introspection The investigation of consciousness by systematic, detached observation of one's mental experiences.

Phenomenological Method An investigator collects descriptions of a particular type of experience from a diverse group of subjects, then tries to identify the essential qualities that distinguish it from all other types of experiences.

CHAPTER 22

Testing Data for Statistical Significance

By the time you finish reading this chapter, you should be able to answer the following questions:

On what grounds have critics argued against the use of tests of statistical significance for evaluating the results of experiments?

One critic wrote, "If rejection of the null hypothesis were the real intention in psychological experiments, there usually would be no need to gather data." How can that be?

Why do critics contend that much more than 5 percent of published articles contain Type I errors?

Increased experimental sophistication has opposite effects on theory testing in physics and psychology. Why?

What three types of alternatives are there to significance testing?

How can the point-biserial correlation coefficient be used to analyze the results of two-group experiments?

How can experiments be designed to yield equations that relate IVs and DVs?

Several methods of testing for statistical significance have been presented throughout this book; they are standard for research methods books and courses. But for more than a quarter of a century, methodologists have argued that reliance on statistical significance testing is a mistake. For example, Paul Meehl (1946), on the occasion of receiving a Distinguished Scientist Award from the American Psychological Association, said: "I believe that the almost universal reliance on merely refuting the null hypothesis as the standard method for corroborating substantive theories . . . is a terrible mistake, is basically unsound, poor scientific strategy, and one of the worst things that ever happened in the history of psychology." (The null hypothesis is the hypothesis that no relationship exists between variables. Scientists use tests of statistical significance in the hope they will show the null hypothesis to be wrong.)

Below are quotes from other eminent methodologists. Taken cumulatively, they may persuade readers that a problem exists. Following the arguments against significance testing, I'll discuss three types of solutions.

William Rozeboom (1960): The thesis to be advanced is that despite the awesome pre-eminence this method has attained in our experimental journals and textbooks of applied statistics, it is based upon a fundamental misunderstanding of the nature of rational inference, and is seldom if ever appropriate to the aims of scientific research. This is not a particularly original view—traditional null-hypothesis procedure has already been superseded in modern statistical theory by a variety of more satisfactory inferential techniques. But the perceptual defenses of psychologists are particularly efficient when dealing with matters of methodology, and so the statistical folkways of a more primitive past continue to dominate the local scene.

David Lykken (1968): The moral of this story is that the finding of statistical significance is perhaps the least important attribute of a good experiment; it is *never* a sufficient condition for concluding that a theory has been corroborated, that a useful empirical fact has been established with reasonable confidence—or that an experimental report ought to be published.

David Bakan (1966): The major point of this paper is that the test of significance does not provide the information concerning psychological phenomena characteristically attributed to it; and that, furthermore, a great deal of mischief has been associated with its use.

Jum Nunnally (1960): "The point of view taken here is that if the null hypothesis is not rejected, it usually is because the N is too small. If enough data are gathered, the hypothesis will generally be rejected. If rejection of the null hypothesis were the real intention in psychological experiments, there usually would be no need to gather data."

Nunnally's position summarizes the relationship discussed on pages 205–208 between power, sample size, size of effect, and significance level. His point is that with a significance level set at .05 (or .01 or any other value), and an effect size that deviates even minutely from zero, use of a large enough sample ensures that power (the likelihood that the null hypothesis will be rejected) will approach one. And as Bakan demonstrated, virtually all nonrandomly selected groups differ somewhat. He collected data on 60,000 subjects and divided the information in arbitrary ways: east vs. west of the Mississippi River, Maine vs. the rest of the country, North vs. South, etc. Every comparison was statistically significant.

The conventional setting for a Type I error is .05, which implies that scientific articles incorrectly reject the null hypothesis only 5 percent of the time. McNemar (1960) argued that publishing practices ensure that the actual error rate is much higher. The reasons are (a) journal editors reject papers without significant results; (b) scientists, knowing editorial policies, don't submit papers with nonsignificant results; (c) scientists select out their significant findings for inclusion in reports; and (d) theorists discard data that don't fit their theories.

Please review the discussion of power on pages 205–206. I noted there that if a coin with a .60 bias toward heads is tossed six times, statistical significance is reached only if the coin lands on heads each time. That probability is .047. Now suppose that thousands of people made six tosses, but journals published only those cases in which six heads had occurred. Readers would be misled about the strength of the bias.

Cohen (1962) analyzed a year's worth of articles in the *Journal of Abnormal and Social Psychology* and concluded that such a situation exists. He reported that the statistical tests had such weak power that they should rarely have led to significant results even if there had been strong relationships between variables.

The Effects of Statistical Significance Testing on Theory

The formulation of a useful theory probably stands at the pinnacle of scientific achievements. Consider:
Pasteur: "Only theory can bring forth and develop the spirit of invention."
Newton: "No great discovery is ever made without a bold guess."
Poincare: "Science is built up with facts, as a house is with stones. But a collection of facts is no more a science than a heap of stones is a house."
Conant: "The history of science demonstrates beyond doubt that the really revolutionary and significant advances come not from empiricism but from new theories."

Popper (1962) urged scientists to boldly propose theories, but then to test them as ruthlessly as possible. Because of the central role of theories in the well-developed sciences, research to test them often has enduring significance. But partly as a result of the popular statistical models, social science theories are poorly developed; and research to test the theories, based largely on the statistical models, is often trivial and quickly forgotten. Meehl (1967) argued persuasively that a crucial difference separates theory testing in physics and psychology. He noted that improvements in technology increase the likelihood that incorrect physics theories will be rejected; but improvements have the reverse effect in the social sciences. In his words,

> In the physical sciences, the usual result of an improvement in experimental design, instrumentation, or numerical mass of data, is to increase the difficulty of the 'observational hurdle' which the physical theory of interest must successfully surmount; whereas, in psychology and some of the allied behavioral sciences, the usual effect of such an improvement in experimental precision is to provide an easier hurdle for the theory to surmount.

The reason is that physical scientists test mathematical relationships between variables, so that specification of the independent variable leads to a narrow range of acceptable values for the dependent variable. As experimental precision increases, the range of DV values that support the theory is reduced.

In most psychology experiments, however, predictions are of the sort that a group treated one way will differ from a group treated otherwise. But since all treatments have some effect (as discussed previously), the null hypothesis (prediction that the groups will be the same) is always false. As a result, increased precision increases the likelihood that a real though trivial difference will be detected. And theories that predict differences in a particular direction, even if they are without foundation, will be confirmed 50 percent of the time if the measuring device is sensitive enough. As a result, to quote Meehl again on how psychologists evaluate theories, "I am saying that the whole business is so radically defective as to be scientifically almost pointless."

Dar (1987) pursued the matter further. He argued that the mere establishment of differences between groups does little to strengthen a theory, because a huge number of plausible alternatives may account for them. On the other hand, failure to find a predicted difference does not mean that a theory must be abandoned; many factors can be imagined that would diminish small differences. But if it is so easy to come up with competing explanations for successful predictions and with after-the-fact explanations for failing ones, why should the theory be taken seriously?

Dar wrote that statistical significance testing has *replaced* good theory building in psychology. He gave three reasons:

1. Graduate students in psychology receive considerable training in statistics, little in philosophy of science or principles of scientific methodology. So when they start research on their own, the graduates feel satisfied with results significant at the .05 level: the criterion for both successful predictions and journal publications. They feel no pressure to develop theories with high internal consistency or explanatory power.

2. Many psychologists have the illusion that statistically significant results provide strong proof that the research hypothesis tested is true. But it is an illusion. See, for example, Box 4.2 on page 55.
3. "High-tech" statistical tests create a second illusion: of scientific respectability. Then, sophisticated tests with computations to several decimal places are used as substitutes for strong theories.

Solutions

Three types of solutions are discussed below: measuring effect size, finding confidence limits, and seeking functional relationships.

Measure Effect Size In Addition to Significance Level

The size of effect of an IV on a DV should be computed routinely and conveys more useful information than does mere significance testing. Eta, defined on page 212, measures effect size. When only two groups are involved, the point-biserial correlation coefficient (see p. 194) can measure both statistical significance and effect size with a single set of calculations. The point-biserial is used when one variable is measured on an interval scale and the other is dichotomous. In a two-group experiment, subjects are assigned to dichotomous groups (experimental and control); so if an interval scale is used for the dependent variable scores, the point-biserial is appropriate. This application was advocated by Nunnally (1960).

1. Compute a point-biserial correlation coefficient from experiment 1, Table 14.4, page 210. Hint: Assign all the subjects in group A to category 1 and all in group B to category 0—or the other way around—it makes no difference as long as you're consistent. Is the correlation significant at the .05 level?

2. In chapter 13, it was stated that correlation does not imply causation. Does correlation imply causation when the point-biserial is used as it was in question 1?

Find Confidence Limits

Significance tests provide unambiguous and rigid criteria for decision-making, for example, if the significance level is .05 act as though the treatment has had an effect, and if the level is .06 act as if it hasn't. Alternatively, researchers can find confidence limits for their data and present results in the following form: "The group that received training averaged 16.4 trials and the control group 24.7, for a difference of 8.3. We can be 95 percent sure that training in comparable subjects will produce a difference between 6.0 and 10.6; we can be 99 percent sure that training will produce a difference between 4.2 and 12.4." For correlational data, rather than concluding that the correlation equals .4, significant at the $p = .05$ level, the researcher would say that she is 95 percent sure that the true correlation is between .25 and .55.

In the next section I show how to calculate 95 percent and 99 percent confidence limits for proportions, experiments with two independent groups, and correlational data. Statistics books show how to calculate limits for other data.

Calculating Confidence Limits

Confidence Limits for Proportions

Suppose that 200 out of 1,000 randomly selected people are left-handed. Thus, p (proportion of left-handers in the sample) $= .2$ and q (proportion of right-handers) $= .8$. To calculate 95 percent confidence limits for the true proportion of left-handers in the population, use the formula

$$\text{true proportion} = p \pm 1.96\sqrt{pq/N} + 1/2N$$
$$= .2 \pm 1.96\sqrt{(.2)(.8)/1000} + 1/2000$$
$$= .2 \pm 1.96\sqrt{.00016} + .0005$$
$$= .2 \pm .025$$

So the investigator could be 95 percent sure that the proportion of left-handers in the population is between 0.175 and 0.225. (For 99 percent limits, replace 1.96 by 2.58.)

Confidence Limits for Experiments with Two Independent Groups

When The Number of Subjects in Each Group (N) is Less than 15 The following directions are for computing confidence limits for groups of equal size. For unequal groups, consult a statistics book. Consider the data from Table 14.4, experiment 1, on page 210. The difference between the group means is 3.0. To get the 95 percent confidence limits for the true difference:

1. Calculate the variances for each group. The sum of squares of group A $= 808 - (80^2/8) = 8$; so the variance of A $= SS_A/(N - 1) = 8/7 = 1.14$. And $SS_B = 396 - (56^2/8) = 4$; so the variance of B $= 4/7 = .57$.
2. Find $\sqrt{(\text{variance of A} + \text{variance of B})/N}$
$= \sqrt{(1.14 + .57)/8} = .46$. Call this quantity M. (Note that N is not the total number of subjects, but the number per group.)
3. Turn to appendix C and find the value for $2N - 2$ degrees of freedom. For 14 df, the value is 2.145.

4. The 95 percent confidence limits are given by the difference between the means ± 2.145M.

$$= 3 \pm 2.145 (.46)$$

$$= 3 \pm .99$$

So the investigator could be 95 percent sure that the true difference between the means is between 2.01 and 3.99.

5. For 14 df, the 99 percent confidence limits are given by the difference between the means ± 2.977.

When the Number of Subjects in Each Group is 15 or More

1. Calculate the variances for each group.
2. Find $\sqrt{(\text{variance of group A} + \text{variance of group B})/N}$. Call this quantity M.
3. The 95 percent confidence limits are given by the difference between the means ± 1.96M. For 99 percent limits, substitute 2.58 for 1.96.

Example

Suppose that 60 subjects are randomly assigned to experimental and control groups. The means and variances are 16.4 and 36.0 for the experimentals and 13.8 and 38.4 for the controls. The 95 percent confidence limits are given by $2.6 \pm \sqrt{(36.0 + 38.4/30} \times (1.96) - 2.6 \pm (1.58)(1.96) = 2.60 \pm 3.10$

Confidence Limits for Correlations

Suppose, as was found from Table 13.3, page 187, that $r = .704$ with an N of six pairs of scores.

1. Use appendix F to convert r to z. From the appendix, $z = 0.875$ (interpolation is necessary to get the precise z value). (In actual practice, N should be at least 10 for the Fisher conversion.)
2. Find $1/\sqrt{N-3} = 1/1.73 = 0.58$
3. The 95 percent confidence limits for z are $.875 \pm 1.96(0.58) = 0.262$ to 2.012.
4. Use appendix F to convert the zs to rs. The 95 percent confidence limits for r are .256 to .965.
5. For 99 percent limits, replace 1.96 by 2.58.

Seek Functional Relationships

Traditional analyses of two-group experiments don't indicate the functional relationship between IV and DV. That is, they show only that the IV has or hasn't had an effect, but not how the DV changes throughout the range of IV values. Equations that show how one variable changes with another are a staple in physics ($E = mc^2$, $F = ma$), and in this respect at least, psychologists should try to emulate physicists. To quote Meehl again, "It is always more valuable to show approximate agreement of observations with a theoretically predicted numerical point value, rank order, or function form, than it is to compute a 'precise probability' that something merely differs from something else."

A method for expressing functional relationships between variables is available for social scientists to try: Use a great many values of the IV, over a wide range. Randomly assign a different value to each subject (or, to obtain more stable measures, use two or three subjects for each value), and submit the data to regression and correlation analyses. The regression equation will give estimates of the effects of all potential values of the IV between the extremes of those tested. The correlation will indicate effect size.

Experimenters who use the traditional two-independent group design may lack sufficient information to choose optimal levels of an IV. Multiple-group designs, with each group receiving a different value of the IV, lessen the problem but require more subjects; and subjects may be scarce or expensive or take a great deal of time to run. So experimenters rarely administer more than five levels of an IV. The suggested approach would greatly reduce the problem of sub-optimal IV levels. It would be useful for people involved in applied research, because they would be able to locate the precise point at which increases in IV intensity begin to have diminishing returns. And it would help theorists, because they could both formulate and test their theories more precisely. Many variations are possible, such as single-subject experiments, concurrent administration of two or more IVs, and curvilinear analyses. Consult a good statistics book for procedures.

As a test of the workability of the regression method, I did a single-subject experiment on myself. Every day, for 72 days, I took my pulse for 30 seconds, then ran in place for a randomly determined period and immediately afterwards took another 30 second pulse. The running periods ranged from 5 to 120 seconds, in 5 second increments, and were assigned 3 times each over the course of the experiment. The DV scores, given in Table 22.1, were changes in pulse rate from before to after the running. The numbers in parentheses are transformed scores—because all the running periods were multiples of 5, I divided by 5 to make the numbers easier to work with.

The regression equation, calculated according to the procedures described on pages 183–184, is $Y = 7.87 - 0.46X$. The correlation is .81, significant at the .05 level. It shows that almost 66 percent of the variance in change of pulse rate is attributable to the length of the running period. Note: Refer to the answer to question 2 to see why results can be interpreted causally.

Table 22.1 *Effects of Jogging in Place on Pulse Rate*

Seconds of Running	Change in Pulse Rate (average over three trials)
5 (1)	3.2
10 (2)	6.2
15 (3)	11.2
20 (4)	12.0
25 (5)	10.6
30 (6)	13.3
35 (7)	13.2
40 (8)	15.1
45 (9)	11.1
50 (10)	12.3
55 (11)	12.5
60 (12)	13.5
65 (13)	9.9
70 (14)	11.9
75 (15)	15.2
80 (16)	12.6
85 (17)	17.5
90 (18)	19.3
95 (19)	17.2
100 (20)	18.7
105 (21)	17.3
110 (22)	20.8
115 (23)	15.5
120 (24)	16.7

Summary

Critics of traditional statistical significance testing contend the following:

1. Significance tests provide no information about sizes of effects. But effect size is more important than statistical significance.
2. If large enough samples are used, statistical significance is guaranteed; thus, it is trivial.
3. Partly because of significance testing, theories in social science lead to predictions of the form "Group A will outperform group B." Such statements are weak and cannot be severely tested.

Three types of alternatives were offered to significance testing: calculating effect sizes, calculating confidence limits, and using regression equations to express functional relationships between variables.

The point-biserial correlation coefficient can be used to test the statistical significance of two-group experiments. An advantage of the procedure is that effect size is found at the same time. All members of one group are given X scores of 0, and members of the other group receive X scores of 1. The DV scores are the Ys.

If each subject is assigned a different value of the IV (or there are two or three subjects per IV), experimenters can use regression and correlational techniques to establish a functional relationship between IV and DV over a wide range of values.

Answers

1. Assign all the subjects in group A an X value of 0, all the subjects in B a value of 1. Then redo the table, as follows:

X	Y	Y^2
0	10	100
0	11	121
0	9	81
0	12	144
0	9	81
0	9	81
0	10	100
0	10	100
1	7	49
1	6	36
1	8	64
1	8	64
1	6	36
1	7	49
1	7	49
1	7	49
		1204

Compute the point-biserial according to the procedures given on page 194. It equals .87. To test its significance, compute

$$t = r\sqrt{\frac{N-2}{1-r^2}}$$

$$= .87\sqrt{57.6}$$

$$= 6.57$$

From appendix C, with 14 df, the critical value for t at the .05 level is 2.145. So, the results are significant.

2. Yes, because subjects were randomly assigned to groups.

PART VI

Drawing Conclusions

CHAPTER 23

Writing Research Reports

By the time you finish reading this chapter, you should be able to answer the following questions:

What did physicist John Ziman mean when he called science public knowledge?

How does scientific writing differ from other forms of writing?

What structure should all original research articles have?

"Work, finish, publish."

Michael Faraday, chemist
(1791–1867)

Theoretical physicist John Ziman defined science as "public knowledge." Ziman wrote (1968):

> An investigation is by no means completed when the last pointer reading has been noted down, the last computation printed out and agreement between theory and experiment confirmed to the umpteenth decimal place. The form in which it is presented to the scientific community, the "paper" in which it is first reported, the subsequent criticisms and citations from other authors and the eventual place that it occupies in the minds of a subsequent generation—these are all quite as much part of its life as the germ of the idea from which it originated or the carefully designed apparatus in which the hypothesis was tested and found to be good.

Whether or not you agree with Ziman, be aware that rewards to scientists depend greatly on others' evaluations of their written reports. So do rewards to students in science research classes.

All Good Writing Shares Certain Characteristics

Good writers have interesting things to say and say them in interesting ways. In addition, good writers are attentive to various problems that can impair the quality of communication. Some of the most troublesome are organized below as a checklist.

Checklist for Writing

- Is the writing free of errors of spelling, punctuation, and capitalization?
- Is the writing well-organized?

    Put everything on the same subject in the same place. Discuss a topic thoroughly and then don't bring it up again. (Each scientific journal has a standard format, which ensures that published articles are organized.)

- Have you shown relationships clearly?

    Link sentences and paragraphs logically. Make sure that each sentence clearly explains, amplifies, or limits the statement it follows; or clearly introduces a new subject. Make liberal use of transitional words and phrases (first, soon afterward, then, moreover, although, nevertheless, but, on the other hand).

- Do subjects and verbs agree?

    Note that the words data, criteria, and phenomena are plural, the singular forms being datum, criterion, and phenomenon. So, write "The data are accurate," not "The data is accurate."

- Have you used tenses correctly?

    Use the present tense to define terms and state laws and hypotheses.

"Aggression is defined as . . ."
"Frustration leads to aggression."

    Use the past tense when referring to work already done, including your own data collection procedures.

"I injected each rat with . . ."

- Have you used the active voice? (The active voice, in which the subject of a sentence performs the action of the verb, is preferable in most cases to the passive voice, in which the subject receives the action.)

NOT: Each rat was injected with . . .
BUT: I injected each rat with . . .

- Have you used precise words?

    Strive for as much precision in your writing as you'd demand from a friend who had arranged a blind date for you. You'd want specifics, not words like "nice," "interesting," and "good personality."

- Have you avoided sexist language? The sentence "Each student must inject his own rat," is sexist. To avoid sexism, try the strategies below.

Rephrase sentences to use plural pronouns.

    Students must inject their own rats.

Eliminate pronouns.

    Each student must inject a rat.

Use "we" or "you."

    We must inject our own rats.
    You must inject your own rat.

Use "he or she" or "his or her."

    Each student must inject his or her own rat.

Alternate pronouns.

    Sometimes use "he" and "him" and sometimes use "she" and "her."

- Have you placed all modifiers next to the word or phrase they modify? (A word, phrase, or clause that gives information about another part of the sentence is called a modifier: in the phrase "young boy," "young" modifies "boy;" in "for the most part friendly," "for the most part" modifies "friendly.")

    The word "only" can modify the sentence "I added the value of x in the equation." But, as is true of all modifiers, where it is placed determines its meaning.

"Only I added . . ."	(Neither you nor she added.)
"I only added . . ."	(I didn't subtract or multiply.)
". . . only the value . . ."	(I didn't add the cube root of x.)
". . . the only value . . ."	(There was just one value of x.)
". . . value of only x . . ."	(I didn't add the value of y.)
". . . x only in the . . ."	(I didn't add x in the grocery list.)
". . . only equation . . ."	(There was only one equation.)

- When you use words like "it," "this," "which," "that," and "these," are the referents clear?

    Two interpretations are possible for the sentence "We warned her that the analysis would take longer than she thought, but it didn't bother her."

    The warning didn't bother her.

    The time spent on analysis didn't bother her.

    Possible misinterpretation can be corrected by replacing "it" with the proper referent. Write either

    . . . , but the warning . . .

    or

    . . . , but the time she spent. . . .

Qualities Unique to Science Writing

There are a few important differences between scientific and other forms of writing.

1. Scientific articles must be concise. When applied to a novel, the adjective 'sprawling' is generally favorable. But journals have tight budgets, and space is expensive. Also, good scientists want to get through their professional reading rapidly; they prefer solving problems to reading about the successful solutions of others.
2. Scientists use a technical vocabulary, which makes it easier for them to be concise.
3. Scientists try to write clearly and without ambiguity. Poets often write on several levels and make obscure references; and James Joyce, a magnificent novelist, would be nobody's nominee for paragon of clarity. Referring to his novel *Ulysses,* Joyce said, "I've put in so many enigmas and puzzles that it will keep the professors busy for centuries arguing over what I meant, and that's the only way of insuring one's immortality." But scientists must be clear.
4. Related to clarity are detail and specificity. Scientists insist on both, so they define key terms unambiguously. They operationally define their independent and dependent variables instead of using vague terms like "heavy," and "anxious;" they give details.

In both my Research Methods and Science Writing classes, I ask students to break up into groups of five or six and define among themselves a countable set of objects. These can be chairs in the room, doors in the hallway, bathrooms, light fixtures, and so forth. Each group is told to define carefully so that all members are certain about what is to be counted. Then they count individually, with no further talking. In my experience, in only about 5 percent of the groups do all the members give identical answers. Some people simply miscount and report, say, 47 chairs instead of 48. Errors of this sort point to the importance of always double-checking data and calculations. The second type of error is more frequent and caused by failure to define with enough specificity. While everybody else in a group counts only classroom doors, some maverick lumps them with bathroom and hallway doors.

Methods sections of scientific articles are much more intricate than simple counting exercises. In addition, scientists don't meet in groups to clarify problems before writing; and they don't typically meet afterwards to find reasons for discrepancies. So, if science is to be "public knowledge," elaborate care must be taken.

5. In contrast to poets and novelists, who try to convey feelings and impressions, scientists write about facts and deductions and inductions from the facts. The facts must be verifiable. Ogburn (1947) gave as an example the statement, "The population of India is increasing to ominous proportions." There is no common measure of "ominous," so the statement cannot be verified. Ogburn's example of a verifiable statement is, "In 1940, India's population increased by 5,000,000. At this rate a population equal to the total population of the United States in 1890 would be added in a single decade to her population."

6. Science writers should not select facts in a biased way to create particular impressions. Ogburn gave two sets of true statements that could be presented about Neanderthals. Both are slanted. They give very different impressions, neither one fair. Science writers should avoid slanting.

   1. Thus the ordinary attitudes characteristic of Homo neanderthalensis would be quite different from our own and quite ungainly. The heavy head, the enormous development of the face, and the backward position of the foramen magnum, through which the spinal cord connects with the brain, would tend to throw the upper part of the body forward and this tendency, with lesser curvature of the neck, the heavy shoulders, and the flattened form of the head, would give this portion of the body a more or less anthropoid aspect.
   2. Unlike the anthropoids the limbs of Neanderthal man were short in comparison with the trunk. His chin was stronger than modern man's, which is decadent with its indentation below the teeth. His vision was

better protected than that of present-day man by virtue of the large brow which acted as an eyeshade over his deep-set eyes, of survival value to a hunter. His broad shoulders and big chest would have made him welcome on a college athletic team. His brain case averaged 1,554 cubic centimeters and was bigger than the brain case of the average college student, which measures about 1,500 cc.

Writing for Scientific Journals

All good writers analyze their audiences. Scientists must choose an appropriate forum for their research proposals and reports from among thousands of publications. *Psychological Abstracts* alone reviews about 1,000 journals that differ in subject matter, style of referencing, acceptable abbreviations, preferred methodology, and editorial standards. So experienced scientists generally conceive their research with specific journals in mind. A psychologist who tests the effects of a new drug on aggression in rats will probably fare better if she submits her findings to the *Journal of Experimental Psychology* or *Psychopharmacology* than to *Psychoanalytic Quarterly*.

Many social science journals follow guidelines set forth in the American Psychological Association's *Publication Manual,* obtainable for a small fee from the APA, 1200 17th Street, N.W., Washington, D.C., 20036. The guidelines specify that each manuscript shall be comprised of several distinct parts, and these are described below in standard journal order. Some journals, like *Science,* do not name distinct parts but require similar organization. (Much of the material below was abstracted from the APA's *Publication Manual.*)

The Parts of a Manuscript

Title   The title is the first and sometimes only part of a journal article that readers see. Reference works such as *Index Medicus* and some computer retrieval systems list only by titles, so articles with uninformative ones may be overlooked. To help indexers, include names of major variables in the title.

Geralyn Rodriguez Clucas' excellent paper, which she submitted for her experimental psychology class at Cal State, Hayward, is reprinted as appendix B. Note that her title contains the key terms "Effects of Knowledge of Results," "Delay of Knowledge," and "Muller-Lyer Illusion."

Abstract   The abstract is a brief (about 100 to 175 words), self-contained summary of the research problem, subjects, methods, results, and conclusions. Do not put information or conclusions in the abstract that are not in the main body of the paper. Abstracts, like titles, are used for indexing and information retrieval.

Geralyn's abstract tells who her subjects were and how they were chosen. She describes the procedures used with each of her three groups and how the dependent variable was measured. She reports her conclusions.

Introduction The introduction allows the author to present a rationale for the study. Describe the research problem and how you designed the study to answer it. Show how your work relates to what is already known about the problem, which generally involves citing pertinent references. Close your introduction by defining your variables, formally stating your hypothesis, and explaining why you anticipate that your hypothesis will be supported.

Geralyn's introduction cites previous research showing that knowledge of results (KR) improves performance in many situations. She then indicates her two purposes in the present research—to examine how KR affects performance on a task involving a powerful visual illusion, and the effects of delaying KR. She concludes with the hypotheses that KR would improve performance and delay of KR would make little difference.

Method The method section describes, in details that must often seem excessive to casual readers, how the research was conducted. The detail is necessary: Your goal should be to enable readers to replicate your study if they so desire. Use labeled subsections in the method section for describing subjects, apparatus, and procedures; use additional subsections if necessary to describe complex material such as a specialized design.

Subjects

State who the subjects were and how you selected them. Include relevant material about age, gender, education, and so forth. Tell how many subjects failed to complete the study and their reasons for dropping out. If you used animals, give the same type of information and also include genus, species, strain, and supplier.

Apparatus

Describe the materials used. Identify specialized equipment by the manufacturer's name and the model number.

Procedures

Summarize each step in the execution of the research. Include a summary of instructions to the participants, and tell how you formed groups and manipulated conditions. Describe special control features of the design.

Geralyn should have given more information about her subjects: the numbers of men and women, how they came to volunteer, and whether they were paid. She does a fine job with the apparatus (materials) and procedure subsections and includes a useful design subsection.

Results The results should summarize the data and how you treated them. Mention all relevant results including those that did not support your hypothesis, but don't discuss the implications. In reporting statistical tests, include information on the obtained magnitude of the test, the probability level, and the direction of the effect. Assume that your reader has professional knowledge of statistics. Use figures and tables as necessary.

Geralyn presents a clear table of her results. Under the heading *Variable errors, Trials 9–24,* she interprets some of the results. She should have presented that material in the discussion section.

Discussion   Open the discussion section with a statement on the support or nonsupport of your hypothesis. Show the status of the hypothesis in light of the results. Examine, interpret, qualify, and draw inferences from the results. You can speculate, but briefly, and only if the speculation is related closely to your data or follows logically from theory.

Geralyn related her findings to those from previous research. She suggested an alternative hypothesis for her findings.

References   List all the references that you cited in the text, and do not list any others.

Rosnow and Rosnow (1986) wrote a short guide for writers of psychology papers. They suggested several questions to ask while planning various sections. These, along with a few additional questions from the APA guidelines, have been compiled as Table 23.1.

Table 23.1   *Questions to Ask When Planning the Different Sections of a Research Paper*

Section	Questions
Abstract	What was the objective or purpose of the research study?
	What principal method was used?
	Who were the research participants?
	What were the major results?
	What was the central conclusion?
Introduction	What was the purpose of the study?
	What is the rationale or logical link between the problem and the research design?
	What are the theoretical implications of the study and its relationships to previous work in the area?
	What terms need to be defined?
	How does the study build on, or derive from, other studies?
	What was the hypothesis or expectation?
	What results do I expect, and why do I expect them?
Results	What did I find?
	How can I say what I found in a careful, detailed way?
	Is what I am planning to say precise and to the point?
	Am I being overly or misleadingly exact?
	Will what I have said be clear to the reader?
	Have I left out anything of importance?
Discussion	What have I contributed here?
	How do my results relate to the purpose of this study?
	Did I make any serendipitous findings of interest?
	Are there larger implications in these findings?
	How valid and generalizable are my findings?

Summary

To be successful within the scientific community, researchers must write up their results. Good science writing, like all good writing, is free of grammatical, spelling, and punctuation errors; is well-organized; and shows relationships clearly. Science writing, unlike some other forms, is concise, clear, detailed, and specific. Scientists should write in an unbiased way about verifiable facts and deductions and inductions from the facts.

Many scientific manuscripts are comprised of the following parts, in the order given below:

Title: The title should include the names of all major variables.

Abstract: The abstract is a 100 to 175 word summary.

Introduction: This provides background and rationale for the study.

Methods: This tells how the subjects were selected, who they are, and whether or not any dropped out before the end of the study; it describes materials and equipment; and it summarizes each step in executing the research.

Results: The results section summarizes the data and tells how they were analyzed.

Discussion: This section is for examining, interpreting, qualifying, and drawing inferences from the results.

References: This is a listing of all references cited in the articles.

CHAPTER 24

Making Sense of Groups of Studies

By the time you finish reading this chapter, you should be able to answer the following questions:

Why have traditional reviews of the scientific literature been inefficient?

What are the key steps in a systematic review?

How can conflicting results between studies be used to generate hypotheses for additional research?

How can the results of studies be combined to provide a single conclusion about the size of effect of a treatment?

Prior to designing their own projects, most scientists read the literature carefully. They learn what has already been done, find out about pitfalls and useful techniques, and clarify their research questions. Sometimes, their findings are considered valuable enough to write about and publish in journals such as *Psychological Bulletin*.

One of the enduring myths about science is that a single study is often enough to justify important and far-reaching conclusions. But as indicated in previous chapters, different studies on the same problem frequently yield dissimilar and even conflicting results. In such cases, a review of the literature that evaluates and organizes previous work may be more valuable than yet another experiment. Consider: Pflaum, Walberg, Karegianes, and Rasher (1980) reviewed 97 studies of methods for teaching children to read; and DerSimonian and Laird (1983) reviewed 36 studies on the benefits of receiving coaching for the Scholastic Aptitude Test. Despite presenting no new data, both sets of authors were able to resolve apparently conflicting findings.

In the past, literature reviews have been conducted unsystematically, with studies combined in idiosyncratic ways. This resulted in inefficiency and unsound conclusions, and distinguished scientists who reviewed the same evidence sometimes disagreed strongly about its collective meaning (Kamin, 1978; Munsinger, 1974, 1978). Fortunately, meta-analysis, a group of techniques developed in recent years for combining studies, can greatly improve the quality of reviews. For example, Cooper and Rosenthal (1980) asked university faculty and graduate students to review seven studies that collectively showed a strong relationship between a subject's gender and his or her task persistence. Reviewers taught meta-analytic procedures were much more likely to detect the effect that those who used their normal criteria for literature reviews.

Light (1984) indicated several benefits of meta-analysis:

1. Combining the results of studies systematically increases the number of subjects on which data has been collected. This increases statistical power (see pp. 205–208).
2. Each study is conducted with a single set of subjects in a particular setting. Reviewers who compare studies with different subjects and settings may be able to isolate key features that make treatments effective, identify particularly sensitive and insensitive subgroups, and learn how robust the treatments are. Such comparisons also provide tests of external validity.
3. Meta-analysis can explain conflicting results.
4. Meta-analysis can relate results to research designs.

Two excellent sources for the discussion of meta-analytic techniques are Rosenthal (1984) and Light and Pillemer (1984).

Organize a Reviewing Strategy

Formulate a Precise Question

Decide exactly what question you want the reviewer to answer. The most common questions are:

What is the Average Effect of a Treatment? Although an answer to this question is often useful, especially when most reviewed studies yield similar findings, additional information is generally desirable; researchers will want to know if the treatment works better with some groups than with others. And when findings conflict, astute researchers seek out reasons; this entails going beyond average effects.

Under What Conditions is a Treatment Particularly Effective or Ineffective? Effectiveness may depend on many factors. A heart transplant, for example, may work wonders on someone with a defective heart; but it won't increase the life expectancy of a healthy 20-year-old. In the social sciences, interactions between variables occur routinely (see p. 232) and are often subtle. Systematic reviews are more likely than individual studies to uncover them.

Will It Work Here? Some research is designed to influence therapeutic practices. A reviewer interested in implementing a treatment in a particular place will watch for studies conducted under circumstances similar to his own.

Decide Whether the Review is to Test a Specific Hypothesis or is Exploratory

Reviewers testing a specific hypothesis should focus on studies that bear on it. But when conducting exploratory research, they don't know beforehand what they will find, so they should examine a broad range of studies. This creates a problem: As more and more potential relationships are examined, the likelihood increases that some will appear significant because of chance factors. To minimize this possibility, Light and Pillemer suggested that the full collection of studies be randomly divided into two groups. The first group should be used to explore relationships among variables and generate hypotheses; the remaining studies can then be used as though new, to rigorously test the hypotheses.

Decide How Studies Should Be Selected

Try to locate every available study, both published and unpublished. If some are excluded because seriously flawed, indicate why. (If only published studies are included, the size of the effect is likely to appear greater than it actually is. The reason is that research findings not reaching statistical significance are less likely to be submitted for publication and less likely to be accepted if submitted.) Finding all the studies is another matter. Use dissertation abstracts in addition to the sources recommended in chapter 3. Computer retrieval services help a great deal and may be essential.

Express Information Quantitatively

Before studies can be compared, their outcomes must be expressed in similar units. Several formulae are available for expressing the size of an effect. The simplest method is to compute the correlation, r, between the independent and dependent variables (see p. 185). Investigators do not routinely provide estimates of effect size. (As discussed in chapter 14, the practice should be changed.) But, as shown by Rosenthal (1983), effect sizes can be derived from tests of statistical significance that typically are given. Of the several formulae available, the most general is Effect Size $= z/$square root of N where N is the number of subjects in a study and z, the standard normal deviate, is found from looking up the reported probability level of the study in appendix E. Thus, if the reported probability level is $p = .10$ (one-tailed) with 100 subjects, then effect size $= 1.28/10 = .128$. Rosenthal gave several formulae for computing effect size directly from data. Once effect size has been calculated, the results from individual studies can be combined to answer the following types of questions:

- Are the effect sizes of the studies essentially similar, or is at least one of them significantly different from the others?
- What is the combined effect size?
- If the studies vary along some dimension, such as year of publication, or sex or number of subjects, does the variation account for differences in results between the studies?
- Are the combined results statistically significant?

Methods for answering the first two questions are described below, with examples taken from Rosenthal. For other methods, see his book.

Are the Effect Sizes Similar?

Let effect size be symbolized by r and the number of subjects by N. Suppose that four studies yield effect sizes of $r = .70, .45, .10,$ and $-.15$ with Ns of 30, 45, 20, and 25, respectively.

1. Use appendix F to convert the rs to Fisher zs, as in Table 24.1. (The many uses of this transformation are explained in statistics books; for the present purpose, knowing how to use appendix F is enough.)
2. For each study, take $N - 3$ and multiply this number by r. The sum of these products, divided by the sum of all the $N - 3$s, is called the weighted mean z (z). In the example, $(z) = \{27(.87) + 42(.48) + 17(.10) + 22(-.15)\}/\{27 + 42 + 17 + 22\} = 42.05/108 = .39$.

Table 24.1 Results from Four Fictitious Studies

Study	r	Fisher z
1	.70	.87
2	.45	.48
3	.10	.10
4	−.15	−.15

Table 24.2 Calculating Whether Effect Sizes are Similar

Study	N − 3	Fisher z	(z)	(N − 3) (Fisher z − z)
1	27	.87	.39	27 × .48² = 6.22
2	42	.48	.39	42 × .09² = 0.34
3	17	.10	.39	17 × −.29² = 1.39
4	22	−.15	.39	22 × −.54² = 6.42

3. For each study, subtract (z) from the Fischer z and square the result. Multiply the answer by $N - 3$. See Table 24.2.
4. Get the sum of the $(N - 3)$(Fisher $z - (z)$).

$$6.22 + 0.34 + 1.39 + 6.42 = 14.37 = \chi^2$$

5. Use appendix G for χ^2 with df = the number of studies minus 1. The appendix shows that the results are statistically significant at $p < .01$. The conclusion is that the effect sizes of the four studies are different.

What is the Combined Effect Size?

Suppose that four studies yield effect sizes of r = .70, .45, .10, and −.15, respectively.

1. Convert the rs to Fisher zs. The corresponding Fisher zs are .87, .48, .10, and −.15. Get the sum of these.
$.87 + .48 + .10 + (-.15) = 1.30$.
2. Divide by the number of studies to find the mean Fisher z.
$1.30/4 = .32$
3. Convert the Fisher z to an r. From appendix E, a Fisher z of .32 = an r of .31. The average effect size is .31. (Notes: As is true of all averages, a single large score can distort the results. Also, effect size is likely to be overestimated, because studies that yield small effects are rarely published.)

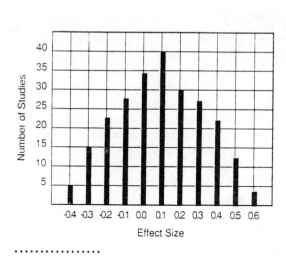

Figure 24.1

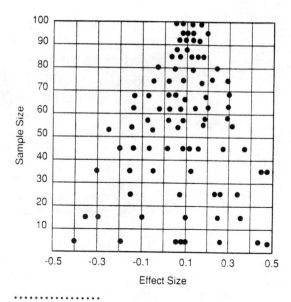

Figure 24.2

Use Visual Displays

Visual displays often demonstrate order in data that might otherwise appear chaotic. For example, tests of a treatment that produces an average effect size of 0.1 will probably cluster around an average of 0.1, but some will show no effect and some a negative effect. A display, as in Figure 24.1, will make the pattern clear. The display is a frequency distribution with effect size on the abscissa and number of studies at each effect size on the ordinate.

In Figure 24.2, effect size is plotted on the abscissa and sample size on the ordinate. Larger samples will usually approximate the true population mean more closely than will smaller samples. Then, if different studies measure the same underlying variable, the graph will resemble a funnel. At the top, the large sample studies will cluster tightly around the population mean; and at the bottom, the small sample studies will spread out. So, funnel displays can help reviewers decide if studies all survey the same population; they can also indicate whether well-done studies that failed to reach statistical significance have been omitted from consideration (see Light and Pillemer); and if population estimates have changed over time. (Compare plots of effect size versus dates of publication.)

Evaluate the Findings

Determine the Population to Which Results Can Be Generalized

The issues involved in generalizing from individual studies are discussed on pages 48–50. When groups of studies are reviewed, an additional complication arises that can be illustrated with an example from Light and Pillemer. Suppose a review of job-training programs showed an inverse relationship between hours of training and income for the trainees. This would seem to imply that the programs are harmful. But suppose further that, within each study, increased training was associated with higher salaries. The conflict would be resolved with the realization that programs providing the most training are based in the poorest neighborhoods, where job prospects are dimmest.

The above example shows that relationships observed in individual and group studies may differ, and the former may represent the truth more closely. The reverse may also be true: A small program that places all its trainees might, if implemented more widely, fail because the market is unable to accommodate more workers in the particular field. Nothing in the single study's research design would suggest the saturation problem; only by reviewing many job training studies might the issue be identified.

Relate Outcomes to Study Characteristics

If results from different studies conflict, reviewers have an excellent opportunity to extend knowledge about the phenomenon in question; they can see if outcomes vary systematically with specific attributes of studies. Rather than mechanically averaging effect sizes, reviewers can form subgroups of studies based on criteria such as date of publication, type of subject, duration of treatment, and so forth. They can then compute the average effect size for each subgroup. Reviewers should look for studies with extreme outcomes (outliers) and develop hypotheses that account for differences between the best and worst studies. These can be tested either by using data from the less extreme studies or by conducting new studies.

Light and Pillemer suggested several attributes that should be considered routinely when examining human service programs. Adapted slightly, they apply to any body of research:

1. Program characteristics. Even though programs have similar labels, they may differ in times of availability or types of services or numbers of participants.
2. Setting characteristics. Where a program is located and who administers it may affect outcomes.
3. Participant characteristics. Only certain types of people may benefit from a program. For example, only poor children show consistent cognitive benefits from daycare programs (Belsky & Steinberg, 1978).

4. Research design. Results may differ because some researchers measure short-term outcomes and others measure long-term outcomes. New and old studies may differ because of advances in instrumentation and data analysis.
  5. Analysis techniques. Researchers interested in the correlation between SAT scores and grade point averages might proceed in at least two different ways. Some might compute overall grade point averages for each of several schools, the average SAT scores of students at the schools, and the correlation between school GPA and student SAT scores. Other researchers might analyze data from a single school to see if students with high GPAs also score high on the SAT. The conclusions might differ (perhaps obviously, under the circumstances; but when subtle differences in research design are embedded in lengthy reports, they may be missed; meta-analysts deliberately investigate relationships between statistical procedures and conclusions).

Summary

Prior to conducting research, scientists typically review the literature in their field. Systematic reviews, in which objective techniques are used for combining results from different studies, have other uses as well. They expand the data base, help uncover the conditions under which treatments work, resolve conflicting findings, and relate results to research designs. To conduct a systematic review:

  1. Organize a reviewing strategy.
      a. Formulate precise questions.
      b. Decide whether the review is to test a specific hypothesis or is exploratory.
      c. Decide how studies should be selected for inclusion and exclusion.
  2. Use visual displays.
  3. Express results quantitatively. Find the average effect size of a group of studies, and test to see if the effect sizes from different studies differ significantly.
  4. Relate outcomes to characteristics of studies. Determine the population to which results can be generalized.

Key Terms

Effect Size The strength of an experimental treatment.

Meta-analysis A group of techniques for systematically analyzing the results of sets of studies on a single topic.

APPENDIX A

Bibliographic Retrieval for the Social and Behavioral Scientist

MaryLu C. Rosenthal

The library's role in helping the social or behavioral scientist find bibliographic material is explained, and methods for the retrieval of bibliographic information are given. A table giving titles of relevant abstracts and indexes also provides time spans covered, frequencies of publication, disciplines for which each is useful, and special features.

The purposes of this paper are to suggest a broad range of databases and printed abstracts and indexes that may aid the social and behavioral scientist in retrieving all published and unpublished studies on a given social and behavioral science research topic, and to acquaint the researcher with the way the research library and librarian can be of assistance in the retrieval effort.

"By-hand" searches of printed indexes and abstracts are still very much in use for many kinds of literature searches, particularly if the researcher is geographically located in an area where machine-readable databases are unavailable. If we are interested in a quick, comprehensive retrieval of citations on a given topic of research, however, the printout from machine-readable databases is the quickest and most comprehensive single searching procedure. For this the researcher will want to contact the college, university, or private research library, or one of the private information/library research companies found listed in the telephone directory yellow pages.

Research library staffs providing on-line bibliographic retrieval keep themselves currently aware of the status of bibliographic databases. There are comprehensive directories, vendors' newsletters and bulletins, monthly updates concerning changes, additions, and deletions in a database, and reviews about new databases found in the library science literature. All of these enable the reference librarian to know what kinds of publications are indexed and/or abstracted and included in a given database, and to assist the researcher in selecting the appropriate databases for the kinds of literature needed.

Ordinarily, researchers know which journals they usually use for their work; they also know where those journals are indexed and abstracted in the usual printed form. The likely next step is simply to seek those on-line databases that are the equivalent to the printed indexes and abstracts. For some purposes, however, it is desirable for the researcher to delve into some of the less frequently used indexes and abstracts in order to find the more elusive studies from the more obscure sources in the worldwide social and behavioral sciences community. For example, meta-analysis is a growing area of interest in which the researcher analyzes the results of many studies testing a single hypothesis. In order to find the studies meeting this criterion, the scientist can search a number of different databases that will yield nearly every published or unpublished study testing the hypothesis in question.

Before any on-line bibliographic search is done, the researcher and reference librarian use thesauri and vocabulary-term lists provided for each database; the process of choosing the correct key words and descriptors is an important one, for it will facilitate the retrieval of the desired citations and the exclusion of unwanted citations, thereby saving time and costs. Careful planning of the search strategy ahead of time will make it unnecessary for the researcher to be present when the search is being done. Some scientists, however, do like to take the time to sit at the video display unit and to watch the search in order to learn immediately if they are retrieving the kinds of citations they want. If they are not receiving them, changes in the strategy may be made immediately when the expert on the subject is available to suggest alternative terminology to be used.

Table 1, which appears at the end of the paper, presents the names of the databases, their acronyms (if any), and the names of their corresponding printed sources, if there are any. (The researcher should be aware that there is almost never an exact correspondence from a printed to a computer-readable source; even though there may be the same number of citations in each, the database may contain additional information in many of its citations.) Any important printed source with a longer time span than its corresponding database is also listed as a separate entry. The list further contains a checklist of disciplines served by each database or printed source (we have defined the "social and behavioral sciences" area somewhat broadly to include economics and business and management), the time span each covers, the frequency with which each is updated, and the countries of publication of the materials indexed or abstracted for each database. When relevant, there are included special features or aspects that might identify it as having a possible usefulness to a specific discipline within the social and behavioral sciences. The following time span abbreviations are used in Table 1:

a	= annually	m	= monthly
bi-a	= biannually	q	= quarterly
bi-m	= bimonthly	s-a	= semiannually
bi-w	= biweekly	w	= weekly
cont.	= continuous	yr	= year
d	= daily		

After the researcher has acquired the citations and abstracts from the desired databases or printed sources, the next step is actually to obtain the full texts of the journal articles, reports, etc. The library will request photocopies of journal articles and government documents to which it does not subscribe, and it will obtain books it does not own through interlibrary loan. The reference department will also help furnish the addresses of institutions and/or authors to whom the researcher may write for unpublished materials. Some materials may be in microform, in which case they will have to be used in the library. The cost of photocopying is usually passed on to the researcher, but the interlibrary loan of books or microforms is usually free.

As the researcher accumulates the studies on the particular topic of interest, he or she may eventually discover that the articles, regardless of outcome, have the same bibliographical references. At that point there may be nothing more to find on the topic and the search may have come to its end.

References

BRS Bulletin (monthly newsletter). Latham, N.Y.: Bibliographic Retrieval Services.

Chronolog (monthly newsletter). Palo Alto, Calif.: Dialog Information Services, Inc.

Directory of Online Databases (published quarterly). Santa Monica, Calif.: Cuadra Associates, Inc.

Directory of Online Information Resources (1983). Kensington, Md.: CSG Press.

Edelhart, Mike, and Davies, Owen (1983). *Omni Online Database Directory*. New York: Macmillan.

Hall, James L., and Brown, Marjorie J. (1983). *Online Bibliographic Databases: An International Directory* (2nd ed.). London: Aslib.

Information Industry Market Place; an International Directory of Information Products and Services (1983). New York: R. R. Bowker.

Schmittroth, John, Jr. (1983). *Abstracting and Indexing Services Directory* (1st ed.). Detroit, Mich.: Gale Research Co.

Schmittroth, John, Jr. (1985–86). *Encyclopedia of Information Systems and Services* (6th ed.). Detroit, Mich.: Gale Research Co.

Searchlight (monthly newsletter). Santa Monica, Calif.: SDC Information Services.

Sheehy, Eugene P., ed. (1976). *Guide to Reference Books* (9th ed.). Chicago, Ill.: American Library Association.

Sheehy, Eugene P., ed. (1980). *Guide to Reference Books* (9th ed. suppl.). Chicago, Ill.: American Library Association.

Williams, Martha E., ed. (1982). *Computer-Readable Databases: A Directory and Data Sourcebook*. White Plains, N.Y.: Knowledge Industry Publications, Inc.

Received April 23, 1985

From MaryLu Rosenthal, "Bibliographic Retrieval for the Social and Behavioral Scientist" in *Research in Higher Education*, 22:315-333. Copyright © 1985 Agathon Press, New York, NY. Reprinted by permission of Human Sciences Press, Inc.

Table 1 Databases for Ten Disciplines

Title of Database or Printed Abstract or Index	Acronym	Corresponding Print Product (if database)	Discipline										Time Span Covered	Frequency of Updating	Country of Origin of Materials; Special Features
			Alcohol & Drug Studies	Anthropology	Economics	Education	Linguistics	Management & Business	Psychiatry	Psychology	Social Work	Sociology			
ABI/Inform	—	none			x			x					1971–present	m	Aust., Canada, Germany, Netherlands, Switzerland, U.K., U.S.A.
Abstracts for Social Workers	—	print		x	x						x	x	1965–present	q	U.S.A.
Abstracts in Anthropology	—	print		x	x		x						1970–present	q	U.S.A.
Abstracts of Dissertations, Oxford University	—	print		x	x					x		x	1928–1947	—	England
Abstracts of Dissertations Approved for the Ph.D., M.Sc., M. Litt. degrees in the University of Cambridge	—	print								x		x	1925–1959	—	England
Abstracts of Doctoral Dissertations in Anthropology	—	print		x			x					x	1969–present	a	U.S.A.

Table continues

From MaryLu Rosenthal, "Bibliographic Retrieval for the Social and Behavioral Scientist" in *Research in Higher Education*, 22:315-333. Copyright © 1985 Agathon Press, New York, NY. Reprinted by permission of Human Sciences Press, Inc.

Table 1 (Continued)

Title of Database or Printed Abstract or Index	Acronym	Corresponding Print Product (if database)	Alcohol & Drug Studies	Anthropology	Economics	Education	Linguistics	Management & Business	Psychiatry	Psychology	Social Work	Sociology	Time Span Covered	Frequency of Updating	Country of Origin of Materials; Special Features
Administration on Aging Data Base	SCAN	none								x	x	x	1978–present	m	Canada, U.S.A.; a merged database: Source Database, Medline, National Health Planning Information Center
Aggregate File of the Drug Abuse Epidemiology Data Center	DAEDAC	none	x			x				x	x	x	1960–present	—	Australia, Canada, Europe, U.S.A.
Agricultural Economics	AG ECON	Journal articles are found in Bibliography of Agriculture 1970–1980			x			x		x	x	x	1970–present	m	U.S.A., Canada, world-wide; rural sociology, human development
Alcohol Use/Abuse Information File	—	none	x		x	x			x	x	x	x	1955–present; some pre-1955	q	Canada, Europe, U.S.A.; evaluation of treatments, counselor training, therapies
American Doctoral Dissertations	—	print	x		x	x				x	x	x	1955–1964	a	U.S.A.

Table continues

Table 1 *(Continued)*

Title of Database or Printed Abstract or Index	Acronym	Corresponding Print Product (if database)	Alcohol & Drug Studies	Anthropology	Economics	Education	Linguistics	Management & Business	Psychiatry	Psychology	Social Work	Sociology	Time Span Covered	Frequency of Updating	Country of Origin of Materials; Special Features
Animal Behaviour Abstracts	—	Animal Behaviour Abstracts								x			1978–present	q	Worldwide
Australian Education Index	AEI	Australian Education Index				x				x		x	1978–present	q	Australia
Australian Public Affairs Information Service	APAIS	Australian Public Affairs Information Service			x			x		x		x	1978–present	m	Australia
Australian Road Index	ARI	Australian Road Index							x	x		x	1977–present	q	Australia, U.K., U.S.A., Worldwide; human factor aspects of driving, drivers
British Education Index	—	print				x				x			1951–present	3/yr.	U.K.
Centre National de la Recherche Scientifique	CNRSLAB	Annuaires CNRS	x	x						x		x	1979–present	a	France; research project descriptions

Table continues

Table 1 (Continued)

Title of Database or Printed Abstract or Index	Acronym	Corresponding Print Product (if database)	Alcohol & Drug Studies	Anthropology	Economics	Education	Linguistics	Management & Business	Psychiatry	Psychology	Social Work	Sociology	Time Span Covered	Frequency of Updating	Country of Origin of Materials; Special Features
Child Abuse and Neglect	NCCAN	1. Child Abuse and Neglect Programs; 2. Child Abuse and Neglect Research: Projects and Publications; 3. Child Abuse and Neglect Audio-Visual Materials	x			x			x	x	x	x	1967–present	s-a	U.S.A.
Child Development Abstracts and Bibliography	—	print		x	x	x	x		x	x	x	x	1927–present	3/yr.	U.S.A.
Comprehensive Dissertation Abstracts	CDI	Comprehensive Dissertation Abstracts; Dissertation Abstracts International	x	x	x	x	x		x	x	x	x	1861–present	m	Canada, U.S.A.
Computer Retrieval of Information on Scientific Projects (NIH Division of Research Grants)	CRISP	Research Awards Index (Sup. of Documents)				x			x	x		x	1971–present	d	U.S.A.; research projects sponsored by NIH and NIMH jointly

Table continues

Table 1 *(Continued)*

Title of Database or Printed Abstract or Index	Acronym	Corresponding Print Product (if database)	Alcohol & Drug Studies	Anthropology	Economics	Education	Linguistics	Management & Business	Psychiatry	Psychology	Social Work	Sociology	Time Span Covered	Frequency of Updating	Country of Origin of Materials; Special Features
A Computerized London Information Service	ACOMPLINE	Urban Abstracts			x			x		x	x	x	1974–present	bi-m	England, France, Germany; includes papers presented at conferences, theses
Conference Papers Index	CPI	Conference Papers Index								x			1973–present	m	International
Current Index to Journals in Education (part of ERIC database)	CIJE	Current Index to Journals in Education				x	x			x			1969–present	m, q, a	U.S.A.
Current Research Information System	CRIS	none								x	x	x	past two years	q	active research and research terminated for two years; agricultural issues related to the behavioral sciences
Datennachweis Informations-system	—	none				x				x		x	1980–present	—	Germany, Austria, Switzerland
Doctoral Dissertations in Social Work	—	print								x	x	x	1954–present	a	U.S.A.

Table continues

Table 1 (Continued)

Title of Database or Printed Abstract or Index	Acronym	Corresponding Print Product (if database)	Discipline — Alcohol & Drug Studies	Anthropology	Economics	Education	Linguistics	Management & Business	Psychiatry	Psychology	Social Work	Sociology	Time Span Covered	Frequency of Updating	Country of Origin of Materials; Special Features
Drug Info	—	none	x			x			x	x	x	x	1969–present	q	U.S.A., Canada, Europe; unpublished works held by Drug Information Service Center
ERIC Clearinghouse for Early Childhood Education	—	print				x				x	x	x	1948–1973	—	U.S.A.
Economics Abstracts International	—	Key to Economic Science; Economic Titles/Abstracts			x			x					1974–present	m	Worldwide
Education Database	—	Research in Education; Current Index to Journals in Education				x				x			most current 4 years	m	U.S.A.
Education Index	—	print				x				x			1929–present	10/yr.	U.S.A.
Education Research	EDUC	Ontaris Abstracts; Ontaris Printed Index				x				x			1960–present	m	Canadian educational research
Educational Sciences	—	Educational Sciences				x				x	x	x	1972–present	bi-m	Worldwide

Table continues

Table 1 (Continued)

Title of Database or Printed Abstract or Index	Acronym	Corresponding Print Product (if database)	Discipline										Time Span Covered	Frequency of Updating	Country of Origin of Materials; Special Features
			Alcohol & Drug Studies	Anthropology	Economics	Education	Linguistics	Management & Business	Psychiatry	Psychology	Social Work	Sociology			
Environmental Psychology	ENVPSYCH	none								x	x	x	1940–present	cont.	the relationships of people to all environments
EUDISED (European Documentation and Information System for Education)	—	print				x				x	x	x	1981–present	5/yr.	16 European Countries; recently completed and on-going research
Exceptional Child Education Resources	ECER	Exceptional Child Resources				x			x	x			1966–present	m	Canada, U.K., Australia, U.S.A.; handicapped and gifted children
Excerpta Medica (includes the subset DRUGDOC)	EMBASE	Drug Literature Index (partly); Excerpta Medica	x						x	x			1947–present	w	Worldwide
Family Resources Database (a merged file: Inventory of Marriage and the Family, plus National Council on Family Relations	—	Inventory of Marriage and the Family, 1973–						x	x	x	x	x	1970–present	m	includes work in progress
Fonds Quetelet	—	Acquisitions List; Catalogues on Microfiche			x					x		x	1969–present	bi-m	Belgium, France, U.K., U.S.A.

Table continues

Table 1 (Continued)

Title of Database or Printed Abstract or Index	Acronym	Corresponding Print Product (if database)	Discipline											Time Span Covered	Frequency of Updating	Country of Origin of Materials; Special Features
			Alcohol & Drug Studies	Anthropology	Economics	Education	Linguistics	Management & Business	Psychiatry	Psychology	Social Work	Sociology			
French Retrieval Automated Network for Current Information in Social and Human Sciences	FRANCIS	Corresponds with 17 printed sources			x	x				x		x	1972–present	m	France
Government Reports Announcements and Index	—	print	x		x	x		x	x	x		x	1946–present	s-m	U.S.A.
Human Sciences of Health	RESHUS	Human Sciences of Health			x		x	x	x		x	x	1977–present	—	Europe, U.S.A.
Index of Economic Articles in Journals and Collective Volumes	—	print			x								1886–present	a	Worldwide
Information Exchange Centre for Federally Supported Research in Universities	IEC	Directory of Federally Supported Research in Universities	x	x	x	x	x	x	x	x	x	x	1971–present	a	Canada
Information Retrieval System for the Sociology of Leisure and Sport	SIRLS	Sociology of Leisure and Sport Abstracts				x				x	x	x	1950–present	3/yr.	U.S.A., Canada, Europe

Table continues

Table 1 (Continued)

Title of Database or Printed Abstract or Index	Acronym	Corresponding Print Product (if database)	Discipline										Time Span Covered	Frequency of Updating	Country of Origin of Materials; Special Features
			Alcohol & Drug Studies	Anthropology	Economics	Education	Linguistics	Management & Business	Psychiatry	Psychology	Social Work	Sociology			
Information Service on Social Sciences Research	FORIS	none		x	x	x	x	x		x	x	x	1971–present	a	Austria, Germany, Switzerland; completed on-going or planned research
International Bibliography of Social and Cultural Anthropology	—	print		x			x			x		x	1955–present	a	Worldwide
International Bibliography of the Social Sciences	IBSS	International Bibliography of Economics; International Bibliography of Sociology						x		x		x	1979–present	irreg.	Worldwide
Inventory of Marriage and Family Literature	—	International Bibliography of Research in Marriage and the Family, 1900–1972; Inventory of Marriage and Family Literature, 1973–1974; Inventory of Marriage and the Family, 1975–1980							x	x	x	x	1900–present	a	U.S.A., Worldwide

Table continues

Table 1 *(Continued)*

Title of Database or Printed Abstract or Index	Acronym	Corresponding Print Product (if database)	Discipline										Time Span Covered	Frequency of Updating	Country of Origin of Materials; Special Features
			Alcohol & Drug Studies	Anthropology	Economics	Education	Linguistics	Management & Business	Psychiatry	Psychology	Social Work	Sociology			
Iowa Drug Information Service	IDIS	Drug Literature Microfilm file	x						x		x		1966–present	m	Worldwide
Journal of Economic Abstracts	—	print			x								1963–1968	q	U.S.A.
Journal of Economic Literature	—	Journal of Economic Literature; Index of Economic Articles			x			x					1969–present	q	U.S.A.
LIBCON	—	records from MARC		x	x	x		x		x	x	x	1968–present	w	U.S.A., Worldwide
Language and Language Behavior Abstracts	LLBA	Language and Language Behaviour Abstracts		x			x			x	x	x	1965–present	q	Worldwide
Management Contents Data Base	—	Management Contents; Legal Contents			x			x					1974–present	m	Worldwide
Medlars Online	Medline	Index Medicus							x	x			1979–present	m	Worldwide; biomedical literature
National Clearinghouse for Mental Health Information Data Base	NCMHI	Partial coverage by Psychopharmacology Abstracts	x						x	x		x	1965–present	m	France, W. Germany, Russia, U.K., U.S.A.

Table continues

Table 1 (Continued)

Title of Database or Printed Abstract or Index	Acronym	Corresponding Print Product (if database)	Alcohol & Drug Studies	Anthropology	Economics	Education	Linguistics	Management & Business	Psychiatry	Psychology	Social Work	Sociology	Time Span Covered	Frequency of Updating	Country of Origin of Materials; Special Features
National Council on Family Relations	NCFR	will be: Information Guide to Family Literature, Programs and Services	x						x	x	x	x	1970–present	m	U.S.A.; literature is non-journal; work planned, work in progress
National Rehabilitation Information Center Data Base	NARIC	none			x				x	x	x	x	1952–present	m	U.S.A.; reports prepared under grants and proposals
National Technical Information Service Bibliographic Data Base	NTIS	Government Reports Announcements	x					x	x	x	x	x	1964–present	bi-w	U.S.A.; reports from government-sponsored research
Northern Research Information Service	NRIS	none	x			x					x	x	1976–present	a	Canada; current research projects
Office on Smoking and Health Database	OSH	Smoking and Health Bulletin, and Bibliography on Smoking and Health								x			1961–present	irreg.	Worldwide
Pascal 390	—	Bulletin Signaletique				x			x	x		x	1968–present	m	Worldwide
Population Bibliography	—	none			x					x		x	1967–present	bi-m	Worldwide; unpublished papers; family planning

Table continues

Table 1 *(Continued)*

Title of Database or Printed Abstract or Index	Acronym	Corresponding Print Product (if database)	Alcohol & Drug Studies	Anthropology	Economics	Education	Linguistics	Management & Business	Psychiatry	Psychology	Social Work	Sociology	Time Span Covered	Frequency of Updating	Country of Origin of Materials; Special Features
Population Information On Line	POPLINE	none								x	x	x	1968–present	m	Worldwide; family planning
Psychological Abstracts	—	print	x	x		x	x	x	x	x	x	x	1927–present	m	Worldwide
Psychological Abstracts Information Service	Psych INFO	Psychological Abstracts	x			x	x	x	x	x	x	x	1967–present	m	Worldwide
Public Affairs Information Service Bulletin	PAIS	PAIS Bulletin			x			x		x		x	1976–present	m	Worldwide; all from English-language publications
Public Affairs Information Service Foreign Language Index	PAIS FLI	Foreign Language Index			x	x		x		x		x	1972–present	q	Worldwide; European-language publications
Resources in Education (part of ERIC database)	RIE	Resources in Education				x	x			x			1966–present	—	U.S.A.; theses, conference papers
Rural Development Abstracts	RDA	Rural Development Abstracts			x	x		x		x	x	x	1978–present	q	Worldwide
Rural Extension, Education and Training Abstracts	REETA	Rural Extension, Education, and Training Abstracts				x							1978–present	q	Worldwide

Table continues

Table 1 *(Continued)*

Title of Database or Printed Abstract or Index	Acronym	Corresponding Print Product (if database)	Discipline										Time Span Covered	Frequency of Updating	Country of Origin of Materials; Special Features
			Alcohol & Drug Studies	Anthropology	Economics	Education	Linguistics	Management & Business	Psychiatry	Psychology	Social Work	Sociology			
Selective Cooperative Indexing of Management Periodicals	SCIMP	European Index of Management Periodicals			x			x					1978–present	m	France, Germany, U.K., U.S.A.
Smithsonian Science Information Exchange	SSIE	Notice of Research Projects	x	x	x	x	x	x	x	x	x	x	1978–present	m	U.S.A.; current research funded by federal and local government, public and private institutions
Social Science Data Base	—	none			x	x	x	x		x	x	x	most current 3 yrs.	entire database updated every 6 mos.	Worldwide; monographs, theses
Social Science Information System (a merged database: Canadian Social Science Abstracts, Current Index to Journals in Education, Psychological Abstracts, Social Science Journal File	SSIS	partly with Psychological Abstracts, 1975– and CIJE, 1975–	x	x	x	x	x	x	x	x	x	x	1968–present	q	U.S.A., Canada, Worldwide
Social Science Citation Index	SSCI	Social Science Citation Index				x		x		x	x	x	1969–present	w	Worldwide
Social Sciences Index	—	print	x	x	x	x	x	x		x	x	x	1974–present	q	U.K., U.S.A.

Table continues

Appendix A

Table 1 *(Continued)*

Title of Database or Printed Abstract or Index	Acronym	Corresponding Print Product (if database)	Alcohol & Drug Studies	Anthropology	Economics	Education	Linguistics	Management & Business	Psychiatry	Psychology	Social Work	Sociology	Time Span Covered	Frequency of Updating	Country of Origin of Materials; Special Features
Sociological Abstracts	—	print		x			x			x	x	x	1952–present	frequency varies	Worldwide
Sociology	—	Sociology		x						x	x	x	1972–present	8/yr.	Worldwide
Sociology Theses Register	—	print		x						x	x	x	1976–present	a	U.K.
Soziologisches Literatur-Informations system	SOLIS	none		x	x					x	x	x	1976–present	m	Austria, Germany, Switzerland
Spoglio Reviste	—	none			x	x				x	x	x	1979–present	m	Worldwide
Successful Candidates for the Degree of Doctor of Philosophy; Oxford University	—	print		x	x					x	x	x	1940–present	a	England
Titles of Dissertations Approved for the Ph.D., M.Sc., and M. Litt. degrees in the University of Cambridge	—	print								x	x	x	1959–present	a	England
World Agricultural Economics and Rural Sociology Abstracts	—	none		x							x	x	1973–present	m	Worldwide

APPENDIX B

Effects of Knowledge of Results and Delay of Knowledge on the Judgment of the Muller-Lyer Illusion

Geralyn Rodriguez Clucas
California State University, Hayward
Hayward, California 94542

Abstract

This study was designed to investigate the effects of knowledge of results (KR) on judging the relative lengths of the Muller-Lyer figures, and the effects of delaying this information. Subjects were volunteers from the S.F. Bay Area, randomly assigned to three groups—no knowledge (NK), immediate knowledge (IK), or delayed knowledge (DK), and were asked to adjust an apparatus to make two Muller-Lyer lines equal in length. A control group (NK) was given no feedback, while the IK and DK groups received KR immediately and 20 seconds after each response respectively. Performance as measured by the method of average errors was found to improve dramatically only when KR was given, while the NK group showed no significant improvement. Furthermore, no significant difference in performance was found between the IK and DK groups. It was concluded that knowledge of results improved performance by enabling subjects to compensate for the effects of the illusion, and that delay of informational feedback by 20 seconds did not detrimentally affect performance in this task. These conclusions were discussed with respect to previous investigations in these areas.

Effects of Knowledge of Results and Delay of Knowledge on the Judgment of the Muller-Lyer Illusion

The effects of knowledge of results (KR) have received ample documentation, often in the guise of other theoretical formulations. Thorndike's line-drawing experiment (1927) provided KR as incentive, postulating that the word "right" would increase the frequency of correct responses. As Thorndike predicted, the incentive condition showed more significant improvement than the no knowledge condition. Trowbridge & Cason (1932) replicated his procedure, including a "correct" group which received information regarding the direction and magnitude of errors. Subjects receiving specific information exhibited greater improvement than the other groups, outperforming Thorndike's incentive condition, while the no-knowledge group showed no significant improvement. These investigators concluded that the specificity of feedback was a more powerful determinant of improved performance than the incentive value of KR.

The theory that knowledge of results contributes to improved performance has received further empirical support. Knowledge of results has been shown to improve estimation of line length, drawing lines, hitting targets, and performing various motor tasks (Ammons, 1956). In his extensive review of the literature, Ammons concluded that KR almost universally improved performance, enabling informed subjects to learn quicker and with greater proficiency than their non-informed counterparts. Using a two-handed coordination task, Ellwell & Grindley (1938) found great improvement with knowledge of results, no significant improvement in its absence, and a significant deterioration when KR was removed. In agreement with Trowbridge & Cason, these investigators concluded that specific KR provides a "directive effect" which may act in conjunction with, or in the absence of the incentive effect hypothesized by Thorndike. McGuigan's research (1959) examined effects of precision, schedule and delay of KR in a line-drawing procedure. The only significant variable was schedule, with 100% KR eliciting the most improved performance. Furthermore, in agreement with the aforementioned studies, the no knowledge group showed no improvement. These studies all indicate that knowledge of results improves performance, while the absence of KR results in no significant improvement.

The effects of withholding KR in repeated trials were dramatically presented in Judd's study (1905) of the perception of angles. Subjects responded in a systematically erroneous manner in the absence of KR. These habitual modes of response became increasingly hard to modify, even when subjects were given knowledge of results. Seashore & Bavelas (1941) noticed similar effects during an examination of Thorndike's line-drawing data. Without KR, each subject's performance did not improve or deteriorate, but stabilized around an incorrect line length. They concluded that subjects relied upon subjective, often inaccurate conceptions of a correct response in the absence of objective KR. Ammons (1956) agreed with their conclusion, noting that subjects relied upon their own hypotheses of correct response based on the observation of cues, memory, or sometimes irrelevant criteria. He concluded that KR, when presented, interacted with these subjective hypotheses.

One purpose of this investigation is to examine the influence of KR on performance using the powerful Muller-Lyer illusion. The destruction of this illusion has been shown to be a slow process requiring hundreds of comparisons (Judd, 1902). Therefore, objective KR should conflict with subjects' hypotheses concerning the relative lengths of the Muller-Lyer figures. This study seeks to determine how powerful perceptual misinformation interacts with objective and precise knowledge of results.

Secondly, this study will examine the effects of delaying KR. This topic has received much attention, with mixed empirical findings. Greenspoon & Foreman's line-drawing experiment (1956) yielded significant differences between 0, 10, 20 and 30 second delays, with slower rates of learning associated with increasing delay. They concluded that delaying KR did not improve performance as effectively as providing immediate KR. Bilodeau & Ryan's attempt (1960) to replicate their finding proved unsuccessful, however, with no significant difference between the delay groups. McGuigan's previously mentioned study (1959) examined the effects of delay of knowledge, and also failed to find significant differences. He hypothesized that the conflicting findings might have resulted from failures to control confounding independent variables, such as intertrial and post-knowledge of results intervals. In a series of experiments designed to uncover the effects of each interval, (post-knowledge, intertrial, and delay), Bilodeau & Bilodeau (1958) found no significant difference in delays of knowledge up to a week. Furthermore, they found that the intertrial interval manipulation showed a significant difference in amount of learning, with longer intervals detrimental to performance. A line-drawing experiment (Denny, Allard, Hall & Rokeach, 1960) lent support to this hypothesis. These investigators concluded that the intertrial interval, rather than the delay interval, was the critical factor affecting performance. Failures to replicate their findings cast doubt on Greenspoon & Foreman's conclusions, suggesting the need for further investigation with appropriate controls.

The task of this experiment involves making the Muller-Lyer figures equal in length. The purpose of this investigation is two-fold: to determine whether knowledge of results will influence the judgment (not necessarily the perception) of the Muller-Lyer figures, and to determine whether delay of KR will detrimentally affect the performance of this task. It is hypothesized that providing specific KR will compensate for the powerful effects of the Muller-Lyer illusion, enabling subjects to learn to make the correct judgments. Furthermore, in agreement with previously cited studies (Bilodeau & Bilodeau, 1958; Denny et al., 1960), it is hypothesized that control of the intertrial interval and variation of delay interval will produce no significant difference in performance.

Method

Subjects

The subjects were 150 men and women volunteers from the San Francisco Bay Area. They were assigned to one of three conditions—no knowledge (NK), immediate knowledge (IK), or delayed knowledge group (DK), by the method of randomized blocks.

Materials

The apparatus used in this experiment was the Muller-Lyer Illusion Board, employing the well-known Muller-Lyer figures. A three inch line with the oblique lines turned inward at 45° angles was drawn at one edge of the apparatus. A moveable line with oblique lines turned outward at 45° angles extended from one end of the static three inch line. The length of the moveable line was variable and easily adjusted by manipulating the outside flap on which the line was drawn. The variable line could be shortened by moving the flap inwardly, or lengthened by adjusting the flap outwardly. The reverse side of the apparatus contained a scale which indicated the point at which the variable line length equalled the static line (three inches), and measured deviations from this point in 16th's of an inch. The apparatus was also equipped with flaps which enabled the experimenter to selectively expose or conceal the figures and scale from the subjects.

Procedure

All subjects received the same treatment for the first seven trials. Subjects were instructed to adjust the variable line of the apparatus until the figures were equal in length, and hand the apparatus back to the experimenter. Subjects were urged not to talk, or use extraneous smudges to help judge the lengths of the lines. Each adjustment was followed by a timed 50-second intertrial interval, during which no communication or activity was allowed. The experimenter recorded errors of measurement after each trial, carefully keeping the scale, figures, and written observations hidden from the subjects. On successive trials, the apparatus was handed to subjects with the variable line to the right or left, to be moved in or out. Systematically switching the position of the apparatus and the variable line flap provided counterbalancing with regard to movement and position.

During the intertrial interval between Trials 7–8, the experimenter read instructions to each subject, the content varying between groups. Both feedback groups (IK and DK) were informed about the timing and presentation of KR. The NK subjects were read a paraphrased version of the original instructions. Subjects in the NK group continued to perform the same task without feedback for 17 more trials, replicating the conditions of the first seven trials.

In Trials 8–24, the IK and DK groups received knowledge of the direction and magnitude of their errors in 16th's of an inch. Specifically, the experimenter said, "You undershot (overshot) _____ 16th's of an inch," or "You're exactly correct." Subjects in the IK condition were presented KR immediately upon handing the apparatus to the experimenter. The timed 50 second intertrial interval began immediately after KR was presented. Subjects in the delay group, however, were given KR 20 seconds after their adjustments. Immediately after presenting KR, the experimenter commenced timing the remaining 30 seconds of the intertrial interval. Thus, the timed 50 second intertrial interval for the KR groups did not include the presentation of feedback. Therefore, each group received the same 50 seconds of absolute psychological rest afforded by the absence of communication.

Design. Subjects were randomly assigned to one of three conditions—NK (control), DK, or IK, by the method of randomized blocks. For the first seven trials, all groups were treated the same as the NK group, thereby providing within-group controls. The method of average error was used to measure errors of observation. Average constant errors were computed by separating data from Trials 1–8 from Trials 9–24 within each group. Data from each block of trials were compared with data from the corresponding blocks among the other groups. Independent group t tests were performed comparing constant error data by block between groups. Within-group data were computed using direct-difference t tests to determine differences between the two blocks of trials within each group. Variable error data were computed and compared in the same manner as the constant error data.

Results

The means and standard deviations for the three knowledge of results groups are presented in Table 1. For both the constant error and variable error scores, the data for Trials 1–8 will be presented first, followed by the data for Trials 9–24. Finally, the direct-difference data comparing for each group its performance on Trials 1–8 with its performance on Trials 9–24 will be presented.

Table 1 Mean Constant and Variable Errors for the Three KR Groups

	CONSTANT ERRORS		VARIABLE ERRORS	
	Trials 1–8	Trials 9–24	Trials 1–8	Trials 9–24
No Knowledge (NK)	−14.34 (SD 4.20)	−13.30 (SD 3.85)	1.94 (SD 0.86)	1.88 (SD 0.86)
Immediate Knowledge (IK)	−12.80 (SD 5.72)	−2.46 (SD 2.12)	1.88 (SD 0.77)	3.62 (SD 1.51)
Delayed Knowledge (DK)	−13.30 (SD 5.24)	−2.46 (SD 2.97)	1.88 (SD 0.95)	3.48 (SD 1.24)

From Geralyn Rodriguez Clucas, "Effects of Knowledge of Results and Delay of Knowledge on the Judgment of the Muller-Dyer Illusion," paper presented at California State University, Hayward, CA. Reprinted by permission.

Constant Errors, Trials 1-8

As may be seen from the left-hand column of Table 1, when the three independent KR groups were treated identically on Trials 1–8, their performances were not significantly different from one another, $t(98) = 1.52$, 1.08, and 0.45 for the NK vs. IK, NK vs. DK, and IK vs. DK comparisons respectively.

Constant Errors, Trials 9-24

However, when the three groups were given differential feedback on Trials 9–24, the groups receiving KR showed a level of error which was well over 400% less than the mean error found for the NK group (-2.46 in both cases versus -13.30). Both of these differences obviously were highly significant, $t(98) = 17.36$ and 25.60, $p < .001$ in both cases for the IK vs. NK and DK vs. NK comparisons respectively. It is of considerable interest to see that the 20 second delay interval had no effect at all in the potency of the informational feedback, $t(98) = 0.00$ for the IK vs. DK comparison.

Constant Errors, Trials 1-8 vs. Trials 9-24

These same data may be viewed from an alternative vantage point by comparing for each group its performance on the early trials where there was no informational feedback with its performance on Trials 9–24. The NK group which continued to receive no feedback on Trials 9–24 showed no significant improvement in performance, $t(49) = 0.36$. In stark contrast to this finding, as may be seen in Table 1, the two KR groups showed remarkable improvements with the onset of informational feedback, $t(49) = 13.69$ and 15.26 for the IK and DK groups respectively.

Variable Errors, Trials 1-8

The data in Table 1, as may be expected, show almost identical performances for the three KR groups on Trials 1–8, and there obviously were no significant differences among them, $t(98) = 0.36$, 0.33, and 0.00 for the NK vs. IK, NK vs. DK, and IK vs. DK comparisons respectively.

Variable Errors, Trials 9-24

However, as may be seen in the extreme right-hand column of Table 1, the two KR groups showed significantly larger variable errors than the NK group, $t(98) = 7.01$ and 7.42, $p < .001$ in both cases for the IK vs. NK and DK vs. NK comparisons respectively. It may on the surface seem paradoxical that the informational feedback substantially increased the size of the variable errors, while it, at the same time, radically reduced the size of the constant errors. This apparent paradox disappears when it is noted that in order for subjects to improve their accuracy through informational feedback, they must vary their behavior in response to the feedback, thereby increasing their variable errors while, at the very same time, decreasing their constant errors.

Variable Errors, Trials 1-8 vs. 9-24

The same pattern emerged when the performance for each group on Trials 9–24 is compared with its performance prior to the onset of any informational feedback. Both feedback groups showed a significant increase in variable errors on Trials 9–24, $t(49) = 7.24$ and 7.74, $p < .001$ in both cases for the IK and DK groups respectively. No change in variable errors was found for the NK group, $t(49) = 0.34$.

Discussion

The purpose of this investigation was to determine whether KR improves performance on a judgment task using the Muller-Lyer figures and whether delay of this feedback proves detrimental to the performance of this task. It was hypothesized that

KR would prove beneficial, enabling subjects to learn to compensate for the effects of the illusion. Furthermore, it was hypothesized that delay of KR by 20 seconds would not significantly hamper performance in this task.

The present findings lend support to the consensus of previously cited investigations (Ammons, 1956; McGuigan, 1959; Ellwell & Grindley, 1938): informational feedback resulted in remarkable improvement in performance. Consistent with Trowbridge & Cason (1932), information given to IK and DK groups regarding magnitude and direction of error resulted in an astounding reduction in errors of judgment with respect to the NK group and their earlier performances.

It is of considerable interest to notice that the NK group did not significantly improve, but performed with a consistently high constant error. This finding is also in agreement with several studies (Trowbridge & Cason, 1932; Ellwell & Grindley, 1938; McGuigan, 1959). Furthermore, the variable errors were relatively fewer compared to the variable errors of the feedback groups. This finding is consistent with the research of Judd (1905) and Seashore & Bavelas (1941), who noted the stability and consistency of error when subjects are denied KR. The high variable errors of the feedback groups lend indirect support to Ammons' (1956) theory of subjective hypotheses. The variation in response might have resulted from the subjects' conflicts between their hypotheses (perceptions) and objective knowledge of results. Finally, the consistent large errors made by the NK group, as well as the KR groups' large variable errors, testified to the considerable strength of the Muller-Lyer illusion, as originally reported by Judd (1902).

It is suggested that the nature of KR in this task resembles the "directive" effect postulated by Ellwell & Grindley, who suggest that KR serves a directive purpose until a task is learned. The persistent nature of this illusion exemplified by NK's large magnitude and stability of error, as well as the large variable errors of the KR groups suggest that subjects did not necessarily perceive the illusion differently, but learned to compensate for its disturbing effects by varying behavior, if not perception, with regard to the illusion. Since this task involved learning, the informational aspect of KR was perhaps more critical than the incentive value postulated by Thorndike.

It is of considerable interest to note that delay of KR produced no significant differences in performance of this task. This finding, in agreement with Bilodeau & Bilodeau, Denny et al., and McGuigan, has failed to lend credence to Greenspoon & Foreman's conclusions. However, it must be noted that only one 20 second delay of knowledge group was included in this experiment. It is beyond the scope of this investigation to determine the results of using longer delays of knowledge, such as minutes, hours, or days.

Furthermore, it must also be noted that although the intertrial interval remained constant, the post-KR interval was not controlled; this interval was 30 seconds for the DK group, and 50 seconds for the IK group. In light of previously cited studies (Bilodeau & Bilodeau, 1958; Denny et al., 1960), perhaps this uncontrolled variable confounded the possible effects of delay in this experiment. These findings might suggest an alternative hypothesis: control of the intertrial interval may have rendered the true critical variable (intertrial interval) impotent. However, this hypothesis is beyond the scope of this study. On the other hand, it is possible that failure to control the post-KR interval may have diminished the possible effects of delayed KR. For instance, the shorter post-KR interval of the DK group may have counteracted the effects of delay, perhaps by providing a shorter time to remember KR before the next trial. It is evident that the effects of delay of knowledge warrant further investigation, with control of all possible confounding variables.

References

Ammons, R. B. Effects of knowledge of performance: A survey and tentative theoretical formulation. *Journal of General Psychology,* 1932, *7,* 245–260.

Bilodeau, E. A., & Bilodeau, I. M. Variation of temporal intervals among critical events in five studies of knowledge of results. *Journal of Experimental Psychology,* 1958, *55,* 603–612.

Bilodeau, E. A., & Ryan, F. J. A test for interaction of delay of knowledge of results and two types of interpolated activity. *Journal of Experimental Psychology,* 1960, *59,* 414–419.

Denny, M. R., Allard, M., Hall, E., & Rokeach, M. Supplementary Report: Delay of knowledge of results, knowledge of task, and intertrial interval. *Journal of Experimental Psychology,* 1960, *60,* 327.

Ellwell, J. L., & Grindley, G. C. The effect of knowledge of results on learning and performance: I. A coordinated movement of the two hands. *British Journal of Psychology,* 1938, *39,* 39–54.

Greenspoon, J., & Foreman, S. Effect of delay of knowledge of results on learning a motor task. *Journal of Experimental Psychology,* 1956, *51,* 226–228.

Judd, C. H. Practice and its effects on the perception of illusions. *Psychological Review,* 1902, *9,* 27–39.

Judd, C. H. Practice without knowledge of results. *Psychological Review Monograph Supplement,* 1905, *7,* 185–198.

McGuigan, F. J. The effect of precision, delay, and schedule of knowledge of results on performance. *Journal of Experimental Psychology,* 1959, *58,* 79–84.

Seashore, H., & Bavelas, A. The functioning of knowledge of results in Thorndike's line-drawing experiment. *Psychological Review,* 1941, *48,* 155–164.

Thorndike, E. L. The law of effect. *American Journal of Psychology,* 1927, *39,* 212–222.

From Geralyn Rodriguez Clucas, "Effects of Knowledge of Results and Delay of Knowledge on the Judgment of the Muller-Dyer Illusion," paper presented at California State University, Hayward, CA. Reprinted by permission.

APPENDIX C

F Distribution

To read this table, locate the degrees of freedom (df) for both mean-square values of the F ratio. The df for the numerator ($df1$) is located in the top row, the df for the denominator ($df2$) in the column to the far left of the table. Find where the row and column intersect for the dfs in your study. There will be two numbers: the top number gives the minimum F value for the .01 level of significance, the bottom number for the .05 level.*

As an example, if you compute an F ratio of 6.75 with 12 and 4 df, enter the table by the column headed by $df1$ equal 12 and the row headed by $df2$ equal 4. The F value required for significance at the .05 level is 5.91; and for the .01 level, 14.37. So, your results would be significant at the .05 level.

---

*You should decide on the critical probability level before you collect data. In the example, if you had decided that you would call your results significant if the F ratio was at least 5.91, then even if F equalled 20, the results would be significant only at the .05 level and should be reported as such.

F Distribution

Degrees of freedom for greater mean square [numerator]

	1	2	3	4	5	6	7	8	9	10	11	12	14	16	20	24	30	40	50	75	100	200	500	∞
1	161 4,052	200 4,999	216 5,403	225 5,625	230 5,764	234 5,859	237 5,928	239 5,981	241 6,022	242 6,056	243 6,082	244 6,106	245 6,142	246 6,169	248 6,208	249 6,234	250 6,261	251 6,286	252 6,302	253 6,323	253 6,334	254 6,352	254 6,361	254 6,366
2	18.51 98.49	19.00 99.00	19.16 99.17	19.25 99.25	19.30 99.30	19.33 99.33	19.36 99.36	19.37 99.37	19.38 99.39	19.39 99.40	19.40 99.41	19.41 99.42	19.42 99.43	19.43 99.44	19.44 99.45	19.45 99.46	19.46 99.47	19.47 99.48	19.47 99.48	19.48 99.49	19.49 99.49	19.49 99.49	19.50 99.50	19.50 99.50
3	10.13 34.12	9.55 30.82	9.28 29.46	9.12 28.71	9.01 28.24	8.94 27.91	8.88 27.67	8.84 27.49	8.81 27.34	8.78 27.23	8.76 27.13	8.74 27.05	8.71 26.92	8.69 26.83	8.66 26.69	8.64 26.60	8.62 26.50	8.60 26.41	8.58 26.35	8.57 26.27	8.56 26.23	8.54 26.18	8.54 26.14	8.53 26.12
4	7.71 21.20	6.94 18.00	6.59 16.69	6.39 15.98	6.26 15.52	6.16 15.21	6.09 14.98	6.04 14.80	6.00 14.66	5.96 14.54	5.93 14.45	5.91 14.37	5.87 14.24	5.84 14.15	5.80 14.02	5.77 13.93	5.74 13.83	5.71 13.74	5.70 13.69	5.68 13.61	5.66 13.57	5.65 13.52	5.64 13.48	5.63 13.46
5	6.61 16.26	5.79 13.27	5.41 12.06	5.19 11.39	5.05 10.97	4.95 10.67	4.88 10.45	4.82 10.29	4.78 10.15	4.74 10.05	4.70 9.96	4.68 9.89	4.64 9.77	4.60 9.68	4.56 9.55	4.53 9.47	4.50 9.38	4.46 9.29	4.44 9.24	4.42 9.17	4.40 9.13	4.38 9.07	4.37 9.04	4.36 9.02
6	5.99 13.74	5.14 10.92	4.76 9.78	4.53 9.15	4.39 8.75	4.28 8.47	4.21 8.26	4.15 8.10	4.10 7.98	4.06 7.87	4.03 7.79	4.00 7.72	3.96 7.60	3.92 7.52	3.87 7.39	3.84 7.31	3.81 7.23	3.77 7.14	3.75 7.09	3.72 7.02	3.71 6.99	3.69 6.94	3.68 6.90	3.67 6.88
7	5.59 12.25	4.74 9.55	4.35 8.45	4.12 7.85	3.97 7.46	3.87 7.19	3.79 7.00	3.73 6.84	3.68 6.71	3.63 6.62	3.60 6.54	3.57 6.47	3.52 6.35	3.49 6.27	3.44 6.15	3.41 6.07	3.38 5.98	3.34 5.90	3.32 5.85	3.29 5.78	3.28 5.75	3.25 5.70	3.24 5.67	3.23 5.65
8	5.32 11.26	4.46 8.65	4.07 7.59	3.84 7.01	3.69 6.63	3.58 6.37	3.50 6.19	3.44 6.03	3.39 5.91	3.34 5.82	3.31 5.74	3.28 5.67	3.23 5.56	3.20 5.48	3.15 5.36	3.12 5.28	3.08 5.20	3.05 5.11	3.03 5.06	3.00 5.00	2.98 4.96	2.96 4.91	2.94 4.88	2.93 4.86
9	5.12 10.56	4.26 8.02	3.86 6.99	3.63 6.42	3.48 6.06	3.37 5.80	3.29 5.62	3.23 5.47	3.18 5.35	3.13 5.26	3.10 5.18	3.07 5.11	3.02 5.00	2.98 4.92	2.93 4.80	2.90 4.73	2.86 4.64	2.82 4.56	2.80 4.51	2.77 4.45	2.76 4.41	2.73 4.36	2.72 4.33	2.71 4.31
10	4.96 10.04	4.10 7.56	3.71 6.55	3.48 5.99	3.33 5.64	3.22 5.39	3.14 5.21	3.07 5.06	3.02 4.95	2.97 4.85	2.94 4.78	2.91 4.71	2.86 4.60	2.82 4.52	2.77 4.41	2.74 4.33	2.70 4.25	2.67 4.17	2.64 4.12	2.61 4.05	2.59 4.01	2.56 3.96	2.55 3.93	2.54 3.91
11	4.84 9.65	3.98 7.20	3.59 6.22	3.36 5.67	3.20 5.32	3.09 5.07	3.01 4.88	2.95 4.74	2.90 4.63	2.86 4.54	2.82 4.46	2.79 4.40	2.74 4.29	2.70 4.21	2.65 4.10	2.61 4.02	2.57 3.94	2.53 3.86	2.50 3.80	2.47 3.74	2.45 3.70	2.42 3.66	2.41 3.62	2.40 3.60
12	4.75 9.33	3.88 6.93	3.49 5.95	3.26 5.41	3.11 5.06	3.00 4.82	2.92 4.65	2.85 4.50	2.80 4.39	2.76 4.30	2.72 4.22	2.69 4.16	2.64 4.05	2.60 3.98	2.54 3.86	2.50 3.78	2.46 3.70	2.42 3.61	2.40 3.56	2.36 3.49	2.35 3.46	2.32 3.41	2.31 3.38	2.30 3.36
13	4.67 9.07	3.80 6.70	3.41 5.74	3.18 5.20	3.02 4.86	2.92 4.62	2.84 4.44	2.77 4.30	2.72 4.19	2.67 4.10	2.63 4.02	2.60 3.96	2.55 3.85	2.51 3.78	2.46 3.67	2.42 3.59	2.38 3.51	2.34 3.42	2.32 3.37	2.28 3.30	2.26 3.27	2.24 3.21	2.22 3.18	2.21 3.16
14	4.60 8.86	3.74 6.51	3.34 5.56	3.11 5.03	2.96 4.69	2.85 4.46	2.77 4.28	2.70 4.14	2.65 4.03	2.60 3.94	2.56 3.86	2.53 3.80	2.48 3.70	2.44 3.62	2.39 3.51	2.35 3.43	2.31 3.34	2.27 3.26	2.24 3.21	2.21 3.14	2.19 3.11	2.16 3.06	2.14 3.02	2.13 3.00
15	4.54 8.68	3.68 6.36	3.29 5.42	3.06 4.89	2.90 4.56	2.79 4.32	2.70 4.14	2.64 4.00	2.59 3.89	2.55 3.80	2.51 3.73	2.48 3.67	2.43 3.56	2.39 3.48	2.33 3.36	2.29 3.29	2.25 3.20	2.21 3.12	2.18 3.07	2.15 3.00	2.12 2.97	2.10 2.92	2.08 2.89	2.07 2.87
16	4.49 8.53	3.63 6.23	3.24 5.29	3.01 4.77	2.85 4.44	2.74 4.20	2.66 4.03	2.59 3.89	2.54 3.78	2.49 3.69	2.45 3.61	2.42 3.55	2.37 3.45	2.33 3.37	2.28 3.25	2.24 3.18	2.20 3.10	2.16 3.01	2.13 2.96	2.09 2.89	2.07 2.86	2.04 2.80	2.02 2.77	2.01 2.75
17	4.45 8.40	3.59 6.11	3.20 5.18	2.96 4.67	2.81 4.34	2.70 4.10	2.62 3.93	2.55 3.79	2.50 3.68	2.45 3.59	2.41 3.52	2.38 3.45	2.33 3.35	2.29 3.27	2.23 3.16	2.19 3.08	2.15 3.00	2.11 2.92	2.08 2.86	2.04 2.79	2.02 2.76	1.99 2.70	1.97 2.67	1.96 2.65
18	4.41 8.28	3.55 6.01	3.16 5.09	2.93 4.58	2.77 4.25	2.66 4.01	2.58 3.85	2.51 3.71	2.46 3.60	2.41 3.51	2.37 3.44	2.34 3.37	2.29 3.27	2.25 3.19	2.19 3.07	2.15 3.00	2.11 2.91	2.07 2.83	2.04 2.78	2.00 2.71	1.98 2.68	1.95 2.62	1.93 2.59	1.92 2.57
19	4.38 8.18	3.52 5.93	3.13 5.01	2.90 4.50	2.74 4.17	2.63 3.94	2.55 3.77	2.48 3.63	2.43 3.52	2.38 3.43	2.34 3.36	2.31 3.30	2.26 3.19	2.21 3.12	2.15 3.00	2.11 2.92	2.07 2.84	2.02 2.76	2.00 2.70	1.96 2.63	1.94 2.60	1.91 2.54	1.90 2.51	1.88 2.49

Degrees of freedom for lesser mean square [denominator]

F Distribution (continued)

									Degrees of freedom for greater mean square [numerator]												
	$p \leq .05$ / $p \leq .01$	1	2	3	4	5	6	7	8	9	10	11	12	14	16	20	24	30	40	50	75	100	200	500	∞
20		4.35 / 8.10	3.49 / 5.85	3.10 / 4.94	2.87 / 4.43	2.71 / 4.10	2.60 / 3.87	2.52 / 3.71	2.45 / 3.56	2.40 / 3.45	2.35 / 3.37	2.31 / 3.30	2.28 / 3.23	2.23 / 3.13	2.18 / 3.05	2.12 / 2.94	2.08 / 2.86	2.04 / 2.77	1.99 / 2.69	1.96 / 2.63	1.92 / 2.56	1.90 / 2.53	1.87 / 2.47	1.85 / 2.44	1.84 / 2.42
21		4.32 / 8.02	3.47 / 5.78	3.07 / 4.87	2.84 / 4.37	2.68 / 4.04	2.57 / 3.81	2.49 / 3.65	2.42 / 3.51	2.37 / 3.40	2.32 / 3.31	2.28 / 3.24	2.25 / 3.17	2.20 / 3.07	2.15 / 2.99	2.09 / 2.88	2.05 / 2.80	2.00 / 2.72	1.96 / 2.63	1.93 / 2.58	1.89 / 2.51	1.87 / 2.47	1.84 / 2.42	1.82 / 2.38	1.81 / 2.36
22		4.30 / 7.94	3.44 / 5.72	3.05 / 4.82	2.82 / 4.31	2.66 / 3.99	2.55 / 3.76	2.47 / 3.59	2.40 / 3.45	2.35 / 3.35	2.30 / 3.26	2.26 / 3.18	2.23 / 3.12	2.18 / 3.02	2.13 / 2.94	2.07 / 2.83	2.03 / 2.75	1.98 / 2.67	1.93 / 2.58	1.91 / 2.53	1.87 / 2.46	1.84 / 2.42	1.81 / 2.37	1.80 / 2.33	1.78 / 2.31
23		4.28 / 7.88	3.42 / 5.66	3.03 / 4.76	2.80 / 4.26	2.64 / 3.94	2.53 / 3.71	2.45 / 3.54	2.38 / 3.41	2.32 / 3.30	2.28 / 3.21	2.24 / 3.14	2.20 / 3.07	2.14 / 2.97	2.10 / 2.89	2.04 / 2.78	2.00 / 2.70	1.96 / 2.62	1.91 / 2.53	1.88 / 2.48	1.84 / 2.41	1.82 / 2.37	1.79 / 2.32	1.77 / 2.28	1.76 / 2.26
24	0.5 / .01	4.26 / 7.82	3.40 / 5.61	3.01 / 4.72	2.78 / 4.22	2.62 / 3.90	2.51 / 3.67	2.43 / 3.50	2.36 / 3.36	2.30 / 3.25	2.26 / 3.17	2.22 / 3.09	2.18 / 3.03	2.13 / 2.93	2.09 / 2.85	2.02 / 2.74	1.98 / 2.66	1.94 / 2.58	1.89 / 2.49	1.86 / 2.44	1.82 / 2.36	1.80 / 2.33	1.76 / 2.27	1.74 / 2.23	1.73 / 2.21
25		4.24 / 7.77	3.38 / 5.57	2.99 / 4.68	2.76 / 4.18	2.60 / 3.86	2.49 / 3.63	2.41 / 3.46	2.34 / 3.32	2.28 / 3.21	2.24 / 3.13	2.20 / 3.05	2.16 / 2.99	2.11 / 2.89	2.06 / 2.81	2.00 / 2.70	1.96 / 2.62	1.92 / 2.54	1.87 / 2.45	1.84 / 2.40	1.80 / 2.32	1.77 / 2.29	1.74 / 2.23	1.72 / 2.19	1.71 / 2.17
26		4.22 / 7.72	3.37 / 5.53	2.98 / 4.64	2.74 / 4.14	2.59 / 3.82	2.47 / 3.59	2.39 / 3.42	2.32 / 3.29	2.27 / 3.17	2.22 / 3.09	2.18 / 3.02	2.15 / 2.96	2.10 / 2.86	2.05 / 2.77	1.99 / 2.66	1.95 / 2.58	1.90 / 2.50	1.85 / 2.41	1.82 / 2.36	1.78 / 2.28	1.76 / 2.25	1.72 / 2.19	1.70 / 2.15	1.69 / 2.13
27		4.21 / 7.68	3.35 / 5.49	2.96 / 4.60	2.73 / 4.11	2.57 / 3.79	2.46 / 3.56	2.37 / 3.39	2.30 / 3.26	2.25 / 3.14	2.20 / 3.06	2.16 / 2.98	2.13 / 2.93	2.08 / 2.83	2.03 / 2.74	1.97 / 2.63	1.93 / 2.55	1.88 / 2.47	1.84 / 2.38	1.80 / 2.33	1.76 / 2.25	1.74 / 2.21	1.71 / 2.16	1.68 / 2.12	1.67 / 2.10
28		4.20 / 7.64	3.34 / 5.45	2.95 / 4.57	2.71 / 4.07	2.56 / 3.76	2.44 / 3.53	2.36 / 3.36	2.29 / 3.23	2.24 / 3.11	2.19 / 3.03	2.15 / 2.95	2.12 / 2.90	2.06 / 2.80	2.02 / 2.71	1.96 / 2.60	1.91 / 2.52	1.87 / 2.44	1.81 / 2.35	1.78 / 2.30	1.75 / 2.22	1.72 / 2.18	1.69 / 2.13	1.67 / 2.09	1.65 / 2.06
29		4.18 / 7.60	3.33 / 5.42	2.93 / 4.54	2.70 / 4.04	2.54 / 3.73	2.43 / 3.50	2.35 / 3.33	2.28 / 3.20	2.22 / 3.08	2.18 / 3.00	2.14 / 2.92	2.10 / 2.87	2.05 / 2.77	2.00 / 2.68	1.94 / 2.57	1.90 / 2.49	1.85 / 2.41	1.80 / 2.32	1.77 / 2.27	1.73 / 2.19	1.71 / 2.15	1.68 / 2.10	1.65 / 2.06	1.64 / 2.03
30		4.17 / 7.56	3.32 / 5.39	2.92 / 4.51	2.69 / 4.02	2.53 / 3.70	2.42 / 3.47	2.34 / 3.30	2.27 / 3.17	2.21 / 3.06	2.16 / 2.98	2.12 / 2.90	2.09 / 2.84	2.04 / 2.74	1.99 / 2.66	1.93 / 2.55	1.89 / 2.47	1.84 / 2.38	1.79 / 2.29	1.76 / 2.24	1.72 / 2.16	1.69 / 2.13	1.66 / 2.07	1.64 / 2.03	1.62 / 2.01
32		4.15 / 7.50	3.30 / 5.34	2.90 / 4.46	2.67 / 3.97	2.51 / 3.66	2.40 / 3.42	2.32 / 3.25	2.25 / 3.12	2.19 / 3.01	2.14 / 2.94	2.10 / 2.86	2.07 / 2.80	2.02 / 2.70	1.97 / 2.62	1.91 / 2.51	1.86 / 2.42	1.82 / 2.34	1.76 / 2.25	1.74 / 2.20	1.69 / 2.12	1.67 / 2.08	1.64 / 2.02	1.61 / 1.98	1.59 / 1.96
34		4.13 / 7.44	3.28 / 5.29	2.88 / 4.42	2.65 / 3.93	2.49 / 3.61	2.38 / 3.38	2.30 / 3.21	2.23 / 3.08	2.17 / 2.97	2.12 / 2.89	2.08 / 2.82	2.05 / 2.76	2.00 / 2.66	1.95 / 2.58	1.89 / 2.47	1.84 / 2.38	1.80 / 2.30	1.74 / 2.21	1.71 / 2.15	1.67 / 2.08	1.64 / 2.04	1.61 / 1.98	1.59 / 1.94	1.57 / 1.91
36		4.11 / 7.39	3.26 / 5.25	2.86 / 4.38	2.63 / 3.89	2.48 / 3.58	2.36 / 3.35	2.28 / 3.18	2.21 / 3.04	2.15 / 2.94	2.10 / 2.86	2.06 / 2.78	2.03 / 2.72	1.98 / 2.62	1.93 / 2.54	1.87 / 2.43	1.82 / 2.35	1.78 / 2.26	1.72 / 2.17	1.69 / 2.12	1.65 / 2.04	1.62 / 2.00	1.59 / 1.94	1.56 / 1.90	1.55 / 1.87
38		4.10 / 7.35	3.25 / 5.21	2.85 / 4.34	2.62 / 3.86	2.46 / 3.54	2.35 / 3.32	2.26 / 3.15	2.19 / 3.02	2.14 / 2.91	2.09 / 2.82	2.05 / 2.75	2.02 / 2.69	1.96 / 2.59	1.92 / 2.51	1.85 / 2.40	1.80 / 2.32	1.76 / 2.22	1.71 / 2.14	1.67 / 2.08	1.63 / 2.00	1.60 / 1.97	1.57 / 1.90	1.54 / 1.86	1.53 / 1.84
40		4.08 / 7.31	3.23 / 5.18	2.84 / 4.31	2.61 / 3.83	2.45 / 3.51	2.34 / 3.29	2.25 / 3.12	2.18 / 2.99	2.12 / 2.88	2.07 / 2.80	2.04 / 2.73	2.00 / 2.66	1.95 / 2.56	1.90 / 2.49	1.84 / 2.37	1.79 / 2.29	1.74 / 2.20	1.69 / 2.11	1.66 / 2.05	1.61 / 1.97	1.59 / 1.94	1.55 / 1.88	1.53 / 1.84	1.51 / 1.81
42		4.07 / 7.27	3.22 / 5.15	2.83 / 4.29	2.59 / 3.80	2.44 / 3.49	2.32 / 3.26	2.24 / 3.10	2.17 / 2.96	2.11 / 2.86	2.06 / 2.77	2.02 / 2.70	1.99 / 2.64	1.94 / 2.54	1.89 / 2.46	1.82 / 2.35	1.78 / 2.26	1.73 / 2.17	1.68 / 2.08	1.64 / 2.02	1.60 / 1.94	1.57 / 1.91	1.54 / 1.85	1.51 / 1.80	1.49 / 1.78
44		4.06 / 7.24	3.21 / 5.12	2.82 / 4.26	2.58 / 3.78	2.43 / 3.46	2.31 / 3.24	2.23 / 3.07	2.16 / 2.94	2.10 / 2.84	2.05 / 2.75	2.01 / 2.68	1.98 / 2.62	1.92 / 2.52	1.88 / 2.44	1.81 / 2.32	1.76 / 2.24	1.72 / 2.15	1.66 / 2.06	1.63 / 2.00	1.58 / 1.92	1.56 / 1.88	1.52 / 1.82	1.50 / 1.78	1.48 / 1.75

Degrees of freedom for lesser mean square [denominator]

F Distribution (continued)

| | \multicolumn{19}{c}{Degrees of freedom for greater mean square [numerator]} |
	1	2	3	4	5	6	7	8	9	10	11	12	14	16	20	24	30	40	50	75	100	200	500	∞
46	4.05 7.21	3.20 5.10	2.81 4.24	2.57 3.76	2.42 3.44	2.30 3.22	2.22 3.05	2.14 2.92	2.09 2.82	2.04 2.73	2.00 2.66	1.97 2.60	1.91 2.50	1.87 2.42	1.80 2.30	1.75 2.22	1.71 2.13	1.65 2.04	1.62 1.98	1.57 1.90	1.54 1.86	1.51 1.80	1.48 1.76	1.46 1.72
48	4.04 7.19	3.19 5.08	2.80 4.22	2.56 3.74	2.41 3.42	2.30 3.20	2.21 3.04	2.14 2.90	2.08 2.80	2.03 2.71	1.99 2.64	1.96 2.58	1.90 2.48	1.86 2.40	1.79 2.28	1.74 2.20	1.70 2.11	1.64 2.02	1.61 1.96	1.56 1.88	1.53 1.84	1.50 1.78	1.47 1.73	1.45 1.70
50	4.03 7.17	3.18 5.06	2.79 4.20	2.56 3.72	2.40 3.41	2.29 3.18	2.20 3.02	2.13 2.88	2.07 2.78	2.02 2.70	1.98 2.62	1.95 2.56	1.90 2.46	1.85 2.39	1.78 2.26	1.74 2.18	1.69 2.10	1.63 2.00	1.60 1.94	1.55 1.86	1.52 1.82	1.48 1.76	1.46 1.71	1.44 1.68
55	4.02 7.12	3.17 5.01	2.78 4.16	2.54 3.68	2.38 3.37	2.27 3.15	2.18 2.98	2.11 2.85	2.05 2.75	2.00 2.66	1.97 2.59	1.93 2.53	1.88 2.43	1.83 2.35	1.76 2.23	1.72 2.15	1.67 2.06	1.61 1.96	1.58 1.90	1.52 1.82	1.50 1.78	1.46 1.71	1.43 1.66	1.41 1.64
60	4.00 7.08	3.15 4.98	2.76 4.13	2.52 3.65	2.37 3.34	2.25 3.12	2.17 2.95	2.10 2.82	2.04 2.72	1.99 2.63	1.95 2.56	1.92 2.50	1.86 2.40	1.81 2.32	1.75 2.20	1.70 2.12	1.65 2.03	1.59 1.93	1.56 1.87	1.50 1.79	1.48 1.74	1.44 1.68	1.41 1.63	1.39 1.60
65	3.99 7.04	3.14 4.95	2.75 4.10	2.51 3.62	2.36 3.31	2.24 3.09	2.15 2.93	2.08 2.79	2.02 2.70	1.98 2.61	1.94 2.54	1.90 2.47	1.85 2.37	1.80 2.30	1.73 2.18	1.68 2.09	1.63 2.00	1.57 1.90	1.54 1.84	1.49 1.76	1.46 1.71	1.42 1.64	1.39 1.60	1.37 1.56
70	3.98 7.01	3.13 4.92	2.74 4.08	2.50 3.60	2.35 3.29	2.23 3.07	2.14 2.91	2.07 2.77	2.01 2.67	1.97 2.59	1.93 2.51	1.89 2.45	1.84 2.35	1.79 2.28	1.72 2.15	1.67 2.07	1.62 1.98	1.56 1.88	1.53 1.82	1.47 1.74	1.45 1.69	1.40 1.62	1.37 1.56	1.35 1.53
80	3.96 6.96	3.11 4.88	2.72 4.04	2.48 3.56	2.33 3.25	2.21 3.04	2.12 2.87	2.05 2.74	1.99 2.64	1.95 2.55	1.91 2.48	1.88 2.41	1.82 2.32	1.77 2.24	1.70 2.11	1.65 2.03	1.60 1.94	1.54 1.84	1.51 1.78	1.45 1.70	1.42 1.65	1.38 1.57	1.35 1.52	1.32 1.49
100	3.94 6.90	3.09 4.82	2.70 3.98	2.46 3.51	2.30 3.20	2.19 2.99	2.10 2.82	2.03 2.69	1.97 2.59	1.92 2.51	1.88 2.43	1.85 2.36	1.79 2.26	1.75 2.19	1.68 2.06	1.63 1.98	1.57 1.89	1.51 1.79	1.48 1.73	1.42 1.64	1.39 1.59	1.34 1.51	1.30 1.46	1.28 1.43
125	3.92 6.84	3.07 4.78	2.68 3.94	2.44 3.47	2.29 3.17	2.17 2.95	2.08 2.79	2.01 2.65	1.95 2.56	1.90 2.47	1.86 2.40	1.83 2.33	1.77 2.23	1.72 2.15	1.65 2.03	1.60 1.94	1.55 1.85	1.49 1.75	1.45 1.68	1.39 1.59	1.36 1.54	1.31 1.46	1.27 1.40	1.25 1.37
150	3.91 6.81	3.06 4.75	2.67 3.91	2.43 3.44	2.27 3.14	2.16 2.92	2.07 2.76	2.00 2.62	1.94 2.53	1.89 2.44	1.85 2.37	1.82 2.30	1.76 2.20	1.71 2.12	1.64 2.00	1.59 1.91	1.54 1.83	1.47 1.72	1.44 1.66	1.37 1.56	1.34 1.51	1.29 1.43	1.25 1.37	1.22 1.33
200	3.89 6.76	3.04 4.71	2.65 3.88	2.41 3.41	2.26 3.11	2.14 2.90	2.05 2.73	1.98 2.60	1.92 2.50	1.87 2.41	1.83 2.34	1.80 2.28	1.74 2.17	1.69 2.09	1.62 1.97	1.57 1.88	1.52 1.79	1.45 1.69	1.42 1.62	1.35 1.53	1.32 1.48	1.26 1.39	1.22 1.33	1.19 1.28
400	3.86 6.70	3.02 4.66	2.62 3.83	2.39 3.36	2.23 3.06	2.12 2.85	2.03 2.69	1.96 2.55	1.90 2.46	1.85 2.37	1.81 2.29	1.78 2.23	1.72 2.12	1.67 2.04	1.60 1.92	1.54 1.84	1.49 1.74	1.42 1.64	1.38 1.57	1.32 1.47	1.28 1.42	1.22 1.32	1.16 1.24	1.13 1.19
1000	3.85 6.66	3.00 4.62	2.61 3.80	2.38 3.34	2.22 3.04	2.10 2.82	2.02 2.66	1.95 2.53	1.89 2.43	1.84 2.34	1.80 2.26	1.76 2.20	1.70 2.09	1.65 2.01	1.58 1.89	1.53 1.81	1.47 1.71	1.41 1.61	1.36 1.54	1.30 1.44	1.26 1.38	1.19 1.28	1.13 1.19	1.08 1.11
∞	3.84 6.64	2.99 4.60	2.60 3.78	2.37 3.32	2.21 3.02	2.09 2.80	2.01 2.64	1.94 2.51	1.88 2.41	1.83 2.32	1.79 2.24	1.75 2.18	1.69 2.07	1.64 1.99	1.57 1.87	1.52 1.79	1.46 1.69	1.40 1.59	1.35 1.52	1.28 1.41	1.24 1.36	1.17 1.25	1.11 1.15	1.00 1.00

Degrees of freedom for lesser mean square [denominator]

Reprinted by permission from *Statistical Methods*, Eighth Edition, by G. W. Snedecor and W. G. Cochran © 1989 by Iowa State University Press, Ames, Iowa 50010.

APPENDIX D

Table of Critical Values of D (or C) in the Fisher Test†

Totals in right margin		B (or A)†	Level of significance	
			.05	.01
$A + B = 3$	$C + D = 3$	3	0	—
$A + B = 4$	$C + D = 4$	4	0	—
	$C + D = 3$	4	0	—
$A + B = 5$	$C + D = 5$	5	1	0
		4	0	—
	$C + D = 4$	5	1	0
		4	0	—
	$C + D = 3$	5	0	—
	$C + D = 2$	5	0	—
$A + B = 6$	$C + D = 6$	6	2	1
		5	1	0
		4	0	—
	$C + D = 5$	6	1	0
		5	0	—
		4	0	—
	$C + D = 4$	6	1	0
		5	0	—
	$C + D = 3$	6	0	—
		5	0	—
	$C + D = 2$	6	0	—
$A + B = 7$	$C + D = 7$	7	3	1
		6	1	0
		5	0	—
		4	0	—
	$C + D = 6$	7	2	1
		6	1	0
		5	0	—
		4	0	—
	$C + D = 5$	7	2	0
		6	1	0
		5	0	—
	$C + D = 4$	7	1	0
		6	0	—
		5	0	—
	$C + D = 3$	7	0	0
		6	0	—
	$C + D = 2$	7	0	—

† When B is entered in the middle column, the significance levels are for D. When A is used in place of B, the significance levels are for C.

Table of Critical Values of D (or C) in the Fisher Test (Continued)

Totals in right margin		B (or A)†	Level of significance	
			.05	.01
$A + B = 8$	$C + D = 8$	8	4	2
		7	2	1
		6	1	0
		5	0	—
		4	0	—
	$C + D = 7$	8	3	2
		7	2	1
		6	1	0
		5	0	—
	$C + D = 6$	8	2	1
		7	1	0
		6	0	0
		5	0	—
	$C + D = 5$	8	2	1
		7	1	0
		6	0	—
		5	0	—
	$C + D = 4$	8	1	0
		7	0	—
		6	0	—
	$C + D = 3$	8	0	0
		7	0	—
	$C + D = 2$	8	0	—
$A + B = 9$	$C + D = 9$	9	5	3
		8	3	2
		7	2	1
		6	1	0
		5	0	—
		4	0	—
	$C + D = 8$	9	4	3
		8	3	1
		7	2	0
		6	1	0
		5	0	—
	$C + D = 7$	9	3	2
		8	2	1
		7	1	0
		6	0	—
		5	0	—

Table of Critical Values of D (or C) in the Fisher Test (Continued)

Totals in right margin		B (or A)†	Level of significance	
			.05	.01
$A+B=9$	$C+D=6$	9	3	1
		8	2	0
		7	1	0
		6	0	—
		5	0	—
	$C+D=5$	9	2	1
		8	1	0
		7	0	—
		6	0	—
	$C+D=4$	9	1	0
		8	0	0
		7	0	—
		6	0	—
	$C+D=3$	9	1	0
		8	0	—
		7	0	—
	$C+D=2$	9	0	—
$A+B=10$	$C+D=10$	10	6	4
		9	4	3
		8	3	1
		7	2	1
		6	1	0
		5	0	—
		4	0	—
	$C+D=9$	10	5	3
		9	4	2
		8	2	1
		7	1	0
		6	1	0
		5	0	—
	$C+D=8$	10	4	3
		9	3	2
		8	2	1
		7	1	0
		6	0	—
		5	0	—
	$C+D=7$	10	3	2
		9	2	1
		8	1	0
		7	1	0
		6	0	—
		5	0	—

Table of Critical Values of D *(or* C*) in the Fisher Test* (Continued)

Totals in right margin		B (or A)†	Level of significance	
			.05	.01
A + B = 10	C + D = 6	10	3	2
		9	2	1
		8	1	0
		7	0	—
		6	0	—
	C + D = 5	10	2	1
		9	1	0
		8	1	0
		7	0	—
		6	0	—
	C + D = 4	10	1	0
		9	1	0
		8	0	—
		7	0	—
	C + D = 3	10	1	0
		9	0	—
		8	0	—
	C + D = 2	10	0	—
		9	0	—
A + B = 11	C + D = 11	11	7	5
		10	5	3
		9	4	2
		8	3	1
		7	2	0
		6	1	0
		5	0	—
		4	0	—
	C + D = 10	11	6	4
		10	4	3
		9	3	2
		8	2	1
		7	1	0
		6	1	0
		5	0	—
	C + D = 9	11	5	4
		10	4	2
		9	3	1
		8	2	1
		7	1	0
		6	0	—
		5	0	—

Table of Critical Values of D (or C) in the Fisher Test (Continued)

Totals in right margin		B (or A)†	Level of significance	
			.05	.01
$A+B=11$	$C+D=8$	11	4	3
		10	3	2
		9	2	1
		8	1	0
		7	1	0
		6	0	—
		5	0	—
	$C+D=7$	11	4	2
		10	3	1
		9	2	1
		8	1	0
		7	0	—
		6	0	—
	$C+D=6$	11	3	2
		10	2	1
		9	1	0
		8	1	0
		7	0	—
		6	0	—
	$C+D=5$	11	2	1
		10	1	0
		9	1	0
		8	0	—
		7	0	—
	$C+D=4$	11	1	1
		10	1	0
		9	0	—
		8	0	—
	$C+D=3$	11	1	0
		10	0	—
		9	0	—
	$C+D=2$	11	0	—
		10	0	—
$A+B=12$	$C+D=12$	12	8	6
		11	6	4
		10	5	3
		9	4	2
		8	3	1
		7	2	0
		6	1	0
		5	0	—
		4	0	—

Appendix D

Table of Critical Values of D *(or C) in the Fisher Test (Continued)*

Totals in right margin		B (or A)†	Level of significance	
			.05	.01
A + B = 12	C + D = 11	12	7	5
		11	5	4
		10	4	2
		9	3	2
		8	2	1
		7	1	0
		6	1	0
		5	0	—
	C + D = 10	12	6	5
		11	5	3
		10	4	2
		9	3	1
		8	2	0
		7	1	0
		6	0	—
		5	0	—
	C + D = 9	12	5	4
		11	4	3
		10	3	2
		9	2	1
		8	1	0
		7	1	0
		6	0	—
		5	0	—
	C + D = 8	12	5	3
		11	3	2
		10	2	1
		9	2	1
		8	1	0
		7	0	—
		6	0	—
	C + D = 7	12	4	3
		11	3	2
		10	2	1
		9	1	0
		8	1	0
		7	0	—
		6	0	—

Table of Critical Values of D (or C) in the Fisher Test (Continued)

Totals in right margin		B (or A)†	Level of significance	
			.05	.01
$A + B = 12$	$C + D = 6$	12	3	2
		11	2	1
		10	1	0
		9	1	0
		8	0	—
		7	0	—
		6	0	—
	$C + D = 5$	12	2	1
		11	1	1
		10	1	0
		9	0	0
		8	0	—
		7	0	—
	$C + D = 4$	12	2	1
		11	1	0
		10	0	0
		9	0	—
		8	0	—
	$C + D = 3$	12	1	0
		11	0	0
		10	0	—
		9	0	—
	$C + D = 2$	12	0	—
		11	0	—
$A + B = 13$	$C + D = 13$	13	9	7
		12	7	5
		11	6	4
		10	4	3
		9	3	2
		8	2	1
		7	2	0
		6	1	0
		5	0	—
		4	0	—
	$C + D = 12$	13	8	6
		12	6	5
		11	5	3
		10	4	2
		9	3	1
		8	2	1
		7	1	0
		6	1	0
		5	0	—

Table of Critical Values of D (or C) in the Fisher Test (Continued)

Totals in right margin		B (or A)†	Level of significance	
			.05	.01
$A + B = 13$	$C + D = 11$	13	7	5
		12	6	4
		11	4	3
		10	3	2
		9	3	1
		8	2	0
		7	1	0
		6	0	—
		5	0	—
	$C + D = 10$	13	6	5
		12	5	3
		11	4	2
		10	3	1
		9	2	1
		8	1	0
		7	1	0
		6	0	—
		5	0	—
	$C + D = 9$	13	5	4
		12	4	3
		11	3	2
		10	2	1
		9	2	0
		8	1	0
		7	0	—
		6	0	—
		5	0	—
	$C + D = 8$	13	5	3
		12	4	2
		11	3	1
		10	2	1
		9	1	0
		8	1	0
		7	0	—
		6	0	—
	$C + D = 7$	13	4	3
		12	3	2
		11	2	1
		10	1	0
		9	1	0
		8	0	—
		7	0	—
		6	0	—

Table of Critical Values of D (or C) in the Fisher Test (Continued)

Totals in right margin		B (or A)†	Level of significance	
			.05	.01
A + B = 13	C + D = 6	13	3	2
		12	2	1
		11	2	1
		10	1	0
		9	1	0
		8	0	—
		7	0	—
	C + D = 5	13	2	1
		12	2	1
		11	1	0
		10	1	0
		9	0	—
		8	0	—
	C + D = 4	13	2	1
		12	1	0
		11	0	0
		10	0	—
		9	0	—
	C + D = 3	13	1	0
		12	0	0
		11	0	—
		10	0	—
	C + D = 2	13	0	0
		12	0	—
A + B = 14	C + D = 14	14	10	8
		13	8	6
		12	6	5
		11	5	3
		10	4	2
		9	3	2
		8	2	1
		7	1	0
		6	1	0
		5	0	—
		4	0	—

Table of Critical Values of D (or C) in the Fisher Test (Continued)

Totals in right margin		B (or A)†	Level of significance	
			.05	.01
A + B = 14	C + D = 13	14	9	7
		13	7	5
		12	6	4
		11	5	3
		10	4	2
		9	3	1
		8	2	1
		7	1	0
		6	1	—
		5	0	—
	C + D = 12	14	8	6
		13	6	5
		12	5	4
		11	4	3
		10	3	2
		9	2	1
		8	2	0
		7	1	0
		6	0	—
		5	0	—
	C + D = 11	14	7	6
		13	6	4
		12	5	3
		11	4	2
		10	3	1
		9	2	1
		8	1	0
		7	1	0
		6	0	—
		5	0	—
	C + D = 10	14	6	5
		13	5	4
		12	4	3
		11	3	2
		10	2	1
		9	2	0
		8	1	0
		7	0	0
		6	0	—
		5	0	—

Table of Critical Values of D (or C) in the Fisher Test (Continued)

Totals in right margin		B (or A)†	Level of significance	
			.05	.01
$A + B = 14$	$C + D = 9$	14	6	4
		13	4	3
		12	3	2
		11	3	1
		10	2	1
		9	1	0
		8	1	0
		7	0	—
		6	0	—
	$C + D = 8$	14	5	4
		13	4	2
		12	3	2
		11	2	1
		10	2	0
		9	1	0
		8	0	0
		7	0	—
		6	0	—
	$C + D = 7$	14	4	3
		13	3	2
		12	2	1
		11	2	1
		10	1	0
		9	1	0
		8	0	—
		7	0	—
	$C + D = 6$	14	3	2
		13	2	1
		12	2	1
		11	1	0
		10	1	0
		9	0	—
		8	0	—
		7	0	—
	$C + D = 5$	14	2	1
		13	2	1
		12	1	0
		11	1	0
		10	0	—
		9	0	—
		8	0	—

Table of Critical Values of D *(or* C*) in the Fisher Test (Continued)*

Totals in right margin		B (or A)†	Level of significance	
			.05	.01
A + B = 14	C + D = 4	14	2	1
		13	1	0
		12	1	0
		11	0	—
		10	0	—
		9	0	—
	C + D = 3	14	1	0
		13	0	0
		12	0	—
		11	0	—
	C + D = 2	14	0	0
		13	0	—
		12	0	—
A + B = 15	C + D = 15	15	11	9
		14	9	7
		13	7	5
		12	6	4
		11	5	3
		10	4	2
		9	3	1
		8	2	1
		7	1	0
		6	1	0
		5	0	—
		4	0	—
	C + D = 14	15	10	8
		14	8	6
		13	7	5
		12	6	4
		11	5	3
		10	4	2
		9	3	1
		8	2	1
		7	1	0
		6	1	—
		5	0	—

Table of Critical Values of D (or C) in the Fisher Test (Continued)

Totals in right margin		B (or A)†	Level of significance	
			.05	.01
$A + B = 15$	$C + D = 13$	15	9	7
		14	7	6
		13	6	4
		12	5	3
		11	4	2
		10	3	2
		9	2	1
		8	2	0
		7	1	0
		6	0	—
		5	0	—
	$C + D = 12$	15	8	7
		14	7	5
		13	6	4
		12	5	3
		11	4	2
		10	3	1
		9	2	1
		8	1	0
		7	1	0
		6	0	—
		5	0	—
	$C + D = 11$	15	7	6
		14	6	4
		13	5	3
		12	4	2
		11	3	2
		10	2	1
		9	2	0
		8	1	0
		7	1	0
		6	0	—
		5	0	—
	$C + D = 10$	15	6	5
		14	5	4
		13	4	3
		12	3	2
		11	3	1
		10	2	1
		9	1	0
		8	1	0
		7	0	—
		6	0	—

Table of Critical Values of D (or C) in the Fisher Test (Continue

Totals in right margin		B (or A)†	Level of significance	
			.05	.01
$A + B = 15$	$C + D = 9$	15	6	4
		14	5	3
		13	4	2
		12	3	2
		11	2	1
		10	2	0
		9	1	0
		8	1	0
		7	0	—
		6	0	—
	$C + D = 8$	15	5	4
		14	4	3
		13	3	2
		12	2	1
		11	2	1
		10	1	0
		9	1	0
		8	0	—
		7	0	—
		6	0	—
	$C + D = 7$	15	4	3
		14	3	2
		13	2	1
		12	2	1
		11	1	0
		10	1	0
		9	0	—
		8	0	—
		7	0	—
	$C + D = 6$	15	3	2
		14	2	1
		13	2	1
		12	1	0
		11	1	0
		10	0	0
		9	0	—
		8	0	—
	$C + D = 5$	15	2	2
		14	2	1
		13	1	0
		12	1	0
		11	0	0
		10	0	—
		9	0	—

Table of Critical Values of D (or C) in the Fisher Test (Continued)

Totals in right margin		B (or A)†	Level of significance	
			.05	.01
$A + B = 15$	$C + D = 4$	15	2	1
		14	1	0
		13	1	0
		12	0	0
		11	0	—
		10	0	—
	$C + D = 3$	15	1	0
		14	0	0
		13	0	—
		12	0	—
		11	0	—
	$C + D = 2$	15	0	0
		14	0	—
		13	0	—

From D. J. Finney, "The Fisher-Yates Test of Significance in 2 × 2 Contingency Tables" in *Biometrika*, 35:149–154, copyright 1948. Reprinted by permission of the Biometrika Trustees.

APPENDIX E

Table of Probabilities Associated with Values as Extreme as Observed Values of z in the Normal Distribution

The body of the table gives one-tailed probabilities under H_o of z. The left-hand marginal column gives various values of z to one decimal place. The top row gives various values to the second decimal place. Thus, for example, the one-tailed p of $z \geq .11$ or $z \leq -.11$ is $p = .4562$.

z	.00	.01	.02	.03	.04	.05	.06	.07	.08	.09
.0	.5000	.4960	.4920	.4880	.4840	.4801	.4761	.4721	.4681	.4641
.1	.4602	.4562	.4522	.4483	.4443	.4404	.4364	.4325	.4286	.4247
.2	.4207	.4168	.4129	.4090	.4052	.4013	.3974	.3936	.3897	.3859
.3	.3821	.3783	.3745	.3707	.3669	.3632	.3594	.3557	.3520	.3483
.4	.3446	.3409	.3372	.3336	.3300	.3264	.3228	.3192	.3156	.3121
.5	.3085	.3050	.3015	.2981	.2946	.2912	.2877	.2843	.2810	.2776
.6	.2743	.2709	.2676	.2643	.2611	.2578	.2546	.2514	.2483	.2451
.7	.2420	.2389	.2358	.2327	.2296	.2266	.2236	.2206	.2177	.2148
.8	.2119	.2090	.2061	.2033	.2005	.1977	.1949	.1922	.1894	.1867
.9	.1841	.1814	.1788	.1762	.1736	.1711	.1685	.1660	.1635	.1611
1.0	.1587	.1562	.1539	.1515	.1492	.1469	.1446	.1423	.1401	.1379
1.1	.1357	.1335	.1314	.1292	.1271	.1251	.1230	.1210	.1190	.1170
1.2	.1151	.1131	.1112	.1093	.1075	.1056	.1038	.1020	.1003	.0985
1.3	.0968	.0951	.0934	.0918	.0901	.0885	.0869	.0853	.0838	.0823
1.4	.0808	.0793	.0778	.0764	.0749	.0735	.0721	.0708	.0694	.0681
1.5	.0668	.0655	.0643	.0630	.0618	.0606	.0594	.0582	.0571	.0559
1.6	.0548	.0537	.0526	.0516	.0505	.0495	.0485	.0475	.0465	.0455
1.7	.0446	.0436	.0427	.0418	.0409	.0401	.0392	.0384	.0375	.0367
1.8	.0359	.0351	.0344	.0336	.0329	.0322	.0314	.0307	.0301	.0294
1.9	.0287	.0281	.0274	.0268	.0262	.0256	.0250	.0244	.0239	.0233
2.0	.0228	.0222	.0217	.0212	.0207	.0202	.0197	.0192	.0188	.0183
2.1	.0179	.0174	.0170	.0166	.0162	.0158	.0154	.0150	.0146	.0143
2.2	.0139	.0136	.0132	.0129	.0125	.0122	.0119	.0116	.0113	.0110
2.3	.0107	.0104	.0102	.0099	.0096	.0094	.0091	.0089	.0087	.0084
2.4	.0082	.0080	.0078	.0075	.0073	.0071	.0069	.0068	.0066	.0064
2.5	.0062	.0060	.0059	.0057	.0055	.0054	.0052	.0051	.0049	.0048
2.6	.0047	.0045	.0044	.0043	.0041	.0040	.0039	.0038	.0037	.0036
2.7	.0035	.0034	.0033	.0032	.0031	.0030	.0029	.0028	.0027	.0026
2.8	.0026	.0025	.0024	.0023	.0023	.0022	.0021	.0021	.0020	.0019
2.9	.0019	.0018	.0018	.0017	.0016	.0016	.0015	.0015	.0014	.0014
3.0	.0013	.0013	.0013	.0012	.0012	.0011	.0011	.0011	.0010	.0010
3.1	.0010	.0009	.0009	.0009	.0008	.0008	.0008	.0008	.0007	.0007
3.2	.0007									
3.3	.0005									
3.4	.0003									
3.5	.00023									
3.6	.00016									
3.7	.00011									
3.8	.00007									
3.9	.00005									
4.0	.00003									

From Siegel, *Nonparametric Statistics*. Copyright McGraw-Hill Publishing Company, New York, NY. Reprinted by permission.

APPENDIX F

Table of $z = (1/2) \ln (1 + r)/(1 - r)$ to Transform the Correlation Coefficient

r	.00	.01	.02	.03	.04	.05	.06	.07	.08	.09
.0	0.000	0.010	0.020	0.030	0.040	0.050	0.060	0.070	0.080	0.090
.1	.100	.110	.121	.131	.141	.151	.161	.172	.182	.192
.2	.203	.213	.224	.234	.245	.255	.266	.277	.288	.299
.3	.310	.321	.332	.343	.354	.365	.377	.388	.400	.412
.4	.424	.436	.448	.460	.472	.485	.497	.510	.523	.536
.5	.549	.563	.576	.590	.604	.618	.633	.648	.662	.678
.6	.693	.709	.725	.741	.758	.775	.793	.811	.829	.848
.7	.867	.887	.908	.929	.950	.973	.996	1.020	1.045	1.071
.8	1.099	1.127	1.157	1.188	1.221	1.256	1.293	1.333	1.376	1.422

r	.000	.001	.002	.003	.004	.005	.006	.007	.008	.009
.90	1.472	1.478	1.483	1.488	1.494	1.499	1.505	1.510	1.516	1.522
.91	1.528	1.533	1.539	1.545	1.551	1.557	1.564	1.570	1.576	1.583
.92	1.589	1.596	1.602	1.609	1.616	1.623	1.630	1.637	1.644	1.651
.93	1.658	1.666	1.673	1.681	1.689	1.697	1.705	1.713	1.721	1.730
.94	1.738	1.747	1.756	1.764	1.774	1.783	1.792	1.802	1.812	1.822
.95	1.832	1.842	1.853	1.863	1.874	1.886	1.897	1.909	1.921	1.933
.96	1.946	1.959	1.972	1.986	2.000	2.014	2.029	2.044	2.060	2.076
.97	2.092	2.109	2.127	2.146	2.165	2.185	2.205	2.227	2.249	2.273
.98	2.298	2.323	2.351	2.380	2.410	2.443	2.477	2.515	2.555	2.599
.99	2.646	2.700	2.759	2.826	2.903	2.994	3.106	3.250	3.453	3.800

Table of r in Terms of z

z	0.00	0.01	0.02	0.03	0.04	0.05	0.06	0.07	0.08	0.09
0.0	.000	.010	.020	.030	.040	.050	.060	.070	.080	.090
.1	.100	.110	.119	.129	.139	.149	.159	.168	.178	.187
.2	.197	.207	.216	.226	.236	.245	.254	.264	.273	.282
.3	.291	.300	.310	.319	.327	.336	.345	.354	.363	.371
.4	.380	.389	.397	.405	.414	.422	.430	.438	.446	.454
.5	.462	.470	.478	.485	.493	.500	.508	.515	.523	.530
.6	.537	.544	.551	.558	.565	.572	.578	.585	.592	.598
.7	.604	.611	.617	.623	.629	.635	.641	.647	.653	.658
.8	.664	.670	.675	.680	.686	.691	.696	.701	.706	.711
.9	.716	.721	.726	.731	.735	.740	.744	.749	.753	.757
1.0	.762	.766	.770	.774	.778	.782	.786	.790	.793	.797
1.1	.800	.804	.808	.811	.814	.818	.821	.824	.828	.831
1.2	.834	.837	.840	.843	.846	.848	.851	.854	.856	.859
1.3	.862	.864	.867	.869	.872	.874	.876	.879	.881	.883
1.4	.885	.888	.890	.892	.894	.896	.898	.900	.902	.903
1.5	.905	.907	.909	.910	.912	.914	.915	.917	.919	.920
1.6	.922	.923	.925	.926	.928	.929	.930	.932	.933	.934
1.7	.935	.937	.938	.939	.940	.941	.942	.944	.945	.946
1.8	.947	.948	.949	.950	.951	.952	.953	.954	.954	.955
1.9	.956	.957	.958	.959	.960	.960	.961	.962	.963	.963
2.0	.964	.965	.965	.966	.967	.967	.968	.969	.969	.970
2.1	.970	.971	.972	.972	.973	.973	.974	.974	.975	.975
2.2	.976	.976	.977	.977	.978	.978	.978	.979	.979	.980
2.3	.980	.980	.981	.981	.982	.982	.982	.983	.983	.983
2.4	.984	.984	.984	.985	.985	.985	.986	.986	.986	.986
2.5	.987	.987	.987	.987	.988	.988	.988	.988	.989	.989
2.6	.989	.989	.989	.990	.990	.990	.990	.990	.991	.991
2.7	.991	.991	.991	.992	.992	.992	.992	.992	.992	.992
2.8	.993	.993	.993	.993	.993	.993	.993	.994	.994	.994
2.9	.994	.994	.994	.994	.994	.995	.995	.995	.995	.995

$r = (e^{2z} - 1)/(e^{2z} + 1)$.

Reprinted by permission from *Statistical Methods, 7th Edition,* by G. W. Snedecor and W. G. Cochran © 1980 by Iowa State University Press, Ames, Iowa 50010.

APPENDIX G

Chi-Square

To read this table, find the degrees of freedom (df) in the column at the left. The minimum chi-square values for the .05 and .01 levels are given in the middle and right columns. For example, if the chi-square is 15.0 with 20 df, it is not statistically significant; for 20 df, the minimum value for significance at the .05 level is 31.4.

df	.05	.01
1	3.8	6.6
2	6.0	9.2
3	7.8	11.3
4	9.5	13.3
5	11.1	15.1
6	12.6	16.8
7	14.1	18.5
8	15.5	20.1
9	16.9	21.7
10	18.3	23.2
11	19.7	24.7
12	21.0	26.2
13	22.4	27.7
14	23.7	29.1
15	25.0	30.6
16	26.3	32.0
17	27.6	33.4
18	28.9	34.8
19	30.1	36.2
20	31.4	37.6
21	32.7	38.9
22	33.9	40.3
23	35.2	41.6
24	36.4	43.0
25	37.7	44.3
26	38.9	45.6
27	40.1	47.0
28	41.3	48.3
29	42.6	49.6
30	43.8	50.9

From *Biometrika Tables for Statisticians*, Vol. 1, Third Edition, 1966. Reprinted by permission of the Biometrika Trustees.

APPENDIX H

Interval Data Master Sheets

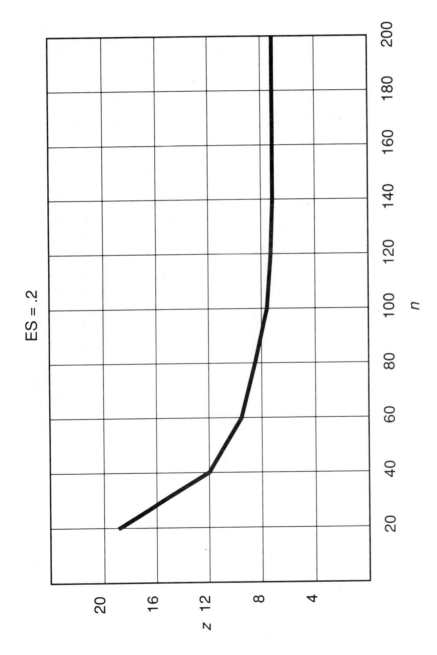

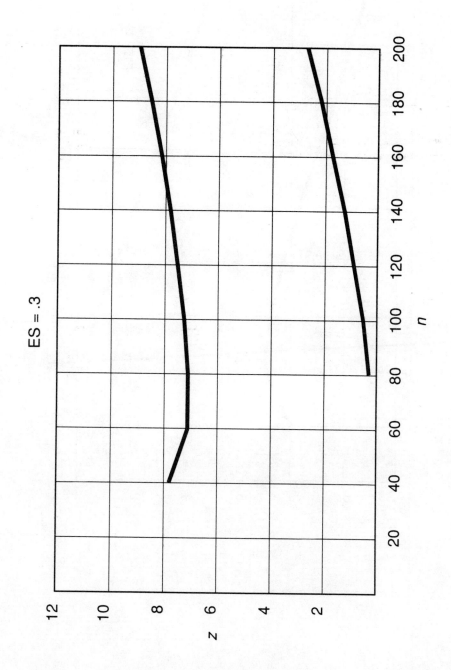

Appendix H

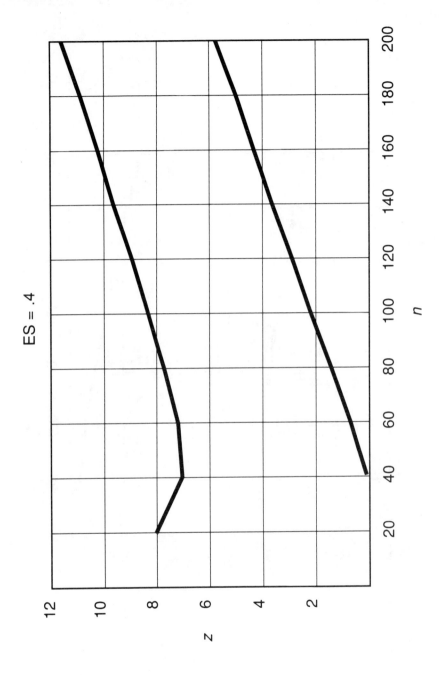

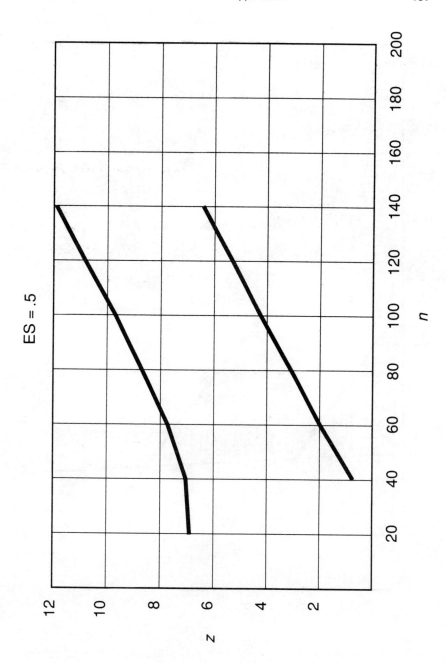

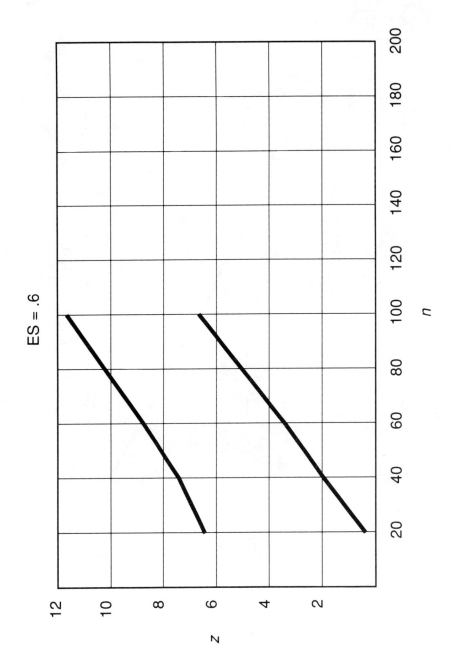

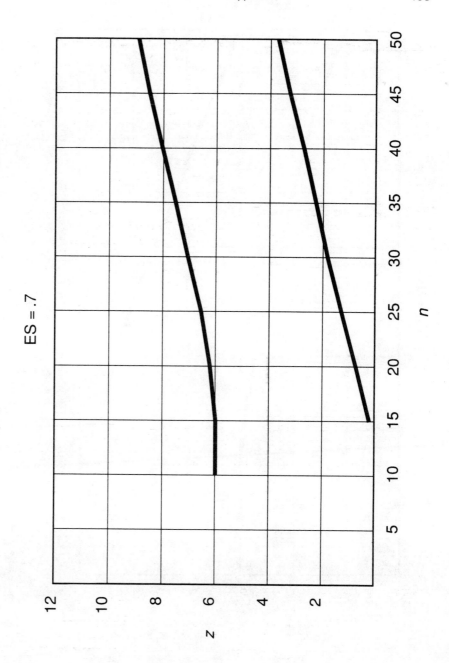

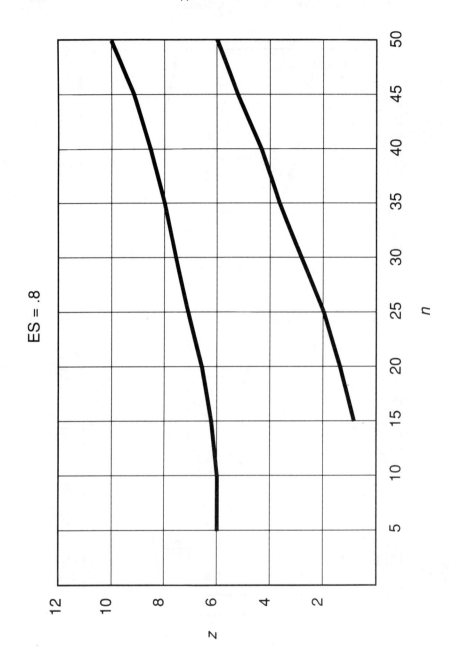

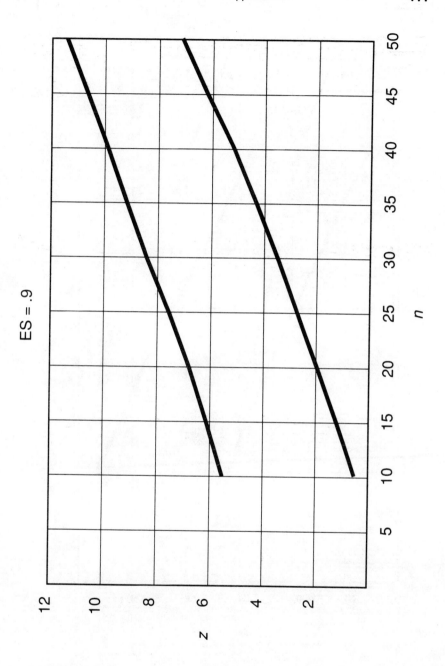

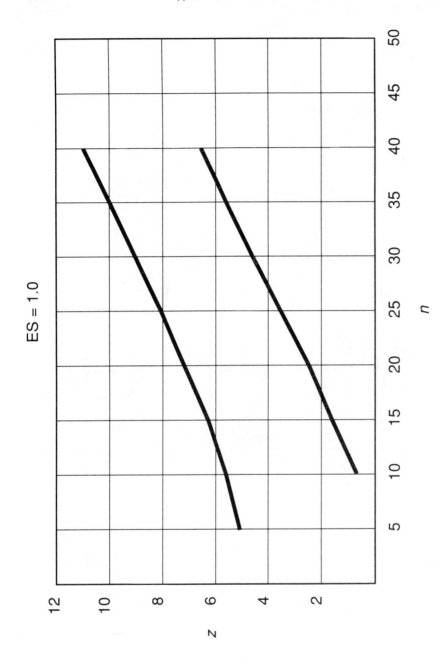

REFERENCES

a

Aderman, D. (1972). Elation, depression, and helping behavior. *Journal of Personality and Social Psychology, 24,* 91–101.

Agras, W. (Ed.). (1977). *Journal of Applied Behavior Analysis, 10,* 97–172. Special issue.

Alcock, J. (1984). *Animal behavior.* Sunderland, MA: Sinauer.

Alper, T. (1974). Achievement motivation in college women: A now-you-see-it-now-you-don't phenomenon. *American Psychologist, 29,* 194–203.

Altman, H., & Evenson, R. (1973). Marijuana use and subsequent psychiatric symptoms: A replication. *Comprehensive Psychiatry, 14,* 415–420.

Altmann, J. (1974). Observational study of behavior: Sampling methods. *Behaviour, 49,* 227–267.

American Marketing Association. (1937). *The technique of marketing.* New York: McGraw-Hill.

Aneshensel, C., Frerichs, R., Clark, V., & Yokopenic, P. (1982). Measuring depression in the community: A comparison of telephone and personal interviews. *Public Opinion Quarterly, 46,* 110–121.

Argyle, M., Bryant, B., & Trower, P. (1974). Social skills training and psychotherapy: A comparative study. *Psychological Medicine, 4,* 435–443.

Armitage, P. (1960). *Sequential medical trials.* Springfield, IL: Charles C. Thomas.

Arnold, D. (1970). Dimensional sampling: An approach for studying a small number of cases. *American Sociologist, 5,* 147–150.

Auerbach, O., Stout, A., Hammond, E., & Garfinkel, L. (1962). Changes in bronchial epithelium in relation to sex, age, residence, smoking, and pneumonia. *New England Journal of Medicine, 267,* 111–119.

b

Babst, D. (1978). Parsimony in designing a drug use survey: A methodology study. *American Journal of Drug and Alcohol Abuse, 5,* 441–454.

Bahrick, H. (1977). Reliability of measurement in investigations of learning and memory. In I. Birnbaum & F. Parker (Eds.), *Alcohol and human memory* (pp. 59–71). New York: Lawrence Erlbaum.

Bakan, D. (1966). The test of significance in psychological research. *Psychological Bulletin, 66,* 423–437.

Bakeman, R., & Gottman, J. (1986). *Observing interaction.* New York: Cambridge University Press.

Baker, E. (1971–1972). Humanistic psychology and scientific method. *Interpersonal Development, 2,* 137–172.

Baker, J. (1966). *The scientific life.* London: George Allen.

Baltes, P., Reese, H., & Nesselroade, J. (1977). *Life-span developmental psychology: Introduction to research methods.* Monterey, CA: Brooks/Cole.

Barber, J. (1985). *The presidential character.* Englewood Cliffs, NJ: Prentice-Hall.

Barker, E. (1971–1972). Humanistic psychology and scientific method. *Interpersonal Development, 2,* 137–172.

Barker, R. (1965). Explorations in ecological psychology. *American Psychologist, 20,* 1–14.

Barker, R., & Wright, H. (1955). *Midwest and its children.* Evanston, IL: Row, Peterson.

Baumrind, D. (1964). Some thoughts on ethics of research: After reading Milgram's "Behavioral study of obedience." *American Psychologist, 19,* 421–423.

Baumrind, D. (1971). Principles of ethical conduct in the treatment of subjects: Reaction to the draft report of the committee on ethical standards in psychological research. *American Psychologist, 26,* 887–896.

Beach, F. (1950). The snark was a boojum. *American Psychologist, 5,* 115–124.

Beauchamp, T., & Faden, R. (Eds.). (1982). *Ethical issues in social science research.* Baltimore: Johns Hopkins University Press.

Beecher, H. (1966). Ethics and clinical research. *New England Journal of Medicine, 24,* 1354–1360.

Belsky, J., & Steinberg, L. (1978). The effects of day care: A critical review. *Child Development, 49,* 929–949.

Belson, W. (1968). Respondent understanding of survey questions. *Polls, 3,* 1–13.

Bennett, C., & Lumsdaine, A. (1975). Social program evaluation: Definitions and issues. In C. Bennett & A. Lumsdaine (Eds.), *Evaluation and experiment* (pp. 1–38). New York: Academic Press.

Bergin, A. (1966). Some implications of psychotherapy research for therapeutic practice. *Journal of Abnormal Psychology, 71,* 235–246.

Bergin, A. (1971). The evaluation of therapeutic outcomes. In A. Bergin & S. Garfield (Eds.), *Handbook of psychotherapy and behavior change.* New York: Wiley.

Bernstein, D. (1973). Situation factors in behavioral fear assessment: A progress report. *Behavior Therapy, 4,* 41–48.

Berrin, K. (1979, October). The art of being Huichol. *Natural History,* pp. 68–75.

Beveridge, W. (1957). *The art of scientific investigation* (2nd ed.). New York: Vintage.

Binder, A., McConnell, D., & Sjoholm, N. (1957). Verbal conditioning as a function of experimenter characteristics. *Journal of Abnormal Social Psychology, 55,* 309–314.

Biology and Gender Study Group. (1988). The importance of feminist critique for contemporary cell biology. *Hypatia, 3,* 61–76.

Blake, J. (1966). Ideal family size among white Americans: A quarter of a century's evidence. *Demography, 3,* 154–173.

Blake, J. (1967). Family size in the 1960's—A baffling fad? *Eugenics Quarterly, 14,* 60–74.

Blake, J. (1967). Income and reproductive motivation. *Population Studies, 21,* 185–206.

Blalock, H. (1984). *Basic dilemmas in the social sciences.* Beverly Hills: Sage Publishing Co.

Blalock, H. (Ed.). (1985). *Causal models in panel and experimental designs.* New York: Aldine.

Block, J. (1976). Issues, problems and pitfalls in assessing sex differences. *Merrill-Palmer Quarterly, 22,* 283–308.

Block, J. (1977). Advancing the psychology of personality: Paradigmatic shift or improving the quality of research. In D. Magnusson & N. Endler (Eds.), *Personality at the crossroads.* Hillsdale, NJ: Erlbaum Associates.

Blumenthal, M., & Ross, H. (1973). *Two experimental studies of traffic laws.* Department of Transportation Reports, HS-800 825 and HS-800 826.

Blurton-Jones, N. (1972). *Ethological studies of child behaviour.* Cambridge, England: Cambridge University Press.

Boice, R. (1971). Laboratizing the wild rat (*Rattus norvegicus*). *Behavior Research Methods and Instrumentation, 3,* 177–182.

Bok, S. (1978). *Lying: Moral choice in public and private life.* New York: Pantheon Books.

Bonjean, C., Hill, R., & McLemore, D. (1967). *Sociological measurement.* San Francisco: Chandler.

Boruch, R. (1987). Conducting social experiments. In H. Cordray & R. Light (Eds.), *Evaluation practice in review.* San Francisco: Jossey-Bass.

Boruch, R., & Cecil, J. (1982). Statistical strategies for preserving privacy in direct inquiry. In J. Sieber (Ed.), *The ethics of social research.* New York: Springer-Verlag.

Boruch, R., McSweeney, A., & Soderstrum, E. (1978). Randomized field experiments for program planning, development, and evaluation. *Evaluation Quarterly, 2,* 655–695.

Bradburn, N. (1982). Question-wording effects in surveys. In R. Hogarth (Ed.), *Question framing and response consistency.* San Francisco: Jossey-Bass.

Brenner, M. (1982). Response effects of "role-restricted" characteristics of the interviewer. In W. Dijkstra & J. van der Zouvven (Eds.), *Response behaviour in the survey-interview.* New York: Academic Press.

Broad, W., & Wade, N. (1982). *Betrayers of the truth.* New York: Simon & Schuster.

Bronowski, J. (1956). *Science and human values.* New York: Harper.

Bruner, J. (1965). The growth of mind. *American Psychologist, 20,* 1007–1017.

Bryant, F., & Wortman, P. (1978). Secondary analysis: The case for data archives. *American Psychologist, 33,* 381–387.

Bryson, M. (1976). The Literary Digest Poll: Making of a statistical myth. *American Statistician, 30,* 184–185.

Burch, P. (1976). *The biology of cancer.* Lancaster, PA: Medical and Technical Publishers.

Bustad, L. (1966). Pigs in the laboratory. *Scientific American, 214,* 94–100.

C

Campbell, D. (1963). From description to experimentation: Interpreting trends as quasi-experiments. In C. Harris (Ed.), *Problems in measuring change.* Madison: University of Wisconsin Press.

Campbell, D. (1974). Unjustified variation and selective retention in scientific discovery. In F. Ayala & T. Dobzhansky (Eds.), *Studies in the philosophy of biology.* Berkeley: University of California Press.

Campbell, D., & Boruch, R. (1975). Making the case for randomized assignment to treatments by considering the alternatives: Six ways in which quasi-experimental evaluations in compensatory education tend to underestimate effects. In C. Bennett & A. Lumsdaine (Eds.), *Evaluation and experiment.* New York: Academic Press.

Campbell, D., & Stanley, J. (1966). *Experimental and quasi-experimental designs for research.* Chicago: Rand McNally.

Campbell, J., Daft, R., & Hulin, C. (1982). *What to study: Generating and developing research questions.* Beverly Hills: Sage Publishing Co.

Carlson, R. (1971). Where is the person in personality research? *Psychological Bulletin, 75,* 203–219.

Cassileth, B., Zupkis, R., & Sutton-Smith, K. (1980). Informed consent—Why are its goals imperfectly realized? *New England Journal of Medicine, 302,* 896–900.

Cederlof, R., Friberg, L., & Lundman, T. (1977). The interactions of smoking, environment and heredity and their implications for disease etiology. A report of epidemiological studies on the Swedish Twin Registries. *Acta Medica Scandanavia, 612,* 1–128.

Chalmers, A. (1976). *What is this thing called science?* Queensland, Australia: University of Queensland Press.

Chalmers, T., Block, J., & Lee, S. (1972). Controlled studies in clinical cancer research. *New England Journal of Medicine, 287,* 75–78.

Chamberlin, T. (1904). The methods of the earth-sciences. *Popular Science Monthly, 66,* 66–75.

Chance, P. (1987, April). Profile: Benjamin Bloom. Master of mastery. *Psychology Today,* pp. 42–46.

Chapanis, A. (1967). The relevance of laboratory studies to practical situations. *Ergonomics, 10,* 557–577.

Chapman, L., & Chapman, J. (1967). The genesis of popular but erroneous psychodiagnostic observations. *Journal of Abnormal Psychology, 72,* 193–204.

Chapman, L., & Chapman, J. (1969). Illusory correlation as an obstacle to the use of valid psychodiagnostic signs. *Journal of Abnormal Psychology, 74,* 271–280.

Cicchini, M. (1976). *Thought content and psychopathology: Themes within the stream of consciousness of depressed and nondepressed subjects.* Unpublished master's thesis, University of Western Australia, Nedlands, Australia.

Clutton-Brock, T., & Harvey, P. (1977). Primate ecology and social organization. *Journal of Zoology, 183,* 1–39.

Cochrane, A. (1971). *Effectiveness and efficiency.* London: Nuffield Provincial Hospitals Trust.

Cohen, D. (1977). *Psychologists on psychology.* New York: Taplinger Publishing Co.

Cohen, J. (1962). The statistical power of abnormal-social psychological research: A review. *Journal of Abnormal Social Psychology, 65,* 145–153.

Cohen, J. (1969). *Statistical power analysis for the behavioral sciences.* New York: Academic Press.

Cohen, M., Liebson, I., Faillace, L., & Speers, W. (1971). Alcoholism: Controlled drinking and incentive for abstinence. *Psychological Reports, 28,* 575–580.

Cohen, W. (1960). Form recognition, spatial orientation, perception of movement in the uniform visual field. In A. Morris & E. Horne (Eds.), *Visual search techniques* (Publication No. 712). Washington, DC: National Academy of Sciences.

Coile, D., & Miller, N. (1984). How radical animal activists try to mislead humane people. *American Psychologist, 39,* 700–701.

Colgrove, M. (1968). Stimulating creative problem solving: Innovative set. *Psychological Reports, 22,* 1205–1211.

Cone, J. (1977). The relevance of reliability and validity for behavioral assessment. *Behavior Therapy, 8,* 411–426.

Conner, R. (1977). Selecting a control group: An analysis of the randomization process in twelve social reform programs. *Evaluation Quarterly, 1,* 195–244.

Converse, P., & Traugott, M. (1986). Assessing the accuracy of polls and surveys. *Science, 234,* 1094–1098.

Cook, L., Toner, J., & Fellows, E. (1954). Effect of diethylaminoethyl 2,2-diphenylpropylacetate-HCl (SKF 525-a) on hexobarbital. *Journal of Pharmacology and Experimental Therapeutics, 111,* 131–141.

Cook, T., & Campbell, D. (1979). *Quasi-experimentation.* Skokie, IL: Rand McNally.

Cooper, H., & Rosenthal, R. (1980). Statistical versus traditional procedures for summarizing research findings. *Psychological Bulletin, 87,* 442–449.

Cooper, J. (1976). Deception and role playing: On telling the good guys from the bad guys. *American Psychologist, 31,* 605–610.

Cowen, E. (1978). Some problems in community program evaluation research. *Journal of Consulting and Clinical Psychology, 46,* 792–805.

Cronbach, L. (1950). Further evidence on response sets and test design. *Educational and Psychological Measurement, 10,* 3–31.

Cronbach, L. (1975). Beyond the two disciplines in scientific psychology. *American Psychologist, 30,* 116–127.

Cronbach, L., Gleser, G., Nanda, H., & Rajaratnam, N. (1972). *The dependability of behavioral measures.* New York: Wiley.

Cronbach, L., & Webb, N. (1975). Between-class and within-class effects in a reported Aptitude X Treatment interaction: Reanalysis of a study by G. L. Anderson. *Journal of Educational Psychology, 67,* 717–724.

Crook, J. (1964). The evolution of social organization and visual communication in the weaver birds (*Ploceinae*). *Behaviour Supplement, 10,* 1–178.

Crowne, D., & Marlowe, D. (1964). *The approval motive.* New York: Wiley.

Cullen, E. (1957). Adaptations in the kittiwake to cliff-nesting. *Ibis, 99,* 275–302.

d

Dabbs, J. (1975). *Physiological and physical activity measures of attitudes.* Paper presented at Navby-Smithsonian Conference on Survey Alternatives, Santa Fe, NM.

Dalbey, W., Nettesheim, P., Griesemer, R., Caton, J., & Guerin, M. (1980). Chronic inhalation of cigarette smoke by F344 rats. *Journal of the National Cancer Institute, 64,* 383–390.

Dancer, D., Braukmann, C., Schumaker, J., Kirigin, K., Willner, A., & Wolf, M. (1978). The training and validation of behavior observation and description skills. *Behavior Modification, 2,* 113–134.

Dar, R. (1987). Another look at Meehl, Lakatos, and the scientific practices of psychologists. *American Psychologist, 42*, 145–151.

Dawis, R. (1987). Scale construction. *Journal of Counseling Psychology, 34*, 481–489.

DeBono, E. (1970). *Lateral thinking: Creativity step by step.* New York: Harper & Row.

DeBono, E. (1972). *PO: A device for successful thinking.* New York: Simon & Schuster.

DeLamater, J. (1982). Response-effects of question content. In W. Dijkstra & J. van der Zouvven (Eds.), *Response behaviour in the survey-interview.* New York: Academic Press.

DeMaio, T. (1980). Refusals: Who, where and why. *Public Opinion Quarterly, 44*, 223–233.

Denenberg, V. (1982). Comparative psychology and single-subject research. In A. Kazdin & A. Tuma (Eds.), *Single-case research designs.* San Francisco: Jossey-Bass.

Denzin, N. (1968). On the ethics of disguised observation. *Social Problems, 15*, 502–506.

DerSimonian, R., & Laird, N. (1983). Evaluating the effect of coaching on SAT scores: A meta-analysis. *Harvard Education Review, 53*, 1–15.

Dewsbury, D. (1975). Diversity and adaptation in rodent copulatory behavior. *Science, 190*, 947–954.

Diener, E., & Crandall, R. (1978). *Ethics in social and behavioral research.* Chicago: University of Chicago Press.

Dillman, D. (1978). *Mail and telephone surveys: The total design method.* New York: Wiley-Interscience.

Doll, R. (1955). Etiology of lung cancer. *Advances in Cancer Research, 3*, 1–50.

Doll, R., & Hill, A. (1954). The mortality of doctors in relation to their smoking habits: A preliminary report. *British Medical Journal, 1*, 1451–1455.

Doll, R., & Hill, A. (1956). Lung cancer and other causes of death in relation to smoking. A second report on the mortality of British doctors. *British Medical Journal, 2*, 1071–1081.

Domer, F. (1971). *Animal experiments in pharmacological analysis.* Springfield, IL: Charles C. Thomas.

Dontenwill, W., & Wiebecke, B. (1966). Tracheal and pulmonary alterations following the inhalation of cigarette smoke by golden hamsters. In L. Severi (Ed.), *Lung tumors in animals* (pp. 519–526). Perugia, Italy: University of Perugia, International Conference on Cancer.

Dukes, W. (1965). N = 1. *Psychological Bulletin, 64*, 74–79.

e

Ebbinghaus, H. (1885). *Uber das Gedachtnis.* Leipzig: Duncker and Humblot.

Edgerton, R. (1965). 'Cultural' vs. 'ecological' factors in the expression of values, attitudes, and personality characteristics. *American Anthropologist, 67*, 442–447.

Edgington, E. (1972). N = 1 experiments: Hypothesis testing. *Canadian Psychologist, 13*, 121–134.

Edwards, A. (1957). *Techniques of attitude scale construction.* New York: Appleton-Century-Crofts.

Einstein, A. (1930). *Living philosophies.* New York: World.

Eisenberg, L. (1977). The social imperatives of medical research. *Science, 198*, 1105–1110.

Ember, M., & Ember, C. (1979). Male-female bonding: A cross-species study of mammals and birds. *Behavioral Science Research, 14*, 37–56.

Epstein, C. (1986). Down's syndrome and Alzheimer's disease: What is the relation? In G. Glenner & R. Wurtman (Eds.), *Advancing frontiers in Alzheimer's disease research.* Austin: University of Texas Press.

Erlenmeyer-Kimling, L., & Jarvik, L. (1963). Genetics and intelligence: A review. *Science, 142*, 1478–1479.

Eron, L., Huesman, L., Lefkowitz, M., & Walder, L. (1972). Does television violence cause aggression? *American Psychologist, 27*, 253–263.

Esecover, H., Malitz, S., & Wilkens, B. (1961). Clinical profiles of paid normal subjects volunteering for hallucinogenic drug studies. *American Journal of Psychiatry, 117*, 910–915.

Evans, C. (1979). An approach to the use of animal models for the study of alcoholism. *British Journal of Alcohol and Alcoholism, 14*, 37–58.

Eysenck, H. (1952). The effects of psychotherapy: An evaluation. *Journal of Consulting Psychology, 16*, 319–324.

Eysenck, H. (1980). *The causes and effects of smoking.* Beverly Hills: Sage Publishing Co.

f

Falk, J. (1969). Conditions producing psychogenic polydipsia in animals. *Annals of the New York Academy of Science, 157*, 569–593.

Falk, J. (1977). Animal model of alcoholism: Critique and progress. In M. Gross (Ed.), *Alcohol intoxication and withdrawal.* New York: Plenum.

Falk, J. (1984). Excessive behavior and drug-taking: Environmental generation and self-control. In P. Levison (Ed.), *Substance abuse, habitual behavior, and self-control* (pp. 81–123). Boulder: Westview Press.

Falk, J. (1985). Schedule-induced determinants of drug-taking. In L. Seiden & R. Balster (Eds.), *Behavioral pharmacology: The current status.* New York: Alan Liss, Inc.

Falk, J. (1986). The formation and function of ritual behavior. In T. Thompson & M. Zeiler (Eds.), *Analysis and integration of behavioral units.* Hillsdale, NJ: Erlbaum Associates.

Faust, D. (1984). *The limits of scientific reasoning.* Minneapolis: University of Minnesota Press.

Feldman, H. (1980). Background and purpose of the ethnographers' policymakers' symposium. In C. Akins & G. Beschner (Eds.), *Ethnography: A research tool for policymakers in the drug and alcohol fields.* Rockville, MD: National Institute of Drug Abuse.

Fenichel, O. (1930). *Ten years of the Berlin Psychoanalytic Institute.* Vienna: International Psychoanalytischer Verlag.

Festinger, L. (1964). Behavioral support for opinion changes. *Public Opinion Quarterly, 28,* 404–417.

Feyerabend, P. (1975). *Against method.* London: NLB.

Fischer, C. (1975). Privacy as a profile of authentic consciousness. *Humanitas, 11,* 27–43.

Fisher, S., & Greenberg, R. (1977). *The scientific credibility of Freud's theories and therapy.* New York: Basic Books.

Foulds, G. (1963). The design of experiments in psychiatry. In P. Sainsbury & N. Kreitman (Eds.), *Methods of psychiatric research.* London: Oxford.

Fox, M. (1986). *Laboratory animal husbandry.* Albany, NY: State University of New York Press.

Freeman, D. (1983). *Margaret Mead and Samoa: The making and unmaking of an anthropological myth.* Cambridge, MA: Harvard University Press.

Freud, A. (1948). *The ego and the mechanisms of defense.* New York: International Universities.

Fries, J., & Loftus, E. (1979). Letter. *Science, 205,* 644–647.

g

Garcia, J., Hankins, W., & Rusiniak, K. (1974). Behavioral regulation of the milieu interne in man and rat. *Science, 185,* 824–831.

Gardner, H. (1983). *Frames of mind: The theory of multiple intelligences.* New York: Basic Books.

Gardos, G. (1974). Are antipsychotic drugs interchangeable? *Journal of Nervous and Mental Disease, 159,* 343–348.

Garfield, P. (1973). Keeping a longitudinal dream record. *Psychotherapy: Theory, Research and Practice, 10,* 223–228.

Garfield, P. (1976). *Creative dreaming.* New York: Ballantine.

Garmezy, N. (1982). The case for the single case in research. In A. Kazdin & A. Tuma (Eds.), *Single-case research designs.* San Francisco: Jossey-Bass.

Gay, C. (1973, November 30). A man collapsed outside a UW building. Others ignore him. What would you do? *University of Washington Daily.*

Gendin, S. (1986). The use of animals in science. In T. Regan (Ed.), *Animal sacrifices.* Philadelphia: Temple University Press.

Gerard, D., Saenger, G., & Wile, R. (1952). The abstinent alcoholic. *Archives of General Psychiatry, 6,* 83–95.

Gergen, K. (1978). Experimentation in social psychology: A reappraisal. *European Journal of Social Psychology, 8,* 507–527.

Geschwind, N. (1984). The biology of cerebral dominance: Implications for cognition. *Cognition, 17,* 193–208.

Gilbert, J., Light, R., & Mosteller, F. (1975). Assessing social innovations: An empirical basis for policy. In C. Bennett & A. Lumsdaine (Eds.), *Evaluation and experiment.* New York: Academic Press.

Gilbert, J., McPeek, B., & Mosteller, F. (1977). Statistics and ethics in surgery and anesthesia. *Science, 198,* 684–689.

Glaser, E. (1977). Letter: Acceptable and unacceptable risks. *British Medical Journal, 2,* 1028–1029.

Glickman, S., & Sroges, R. (1966). Curiosity in zoo animals. *Behaviour, 26,* 151–188.

Goldstein, J. (1972). The effectiveness of manpower programs. A review of research on the impact on the poor. *Studies in public welfare.* Subcommittee on Fiscal Policy, Congress of the U.S. Washington, DC: U.S. Government Printing Office.

Goodman, J. (1977). IQ decline in mentally retarded adults: A matter of fact or methodological flaw. *Journal of Mental Deficiency Research, 21,* 199–203.

Gordon, W. (1961). *Synectics.* New York: Harper.

Graham, S. (1960). The influence of therapist character structure upon Rorschach changes in the course of psychotherapy. *American Psychologist, 15,* 415.

Grant, E. (1965). An ethological description of some schizophrenic patterns of behaviour. *Leeds Symposium on Behavioral Disorders.* Dagenham, Essex, England: May and Baker Ltd.

Grant, E. (1968). An ethological description of non-verbal behavior during interviews. *British Journal of Medical Psychology, 41,* 177–184.

Grant, E. (1969). Human facial expression. *Man, 4,* 525–536.

Gray, B. (1975). *Human subjects in medical experimentation.* New York: Wiley.

Greenberg, J., Pyszcynski, T., Solomon, S., & Steinberg, L. (1988). A reaction to Greenwald, Pratkanis, Leippe, and Baumgardner (1986): Under what conditions does research obstruct theory progress? *Psychological Review, 95,* 566–571.

Greenwald, A., & Pratkanis, A. (1988). On the use of "theory" and the usefulness of theory. *Psychological Review, 95,* 575–579.

Greenwald, A., Pratkanis, A., Leippe, M., & Baumgardner, M. (1986). Under what conditions does theory obstruct research progress? *Psychological Review, 93,* 216–229.

Griffin, D., Webster, F., & Michael, C. (1960). The echolocation of flying insects by bats. *Animal Behavior, 8,* 141–154.

Grunder, T. (1980). On readability of surgical consent forms. *New England Journal of Medicine, 302,* 900–902.

h

Hallberg, G. (1976). A system for the description and classification of movement behaviour. *Ergonomics, 19,* 727–739.

Hammer, E., & Piotrowski, Z. (1953). Hostility as a factor in the clinician's personality as it affects his interpretation of projective drawings. *Journal of Projective Techniques, 17,* 210–216.

Hansel, C. (1966). *ESP: A scientific evaluation*. New York: Scribner's.

Hanson, N. (1958). *Patterns of discovery*. London: Cambridge University Press.

Harlow, H., & Harlow, M. (1962). Social deprivation in monkeys. *Scientific American, 207*, 136–146.

Harris, B. (1979). Whatever happened to Little Albert? *American Psychologist, 34*, 151–160.

Hartmann, E., & Cravens, J. (1973). The effects of long-term administration of psychotropic drugs on human sleep: I. Methodology and the effects of placebo. *Psychopharmacology, 33*, 153–167.

Hartshorne, H., & May, M. (1928). *Studies in the nature of character: Vol. 1. Studies in deceit*. New York: Macmillan.

Hatchett, S., & Schuman, H. (1975). White respondent and race-of-interviewer effects. *Public Opinion Quarterly, 39*, 523–528.

Hauser, P. (1975). *Social statistics in use*. New York: Sage Publishing Co.

Hayes, J. (1982). Issues in protocol analysis. In G. Ungson & D. Braunstein (Eds.), *Decision making: An interdisciplinary approach*. Boston: Kent Publishing Co.

Hefferline, R., Keenan, B., & Harford, R. (1959). Escape and avoidance conditioning in human subjects with and without their observation of the response. *Science, 130*, 1338–1339.

Heise, D. (1975). *Causal analysis*. New York: Wiley.

Hempel, C. (1966). *Philosophy of natural science*. Englewood Cliffs, NJ: Prentice-Hall.

Henshel, R. (1980). Seeking inoperative laws: Toward the deliberate use of unnatural experimentation. In L. Freese (Ed.), *Theoretical methods in sociology*. Pittsburgh: University of Pittsburgh Press.

Herbert, J., & Attridge, C. (1975). A guide for developers and users of observation systems and manuals. *American Educational Research Journal, 12*, 1–20.

Hilgard, E. (1980). Consciousness in contemporary psychology. *Annual Review of Psychology, 31*, 1–26.

Hinde, R. (1970). *Animal behaviour*. New York: McGraw-Hill.

Hite, S. (1987). *The Hite Report: Women and love*. New York: Knopf.

Hodgkin, A., & Huxley, A. (1952). A quantitative description of membrane current and its application to conduction and excitation in nerve. *Journal of Physiology, 117*, 500–544.

Hodgkin, A., & Keynes, R. (1956). Experiments on the injection of substances into squid giant axons by means of a microsyringe. *Journal of Physiology, 131*, 592–616.

Holsti, O. (1969). *Content analysis for the social sciences and humanities*. Reading, MA: Addison-Wesley.

Hopkins, B., & Hermann, J. (1977). Evaluating interobserver reliability of interval data. *Journal of Applied Behavioral Analysis, 10*, 121–126.

Hrdy, S. (1986). Empathy, polyandry, and the myth of the coy female. In R. Bleier (Ed.), *Feminist approaches to science* (pp. 119–146). New York: Pergamon Press.

Humphreys, W. (1968). *Anomalies and scientific theories*. San Francisco: Freemen, Cooper.

Hutt, S., & Hutt, C. (Eds.). (1970). *Behaviour studies in psychiatry*. Oxford: Pergamon Press.

Hyde, J., & Rosenberg, B. (1976). *Half the human experience: The psychology of women*. Lexington, MA: D.C. Heath.

Hyman, H. (1972). *Secondary analysis of sample surveys*. New York: Wiley.

i

Iivanainen, M. (1974). *A study on the origins of mental retardation*. London: Heinemann Medical Publishers.

Isen, A., Daubman, K., & Nowicki, G. (1987). Positive affect facilitates creative problem solving. *Journal of Personality and Social Psychology, 52*, 1122–1131.

Ittelson, W., Proshansky, H., Rivlin, L., & Winkel, G. (1974). *An introduction to environmental psychology*. New York: Holt, Rinehart and Winston.

j

Jaffe, J. (1984). Evaluating drug abuse treatment: A comment on the state of the art. In F. Tims & J. Ludford (Eds.), *Drug abuse treatment evaluation: Strategies, progress, and prospects*. National Institute on Drug Abuse Research Monograph 51. Rockville, MD: National Institute on Drug Abuse.

James, B. (1986). *The Bill James baseball abstract 1986*. New York: Ballantine Books.

James, W. (1892). *Psychology: Briefer course*. New York: Henry Holt.

Jarman, P. (1974). The social organization of antelope in relation to their ecology. *Behaviour, 48*, 215–267.

Jones, J. (1981). *Bad blood: The Tuskegee Syphilis Experiment*. New York: Free Press.

Jones, R., Reid, J., & Patterson, R. (1975). Naturalistic observation in clinical assessment. In P. McReynolds (Ed.), *Advances in psychological assessment* (Vol. 3). San Francisco: Jossey-Bass.

k

Kagan, J. (1986). Rates of change in psychological processes. *Journal of Applied Developmental Psychology, 7*, 125–130.

Kamin, L. (1978). Comment on Munsinger's review of adoption studies. *Psychological Bulletin, 85*, 194–201.

Karpf, D., & Levine, M. (1971). Blank-trial probes and introtacts in human discrimination learning. *Journal of Experimental Psychology, 90*, 51–55.

Kasperson, C. (1978). Psychology of the scientist: XXXVII. Scientific creativity: A relationship with information channels. *Psychological Reports, 42*, 691–694.

Kazdin, A. (1977). Artifact, bias, and complexity of assessment: The ABCs of reliability. *Journal of Applied Behavioral Analysis, 10*, 141–150.

Kazdin, A. (1978). Methodological and interpretive problems of single-case experimental designs. *Journal of Consulting and Clinical Psychology, 46*, 629–642.

Kazdin, A. (1982). Single-case experimental designs in clinical research and practice. In A. Kazdin & A. Tuma (Eds.), *Single-case research designs* (pp. 33–47). San Francisco: Jossey-Bass.

Kazdin, A., & Tuma, A. (Eds.). (1982). *Single-case research designs*. San Francisco: Jossey-Bass.

Keir, G. (1949). The Progressive Matrices as applied to school children. *British Journal of Psychology, statistical section 2*, 140–150.

Kelman, H. (1967). Human use of human subjects: The problem of deception in social psychological experiments. *Psychological Bulletin, 67*, 1–11.

Kelman, H. (1982). Ethical issues in different social science methods. In T. Beauchamp & R. Faden (Eds.), *Ethical issues in social science research*. Baltimore: Johns Hopkins University Press.

Keppel, G., & Saufley, W. (1980). *Introduction to design and analysis*. San Francisco: W. H. Freeman & Co.

Kiewiet, D., & Rivers, D. (1985). The economic basis of Reagan's appeal. In J. Chubb & P. Peterson (Eds.), *The new direction in American politics*. Washington, DC: Brookings Institute.

Kimmel, M. (1987, July). Profile: Joseph Pleck. Real man redux. *Psychology Today*, pp. 48–52.

King, F. (1984, September). Animal research: The case for experimentation. *Psychology Today*, pp. 56–58.

Kinlen, L., Goldblatt, P., Fox, J., & Yudkin, J. (1984). Coffee and pancreatic cancer: Controversy in part explained? *Lancet, 1*, 282–283.

Kinsey, A., Pomeroy, W., & Martin, C. (1948). *Sexual behavior in the human male*. Philadelphia: Saunders.

Kinsey, A., Pomeroy, W., Martin, C., & Gebhard, P. (1953). *Sexual behavior in the human female*. Philadelphia: Saunders.

Kintz, B. (1965). The experimenter effect. *Psychological Bulletin, 63*, 223–232.

Koestler, A. (1964). *The act of creation*. New York: Dell.

Kohlberg, L. (1976). Moral stages and moralization: The cognitive-developmental approach. In T. Likona (Ed.), *Moral development and behavior*. New York: Holt, Rinehart, & Winston.

Kolansky, H., & Moore, W. (1971). Effects of marihuana on adolescents and young adults. *Journal of the American Medical Association, 216*, 486–492.

Kolb, L. (1962). *Drug addiction: A medical problem*. Springfield, IL: Charles C. Thomas.

Kortlandt, A. (1962, May). Chimpanzees in the wild. *Scientific American*, pp. 1–10.

Krebs, J., & Davies, N. (1981). *An introduction to behavioural ecology*. London: Blackwell.

Krebs, J., & Davies, N. (1984). *Behavioural ecology: An evolutionary approach*. London: Blackwell.

Krippendorff, K. (1980). *Content analysis*. Beverly Hills: Sage Publishing Co.

Kunze, M., & Vutuc, C. (1980). Threshold of tar exposure: Analysis of male lung cancer cases and controls. In G. Gori & F. Bock (Eds.), *Banbury Report 3—A safe cigarette* (pp. 29–36). New York: Cold Spring Harbor Lab.

Kurman, P. (1977). Research: An aerial view. *Et Cetera, 34*, 265–276.

l

Lamb, D. (1978). Use of behavioral measures in anxiety research. *Psychological Reports, 43*, 1079–1085.

Lamberg, L. (1988, July). Profile: Rosalind Cartwright. Night pilot. *Psychology Today*, pp. 34–42.

Langley, P., Simon, H., Bradshaw, G., & Zytkow, J. (1987). *Scientific discovery: Computational explorations of the creative processes*. Cambridge, MA: Massachusetts Institute of Technology Press.

Lasagna, L., & von Felsinger, J. (1954). The volunteer subject in research. *Science, 120*, 359–361.

Laudan, L. (1984). *Science and values*. Berkeley: University of California Press.

Lawson, T., & Winstead, D. (1978). Toward a theory of drug use. *British Journal of Addiction, 73*, 149–154.

Lazarsfeld, P. (1948). The use of panels in social research. *Proceed. American Philosophical Society, 2*, 405–410.

Lazarsfeld, P. (1972). Mutual relations over time of two attributes: A review and integration of various approaches. In M. Hammer et al. (Eds.), *Continuities in the language of social research*. New York: Free Press.

Lazarsfeld, P., & Barton, A. (1951). Qualitative measurement in the social sciences. In D. Lerner & H. Lasswell (Eds.), *The policy sciences*. Palo Alto: Stanford University Press.

Leacock, Stephen. (1916). *Further foolishness: Sketches and satires on the follies of the day*. New York: John Lane Co.

Lear, J. (1966, July). Experiments on people—The growing debate. *Saturday Review*, pp. 41–43.

Leavitt, F. (1982). *Drugs and behavior*. New York: Wiley-Interscience.

Lederman, L. (1984, November). The value of fundamental science. *Scientific American*, pp. 40–47.

Lehner, P. (1979). *Handbook of ethological methods*. New York: Garland STPM Press.

Leibovitz, B., & Siegel, B. (1980). Aspects of free radical reactions in biological systems: Aging. *Journal of Gerontology, 35*, 45–56.

Lenneberg, E. (1962). Understanding language without ability to speak: A case study. *Journal of Abnormal and Social Psychology, 65*, 419–425.

Leopold, A. (1978). The act of creation: Creative processes in science. *BioScience, 28*, 436–440.

Levin, H. (1983). *Cost-effectiveness: A primer*. Beverly Hills: Sage Publishing Co.

Levin, H. (1987). Cost-benefit and cost-effectiveness analyses. In H. Cordray & R. Light (Eds.), *Evaluation practice in review*. San Francisco: Jossey-Bass.

Levine, A. (1977). Naturalistic observation: Validity of frequency data. *Psychological Reports, 40*, 1311–1338.

Levine, M. (1988). *Effective problem solving*. Englewood Cliffs, NJ: Prentice-Hall.

Lewis, A. (1984). *The evidence never lies*. New York: Dell.

Lick, J., & Unger, T. (1977). The external validity of behavioral fear assessment. *Behavior Modification, 1,* 283–306.

Lieberman, D. (1979). Behaviorism and the mind. *American Psychologist, 34,* 319–333.

Light, R. (1984). Six evaluation issues that synthesis can resolve better than single studies. In W. Yeaton & P. Wortman (Eds.), *Issues in data synthesis*. San Francisco: Jossey-Bass.

Light, R., & Pillemer, D. (1984). *Summing up*. Cambridge, MA: Harvard University Press.

Linton, M. (1982). Transformations of memory in everyday life. In U. Neisser (Ed.), *Memory observed*. San Francisco: W. H. Freeman & Co.

Lipton, D., & Appel, P. (1984). The State perspective. In F. Tims & J. Ludford (Eds.), *Drug abuse treatment evaluation: Strategies, progress, and prospects*. National Institute on Drug Abuse Research Monograph 51. Rockville, MD: National Institute on Drug Abuse.

Livson, N. (1977, September). *The physically attractive woman at age 40: Precursors in adolescent personality and adult correlates from a longitudinal study*. Paper presented at the International Conference on Love and Attraction, Swansea, Wales.

Lockard, R. (1968). The albino rat: A defensible choice or a bad habit. *American Psychologist, 23,* 734–742.

Lowe, C. (1983). Radical behaviorism and human psychology. In G. Davey (Ed.), *Animal models of human behavior*. New York: Wiley.

Luborsky, L. (1969). Research cannot yet influence clinical practice. *International Journal of Psychiatry, 7,* 135–140.

Lykken, D. (1968). Statistical significance in psychological research. *Psychological Bulletin, 70,* 151–159.

m

MacKay, D. (1988). Under what conditions can theoretical psychology survive and prosper? Integrating the rational and empirical epistemologies. *Psychological Review, 95,* 559–565.

MacKenzie, B. (1984). Explaining race differences in IQ. *American Psychologist, 39,* 1214–1233.

MacMahon, B., Yen, S., Trichopoulos, D., Warren, K., & Nardi, G. (1981). Coffee and cancer of the pancreas. *New England Journal of Medicine, 304,* 630–633.

Mahrer, A. (1988). Discovery-oriented psychotherapy research. *American Psychologist, 43,* 694–702.

Maier, N. (1970). *Problem solving and creativity*. Belmont, CA: Brooks/Cole.

Maini, S., & Nordbeck, B. (1973). Critical moments, the creative process and research motivation. *International Social Science Journal, 25,* 190–204.

Majasan, J. (1972). *College students' achievement as a function of the congruence between their beliefs and their instructor's beliefs*. Unpublished doctoral dissertation, Stanford University, Palo Alto, CA.

Mandler, G., & Sarason, S. (1952). A study of anxiety and learning. *Journal of Abnormal Social Psychology, 47,* 166–173.

Mann, L., & Janis, I. (1968). A follow-up study on the long-term effects of emotional role playing. *Journal of Personality and Social Psychology, 8,* 339–342.

Margolis, J. (1967). Citation indexing and evaluation of scientific papers. *Science, 155,* 1213–1219.

Marler, P. (1970). Birdsong and speech development: Could there be parallels? *American Scientist, 58,* 669–673.

McClearn, G. (1965). Genotype and mouse behaviour. In S. Geerts (Ed.), *Genetics Today, Proceedings of the XI International Congress on Genetics, 3,* 795–805.

McClelland, D. (1961). *The achieving society*. Princeton: Van Nostrand.

McClelland, D., Atkinson, J., Clark, R., & Lowell, E. (1953). *The achievement motive*. New York: Appleton-Century-Crofts.

McGuigan, F. (1963). The experimenter: A neglected stimulus object. *Psychological Bulletin, 60,* 421–428.

McGuire, W. (1973). The Yin and Yang of progress in social psychology: Seven koan. *Journal of Personality and Social Psychology, 28,* 446–456.

McKeachie, W. (1974). The decline and fall of the laws of learning. *Educational Researcher, 3,* 7–11.

McKellar, P. (1949). The emotion of anger in the expression of human aggressiveness. *British Journal of Psychology, 39,* 148–155.

McLellan, A., Luborsky, L., Woody, G., O'Brien, C., & Druley, K. (1983). Predicting response to alcohol and drug abuse treatments. *Archives of General Psychiatry, 40,* 620–625.

McNemar, Q. (1960). At random: Sense and nonsense. *American Psychologist, 15,* 295–300.

Medawar, P. (1969). *Induction and intuition in scientific thought*. Philadelphia: American Philosophical Society.

Medley, D., & Mitzel, H. (1963). Measuring classroom behavior by systematic observation. In N. Gage (Ed.), *Handbook of research in teaching*. Chicago: Rand McNally.

Meehl, P. (1954). *Clinical versus statistical prediction*. Minneapolis: University of Minnesota Press.

Meehl, P. (1967). Theory testing in psychology and physics: A methodological paradox. *Philosophy of Science, 34,* 103–115.

Meehl, P. (1970). Nuisance variables and ex post facto design. In M. Radner & S. Winokur (Eds.), *Analyses of theories and methods of physics and psychology* (pp. 380–412). Minneapolis: University of Minnesota Press.

Meehl, P. (1970). Theory-testing in psychology and physics: A methodological paradox. In D. Morrison & R. Henkel (Eds.), *The significance test controversy*. Chicago: Aldine.

Meehl, P. (1978). Theoretical risks and tabular asterisks: Sir Karl, Sir Ronald, and the slow progress of soft psychology. *Journal of Consulting and Clinical Psychology, 46,* 806–834.

Mello, N. (1972). Behavioral studies of alcoholism. In B. Kissen & H. Begleiter (Eds.), *The biology of alcoholism.* New York: Plenum.

Mello, N., & Mendelson, J. (1972). Drinking patterns during work-contingent and noncontingent alcohol acquisition. *Psychosomatic Medicine, 34,* 139–164.

Merton, R. (1968). The Matthew effect in science. *Science, 159,* 56–63.

Miller, D. (1977). Roles of naturalistic observation in comparative psychology. *American Psychologist, 32,* 211–219.

Miller, N. (1956). Effects of drugs on motivation: The value of using a variety of measures. *Annals of the New York Academy of Science, 65,* 318–333.

Mischel, W. (1973). Toward a cognitive social learning reconceptualization of personality. *Psychological Review, 80,* 252–283.

Mischel, W. (1974). Processes in delay of gratification. In L. Berkowitz (Ed.), *Advances in experimental social psychology.* New York: Academic Press.

Monod, J. (1975). On the molecular theory of evolution. In R. Harre (Ed.), *Problems of scientific revolution* (pp. 11–24). Oxford: Clarendon Press.

Morse, W. (1966). Intermittent reinforcement. In W. Honig (Ed.), *Operant behavior: Areas of research and application.* New York: Appleton-Century-Crofts.

Moser, K., Gadenne, V., & Schröder, J. (1988). Under what conditions does confirmation seeking obstruct scientific progress? *Psychological Review, 95,* 572–574.

Mosteller, F., & Wallace, D. (1964). *Inference and disputed authorship: The Federalist.* Reading, MA: Addison-Wesley.

Motley, M. (1985, September). Slips of the tongue. *Scientific American,* pp. 114–119.

Munsinger, H. (1974). The adopted child's IQ: A critical review. *Psychological Bulletin, 82,* 623–659.

Munsinger, H. (1978). Reply to Kamin. *Psychological Bulletin, 85,* 202–206.

Murphy, K. (1984). Review of Wonderlic Personnel Test. In D. Keyser & R. Sweetland (Eds.), *Test critiques* (Vol. 1). Kansas City: Test Corporation of America.

Murphy, K., & Davidshofer, C. (1988). *Psychological testing.* Englewood Cliffs, NJ: Prentice-Hall.

n

National Academy of Sciences. (1974). *Final report of the panel on manpower training evaluation. The use of Social Security earnings data for assessing the impact of manpower training programs.* Washington, DC: National Academy of Sciences.

National Center for Health Statistics. (1967). [Interview data on chronic conditions compared with information derived from medical records] (PHS Publication No. 1000–2–23). Washington, DC: U.S. Government Printing Office.

National Center for Health Statistics. (1971). [Effect of some experimental interviewing techniques on reporting in the health survey interview] (PHS Publication No. 1000–2–41). Washington, DC: U.S. Government Printing Office.

National Center for Health Statistics. (1972). *Optimum recall period for reporting persons injured in motor vehicle accidents* (DHEW Publication No. HSM 72–1050). Washington, DC: U.S. Government Printing Office.

National Center for Health Statistics. (1977). *A summary of studies of interviewing methodology* (DHEW Publication No. HRA 77–1343). Washington, DC: U.S. Government Printing Office.

Nesselroade, J., & Reese, H. (1973). *Life-span developmental psychology.* New York: Academic Press.

Netting, R. (1977). *Cultural ecology.* Menlo Park, CA: Benjamin/Cummings.

Nietzel, M., & Bernstein, P. (1987). *Introduction to clinical psychology.* Englewood Cliffs, NJ: Prentice-Hall.

Nomura, A., Stemmermann, G., & Heilbrun, L. (1981). Coffee and pancreatic cancer. *Lancet, 2,* 415.

Nunnally, J. (1960). The place of statistics in psychology. *Educational and Psychological Measurement, 20,* 641–650.

o

Ogborne, A., & Smart, R. (1982). Reactions to research: The case of the evaluation of Ontario's detoxification centers. *British Journal of Addiction, 77,* 275–282.

Ogburn, W. (1947). On scientific writing. *American Journal of Sociology, 52,* 383–388.

Olby, R. (1970). Francis Crick, DNA, and the central dogma. *Daedalus, 99,* 938–987.

O'Leary, K., Kent, R., & Kanowitz, J. (1975). Shaping data collection congruent with experimental hypotheses. *Journal of Applied Behavioral Analysis, 8,* 43–51.

Olson, R. (1971). *Science as metaphor.* Belmont, CA: Wadsworth.

Oppenheim, A. (1966). *Questionnaire design and attitude measurement.* New York: Basic Books.

Orians, G. (1969). On the evolution of mating systems in birds and mammals. *American Naturalist, 103,* 589–603.

Orlans, F. (1977). *Animal care from protozoa to small mammals.* Menlo Park, CA: Addison-Wesley.

Orne, M. (1962). On the social psychology of the psychological experiment. *American Psychologist, 17,* 776–783.

Ortega y Gasset, J. (1951). *Revolt of the masses.* London: George Allen.

Osborne, A. (1963). *Applied imagination.* New York: Scribner's.

p

Parry, H., & Crossley, H. (1950). Validity of responses to survey questions. *Public Opinion Quarterly, 14,* 61–80.

Pastore, N. (1949). *The nature-nurture controversy.* New York: King's Crown Press.

Pavlov, I. (1928). *Lectures on conditioned reflexes.* New York: International.

Payer, L. (1988). *Medicine and culture.* New York: Henry Holt.

Pernanen, K. (1974). Validity of survey data on alcohol use. In R. Gibbins, Y. Israel, H. Kalant, R. Popham, W. Schmidt, & R. Smart (Eds.), *Research advances in alcohol and drug problems.* New York: Wiley.

Petzel, T., Johnson, J., & McKillip, J. (1973). Response bias in drug surveys. *Journal of Consulting and Clinical Psychology, 40,* 437–439.

Pflaum, S., Walberg, H., Karegianes, M., & Rasher, S. (1980). Reading instruction: A quantitative analysis. *Educational Researcher, 9,* 12–18.

Phillips, D. (1972). Deathday and birthday: An unexpected connection. In J. Tanur, F. Mosteller, W. Kruskal, R. Link, R. Pieters, & G. Rising (Eds.), *Statistics: A guide to the unknown.* San Francisco: Holden-Day.

Phillips, D. (1982). The impact of fictional television stories on U.S. adult fatalities: New evidence of the effect of mass media on violence. *American Journal of Sociology, 87,* 1340–1359.

Phillips, D. (1983). The impact of mass media violence on U.S. homicides. *American Sociological Review, 48,* 560–568.

Piaget, J. (1965). *The moral judgment of the child.* New York: Free Press.

Piliavin, I., Rodin, J., & Piliavin, J. (1970). Good samaritanism: An underground phenomenon? *Journal of Personality and Social Psychology, 13,* 289–299.

Piliavin, J., & Piliavin, I. (1972). Effects of blood on reactions to a victim. *Journal of Personality and Social Psychology, 23,* 353–361.

Pinney, J. (1979). Preface. In N. Krasnegor (Ed.), *Cigarette smoking as a dependent process* (DHEW Publication No. DM 79–800). Rockville, MD: National Institute on Drug Abuse.

Platt, J. (1964). Strong inference. *Science, 146,* 347–353.

Platt, W., & Baker, R. (1931). The relationship of the scientific "hunch" research. *Journal of Chemical Education, 8,* 1969–2002.

Polsky, R., & McGuire, M. (1979). An ethological analysis of manic-depressive disorder. *Journal of Nervous and Mental Disease, 167,* 56–65.

Polya, G. (1954). *Mathematics and plausible reasoning.* Princeton, NJ: Princeton University Press.

Polya, G. (1957). *How to solve it.* Princeton, NJ: Princeton University Press.

Popper, K. (1962). *Conjectures and refutations.* New York: Basic Books.

Popper, K. (1968). *The logic of scientific discovery.* London: Hutchinson.

Powers, E., Goudy, W., & Keith, P. (1978). Congruence between panel and recall data in longitudinal research. *Public Opinion Quarterly, 42,* 381–389.

Pratt, D. (1980). *Alternatives to pain in experiments on animals.* New York: Argus Archives.

Preble, E., & Casey, J. (1969). Taking care of business: The heroin user's life on the street. *International Journal of the Addictions, 4,* 1–24.

Price, E. (1972). Domestication and early experience effects on escape conditioning in the Norway rat. *Journal of Comparative and Physiological Psychology, 79,* 51–55.

r

Ray, R., & Ray, R. (1976). A systems approach to behavior II: The ecological description and analysis of human behavior dynamics. *Psychological Record, 26,* 147–180.

Razran, G. (1958). Pavlov and Lamarck. *Science, 128,* 758–760.

Regan, T. (Ed.). (1986). *Animal sacrifices.* Philadelphia: Temple University Press.

Reines, B. (1982). *Psychology experiments on animals.* Boston, MA: New England Antivivisection Society.

Richardson, A. (1984). *The experiential dimension of psychology.* Queensland, Australia: University of Queensland Press.

Richter, C. (1954). The effects of domestication and selection on the behavior of the Norway rat. *Journal of the National Cancer Institute, 15,* 727–738.

Rimington, J. (1981). The effect of filters on the incidence of lung cancer in cigarette smokers. *Environmental Research, 24,* 162–166.

Rogers, H., & Layton, B. (1979). Two methods of predicting drug-taking. *International Journal of the Addictions, 14,* 299–310.

Rootman, I., & Smart, R. (1985). A comparison of alcohol, tobacco and drug use as determined from household and school surveys. *Drug and Alcohol Dependence, 16,* 89–94.

Rosenblum, L. (1978). The creation of a behavioral taxonomy. In G. Sackett (Ed.), *Observing behavior* (Vol. 2). Baltimore: University Park Press.

Rosenthal, R. (1963). On the social psychology of the psychological experiment: The experimenter's hypothesis as unintended determinant of experimental results. *American Scientist, 51,* 268–283.

Rosenthal, R. (1977). Biasing effects of experimenters. *Et Cetera, 34,* 253–264.

Rosenthal, R. (1983). Assessing the statistical and social importance of the effects of psychotherapy. *Journal of Consulting and Clinical Psychology, 51,* 4–13.

Rosenthal, R. (1984). *Meta-analytic procedures for social research.* Beverly Hills: Sage Publishing Co.

Rosenthal, R., & Jacobson, L. (1968). *Pygmalion in the classroom.* New York: Holt, Rinehart & Winston.

Rosenthal, R., & Rosnow, R. (1975). *The volunteer subject.* New York: Wiley-Interscience.

Rosnow, R., & Rosnow, M. (1986). *Writing papers in psychology.* Belmont, CA: Wadsworth.

Rossi, P., & Freeman, H. (1985). *Evaluation: A systematic approach.* Beverly Hills: Sage Publishing Co.

Roszak, T. (1972). *Where the wasteland ends*. New York: Doubleday.

Rowan, A. (1984). *Of mice, models, and men*. Albany, NY: State University of New York Press.

Rozeboom, W. (1960). The fallacy of the null-hypothesis significance test. *Psychological Bulletin, 57*, 416–428.

Rugg, D. (1941). Experiment in wording questions, II. *Public Opinion Quarterly, 5*, 91–92.

S

Sackett, G. (Ed.). (1978). *Observing behavior* (Vol. 1). Baltimore: University Park Press.

Sapolsky, A. (1964). An effort at studying Rorschach content symbolism: The frog response. *Journal of Consulting Psychology, 28*, 469–472.

Schaie, K., & Labouvie-Vief, G. (1974). Generational and ontogenetic components of change in adult cognitive behavior. A fourteen-year cross-sequential study. *Journal of Developmental Psychology, 10*, 305–320.

Schewe, C., & Cournoyer, N. (1976). Prepaid vs. promised monetary incentives to questionnaire response: Further evidence. *Public Opinion Quarterly, 40*, 105–107.

Schuman, H., & Presser, S. (1981). *Questions and answers in attitude surveys*. New York: Academic Press.

Schuster, D., & Schuster, L. (1969). Study of stress and sex ratio in humans. *Proceedings of the 77th Annual Convention of the American Psychological Association*, 335–336.

Schuster, D., & Schuster, L. (1969). Theory of stress and sex ratio. *Proceedings of the 77th Annual Convention of the American Psychological Association*, 223–224.

Schwarz, H. (1972). The use of subjective probability methods in estimating demand. In J. Tanur, F. Mosteller, R. Link, R. Pieters, & G. Rising (Eds.), *Statistics: A guide to the unknown*. San Francisco: Holden-Day.

Scott, J., & Franklin, J. (1972). The changing nature of sex references in mass circulation magazines. *Public Opinion Quarterly, 36*, 80–86.

Scudder, S. (1874). A great teacher's method. *Every Saturday*, pp. 369–370.

Seligman, M. (1975). *Helplessness: On depression, development, and death*. San Francisco: Freeman.

Selye, H. (1964). *From dream to discovery*. New York: McGraw-Hill.

Sharman, G. (1973). Adaptations of marsupial pouch young for extra-uterine existence. In C. Austin (Ed.), *The mammalian fetus in vitro*. London: Chapman & Hall.

Sharp, L., & Frankel, J. (1983). Respondent burden: A test of some common assumptions. *Public Opinion Quarterly, 47*, 36–53.

Sheldon, W. (1942). *The varieties of temperament: A psychology of constitutional differences*. New York: Harper.

Sidman, M. (1960). *Tactics of scientific research*. New York: Basic Books.

Siegel, S. (1956). *Nonparametric statistics for the behavioral sciences*. New York: McGraw-Hill.

Sigelman, L. (1981). Question-order effects on presidential popularity. *Public Opinion Quarterly, 45*, 199–207.

Silverman, I., & Margulis, S. (1973). Experiment title as a source of sampling bias in commonly used "subject-pool" procedures. *Canadian Psychologist, 14*, 197–201.

Singer, J. (1975). Navigating the stream of consciousness. *American Psychologist, 30*, 727–738.

Sjoberg, G. (1983). Politics, ethics and evaluation research. In E. Struening & M. Brewer (Eds.), *Handbook of evaluation research*. Beverly Hills: Sage Publishing Co.

Skinner, B. (1956). A case history in scientific method. *American Psychologist, 11*, 221–233.

Skinner, B. (1969). *Contingencies of reinforcement: A theoretical analysis*. New York: Appleton-Century-Crofts.

Smart, R., & Liban, C. (1982). Alcohol consumption as estimated by the informant method: A household survey and sales data. *Journal of Studies on Alcohol, 43*, 1020–1027.

Snedecor, G. (1956). *Statistical methods*. Ames, IA: Iowa State College Press.

Social Science Research Council. (1975). *Basic background items for U.S. household surveys*. Washington, DC: Center for Coordination of Research on Social Indicators.

Speed, G. (1893). Do newspapers now give the news? *Forum, 15*, 705–711.

Sperry, R. (1943). Effects of 180 degree rotation of the retinal field on visuomotor coordination. *Journal of Experimental Zoology, 92*, 263–279.

Stewart, D. (1984). *Secondary research*. Beverly Hills: Sage Publishing Co.

Stone, C. (1939). Copulatory activity in adult male rats following castration and injections of testosterone propionate. *Endocrinology, 24*, 165–174.

Storrs, E., & Williams, R. (1968). A study of monozygous quadruplet armadillos in relation to mammalian inheritance. *Proceedings of the National Academy of Sciences, 60*, 910–914.

Stratton, G. (1897). Vision without inversion of the retinal image. *Psychological Review, 4*, 341–360.

Strupp, H. (1960). Some comments on the future of research in psychotherapy. *Behavioral Science, 5*, 60–71.

Strupp, H., & Hadley, S. (1979). Specific versus nonspecific factors in psychotherapy: A controlled study of outcome. *Archives of General Psychiatry, 10*, 1125–1136.

Stuart, A. (1984). *The ideas of sampling*. New York: Oxford University Press.

Sudman, S., & Bradburn, N. (1974). *Response effects in surveys: A review and synthesis*. Chicago: Aldine.

Sudman, S., & Bradburn, N. (1982). *Asking questions*. San Francisco: Jossey-Bass.

Sudman, S., & Bradburn, N. (1984). Improving mailed questionnaire design. In D. Lockhart (Ed.), *Making effective use of mailed questionnaires.* San Francisco: Jossey-Bass.

Suedfeld, P. (1981). Aloneness as a healing experience. In L. Peplau & D. Perlman (Eds.), *Loneliness: A sourcebook of current theory, research, and therapy.* New York: Wiley-Interscience.

Sutter, A. (1966). The world of the righteous dope fiend. *Issues in Criminology, 2,* 177–222.

t

Taylor, C., Price, P., Richards, J., & Jacobsen, T. (1965). An investigation of the criterion problem for a group of medical general practitioners. *Journal of Applied Psychology, 49,* 399–406.

Taylor, S., & Bogdan, R. (1984). *Introduction to qualitative research methods.* New York: Wiley.

Terman, L. (1948). Kinsey's "Sexual Behavior in the Human Male": Some comments and criticisms. *Psychological Bulletin, 45,* 443–459.

Tetlock, P. (1984). Cognitive style and political belief systems in the British House of Commons. *Journal of Personality and Social Psychology, 46,* 365–375.

Tetlock, P. (1985). Integrative complexity of American and Soviet foreign policy rhetoric: A time-series analysis. *Journal of Personality and Social Psychology, 49,* 1565–1585.

Tetlock, P., Hannum, K., & Michelletti, P. (1984). Stability and change in the complexity of senatorial debate: Testing the cognitive versus rhetorical style hypotheses. *Journal of Personality and Social Psychology, 46,* 979–990.

Thompson, M. (1980). *Benefit-cost analysis for program evaluation.* Beverly Hills: Sage Publishing Co.

Thompson, W., & Olian, S. (1961). Some effects on offspring behavior of maternal adrenalin injection during pregnancy in three inbred mouse strains. *Psychological Reports, 8,* 87–90.

Tims, F. (1982). Assessing treatment: The conduct of evaluation in drug abuse treatment programs (DHHS Publication No. ADM 82-1218). Washington, DC: U.S. Government Printing Office.

Tinbergen, E., & Tinbergen, N. (1972). Early childhood autism: An ethological approach. *Advances in Ethology, 10,* 1–53.

Tinbergen, N. (1963). On aims and methods of ethology. *Zeitschrift fur Tierpsychologie, 20,* 410–433.

Tinbergen, N., Brockhuysen, G., Feekes, F., Houghton, J., Kruuk, H., & Szuk, E. (1962). Eggshell removal by the blackheaded gull, *Larus ridibundus L.:* A behaviour component of camouflage. *Behaviour, 19,* 74–118.

Tokuhata, G. (1972). *Cancer of the lung: Host and environmental interaction.* Harrisburg, PA: Pennsylvania Department of Health.

Toulmin, S. (1961). *Foresight and understanding.* New York: Harper & Row.

Traugott, M. (1987). The importance of persistence in respondent selection for preelection surveys. *Public Opinion Quarterly, 51,* 48–57.

Trivers, R. (1972). Parental investment and sexual selection. In B. Campbell (Ed.), *Sexual selection and the descent of man* (pp. 139–179). Chicago: Aldine.

Trotter, R. (1987, February). Profile: Martin Seligman. Stop blaming yourself. *Psychology Today,* pp. 30–39.

u

Underwood, B., Berenson, J., Cheng, K., Wilson, D., Kulik, J., More, B., & Wenzel, G. (1977). Attention, negative affect, and altruism: An ecological validation. *Personality and Social Psychology Bulletin, 3,* 54–58.

U.S. Department of Health and Human Services. (1982). *The health consequences of smoking.* Washington, DC: U.S. Government Printing Office.

U.S. Public Health Service. (1964). *Smoking and health. Report of the Advisory Committee to the Surgeon General of the Public Health Service.* Washington, DC: U.S. Government Printing Office.

v

Van Gundy, A. (1988). *Techniques of structured problem solving.* New York: Van Nostrand.

Vigderhous, G. (1981). Scheduling telephone interviews: A study of seasonal patterns. *Public Opinion Quarterly, 45,* 250–259.

Viney, L. (1983). Assessment of psychological states through content analysis of verbal communications. *Psychological Bulletin, 94,* 542–563.

von Frisch, K. (1950). *Bees: Their vision, chemical senses, and language.* Ithaca, NY: Cornell University Press.

w

Wagner, R. (1949). The employment interview: A critical review. *Personnel Psychology, 2,* 17–46.

Wahlke, J. (1979). Pre-behavioralism in political science. *American Political Science Review, 73,* 9–31.

Wald, A. (1947). *Sequential analysis.* New York: Wiley.

Warga, C. (1987, August). Profile: Ronald Melzack. Pain's gatekeeper. *Psychology Today,* pp. 50–56.

Warner, K. (1977). *Possible increases in the understating of cigarette consumption.* Paper presented at 105th Annual Meeting of American Public Health Association, Washington, DC.

Warwick, D. (1982). Types of harm in social research. In T. Beauchamp & R. Faden (Eds.), *Ethical issues in social science research.* Baltimore: Johns Hopkins University Press.

Watson, J. (1913). Psychology as the behaviorist views it. *Psychological Review, 20,* 158–177.

Watson, J., & Rayner, R. (1920). Conditioned emotional reactions. *Journal of Experimental Psychology, 3,* 1–14.

Wattenberg, W., & Clifford, C. (1964). Relation of self-concept to beginning achievement in reading. *Child Development, 35,* 461–467.

Watts, A. (1970, February). The drug revolution. *Playboy,* pp. 53–74.

Webb, E., Campbell, D., Schwartz, R., & Sechrest, L. (1966). *Unobtrusive measures: Nonreactive research in the social sciences.* Chicago: Rand McNally.

Webb, E., Campbell, D., Schwartz, R., Sechrest, L., & Grove, J. (1981). *Nonreactive measures in the social sciences* (2nd ed.). Boston: Houghton-Mifflin.

Weiss, B., et al. (1980). Behavioral responses to artificial food colors. *Science, 207,* 1487–1489.

Weiss, S., Jurs, S., LeSage, J., & Iverson, D. (1984). A cost-benefit analysis of smoking cessation program. *Evaluation and Program Planning, 7,* 337–346.

Weller, L., & Livingston, R. (1988). Effect of color of questionnaire on emotional responses. *Journal of General Psychology, 115,* 433–440.

Werth, L., & Flaherty, J. (1986). A phenomenological approach to human deception. In R. Mitchell & N. Thompson (Eds.), *Deception: Perspectives on human and nonhuman deceit.* Albany, NY: State University of New York Press.

Whimbey, A. (1976). *Intelligence can be taught.* New York: Bantam.

Whimbey, A. (1980). Students can learn to be better problem solvers. *Educational Leadership, 37,* 560–565.

White, N. (Ed.). (1974). *Ethology and psychiatry.* Toronto: University of Toronto Press.

White, R. (1966). Misperception as a cause of two world wars. *Journal of Social Issues, 22,* 1–19.

Why scientific fact is sometimes fiction. (1987, February). *The Economist,* pp. 103–104.

Wicker, A. (1985). Getting out of our conceptual ruts. *American Psychologist, 40,* 1094–1103.

Willems, E. (1973). Behavioral ecology and experimental analysis: Courtship is not enough. In J. Nesselroade & H. Reese (Eds.), *Life-span developmental psychology.* New York: Academic Press.

Williams, R. (1956). *Biochemical individuality.* Austin: University of Texas Press.

Williams, S. (1977). Social evolution of a fact. *Bulletin of the British Psychological Society, 30,* 241–246.

Willick, D., & Ashley, R. (1971). Survey question order and political party preferences of college students and their parents. *Public Opinion Quarterly, 35,* 189–199.

Willner, P. (1985). *Depression: A psychobiological synthesis.* New York: Wiley-Interscience.

Windle, C. (1986). Measures: Derivation, analysis, and interpretation. In C. Windle, J. Jacobs, & P. Sherman (Eds.), *Mental health program performance measurement* (DHHS Publication No. ADM 86–1441). Washington, DC: U.S. Government Printing Office.

Windle, C. (1986). An orientation to performance measurement. In C. Windle, J. Jacobs, & P. Sherman (Eds.), *Mental health program performance measurement* (DHHS Publication No. ADM 86–1441). Washington, DC: U.S. Government Printing Office.

Winget, C., & Kramer, M. (1979). *Dimensions of dreaming.* Gainesville, FL: University Presses of Florida.

Wohlwill, J. (1970). Methodology and research strategy in the study of developmental change. In L. Goulet & P. Baltes (Eds.), *Life-span developmental psychology: Research and theory.* New York: Academic Press.

Wolf, S. (1962). Placebos: Problems and pitfalls. *Clinical Pharmacology and Therapeutics, 3,* 254–257.

Wolins, L. (1962). Responsibility for raw data. *American Psychologist, 17,* 657–658.

Women on Words and Images. (1972). *Dick and Jane as victims.* Princeton, NJ.

Worden, M. (1979). Popular and unpopular prevention. *Journal of Drug Issues, 3,* 425–433.

Wright, H. (1960). Observational child study. In P. Mussen (Ed.), *Handbook of research methods in child development.* New York: Wiley.

Wynne-Edwards, V. (1962). *Animal dispersion in relation to social behavior.* London: Oliver & Boyd.

Y

Yukawa, H. (1973). *Creativity and intuition.* Tokyo: Kodansha International Limited.

Z

Ziman, J. (1968). *Public knowledge: The social dimension of science.* Cambridge, England: Cambridge University Press.

Ziman, J. (1980). The proliferation of scientific literature: A natural process. *Science, 208,* 369–371.

Zimbardo, P. (1985). *Psychology and life.* Glenview, IL: Scott, Foresman.

Zimmerman, D., & Clark, D. (1987). *Guide to technical and scientific communication.* New York: Random House.

Zirkle, C. (1958). Pavlov's beliefs. *Science, 128,* 1476.

Zucker, I. (1988). Seasonal affective disorders: Animal models non fingo. *Journal of Biological Rhythms, 3,* 209–223.

Zucker, I., Rusak, B., & King, R. (1976). Neural bases for circadian rhythms in rodent behavior. In A. Reisen & R. Thompson (Eds.), *Advances in psychobiology* (Vol. 3). New York: Wiley.

INDEX

a

ABAB design, 244
Accretion measures, 92
Aderman, D., 221
Agassiz, Louis, 145
Aging, animal models of, 304
Alcock, J., 311, 312
Alcoholism, animal models of human, 303
Altman, H., 176
Altmann, J., 144
Alzheimer's disease, 71
American Political Science Review, 153
American Psychological Association (APA), 111
    guidelines for ethical treatment of animals, 121
    guidelines for ethical treatment of humans, 112–13
    publication manual, 344
Ames bacterial response test, 301
Analogies, value of
    in finding research problems, 29
    in solving research problems, 129–30
Analysis of variance, 210–18
    computational procedure, 211 (box)
    implications of, for experimenters, 212–16
    for more than two independent groups, 230–31 (box)
    summary of, for 2 × 2 factorial study, 239 (table)
Analytic statements, 11
Aneshensel, C., 159
Anger, self-reports on, 325
Animal subjects, 299–314
    bee behavior, 308–9
    bird behavior, 6, 69, 139, 295, 313
    creating models of human conditions using, 301–2
        alcoholism, 303
        learned helplessness, 302
    curiosity in zoo animals, 311, 312
    development of behavior in, 295
    ethics of research on, 120, 121–23
    learning about human systems using, 303–5
    learning how independent variables affect humans by using, 300–301
    polygyny and monogamy in, 66
    rats and mice (*see* Rats and mice, experiments using)
    studying evolutionary aspects in, 307–13
        compared to humans, 313
        ecological aspects, 311–13
        evolutionary history and, 311
        generating hypotheses, 310
    studying unusual properties in, 305–7
Anxiety, sexual, and verbal errors, 70
Appel, P., 287–88
Archives, measurement using, 92–93
Argyle, M., 146
Armadillos, development research on, 306
Armitage, P., 251
Arnold, D., 244
Ashley, R., 169
Association methods, 323–24
Attitudes, writing survey questions about, 165–66
Attridge, C., 147
Auerbach, O., 102

b

Babst, D., 166
Bacon, Francis, 78, 202
Bacterial response, 301
Bahrick, H., 90
Bakeman, R., 143, 144
Baker, J., 4, 132
Baker, R., 132
Baltes, P., 295
Barber, J., 264
Barker, E., 317, 321
Barker, R., 138, 221
Barton, A., 80–81
Baseball, statistics on, 262, 263 (box)
Bateson, William, 71
Bats, research on, 307
Baumgardner, M., 67
Baumrind, D., 111
Baumrind, H., 111
Beach, F., 306
Beauchamp, T., 111
Beecher, H., 110, 111, 217
Bees, research on, 307, 308–9 (box)
Behavior(s)
    assessing, through observation, 139–40
    boundaries of, 68
    developing lists of, for observation, 142–43
    ecological links to, 311–13
    mechanisms of, 69
    methods of recording, 144 (*see also* Scorer and scoring)
    paradoxical, 71–72
    research on development of, 290–96
        advanced and miscellaneous techniques, 294–95
        cross-sectional designs, 291–92
        longitudinal designs, 292–94

studying evolutionary history of, 311
survey questions about, 163–65
survival values of, 69
Behavioral Ecology: An Evolutionary Approach (Clutton-Brock, Harvey), 311
Belsky, J., 354
Belson, W., 163
Bennett, C., 278
Bergin, A., 78, 243
Bernard, Claude, 35, 138
Bernstein, P., 243
Berra, Yogi, 137
Berrin, K., 150
Between-group variability, 229
  sequential analysis, 251
Beveridge, W., 151
Bias in researchers, 47, 144, 150
Binder, A., 241
Binomial formula, 51–53
  calculating, 52 (box)
Biochemical Individuality (Williams), 16
Bioethics, 111
Biological Abstracts, 36
Biology and Gender Study Group, 13
Biorhythm(s), 67, 69, 245, 301
Birds
  bearded tit nesting behavior, 139
  gull nesting behavior, 6, 69, 311
  learning periods of songs, 295
  weaver bird behavior linked to habitat ecology, 313
Blake, Judith, 260
Blalock, H., 191
Block, J., 82, 283
Blocking technique, 241
Bloom, Benjamin, 30
Blumenthal, M., 281
Blurton-Jones, N., 143
Body type and personality, 182
Bogdan, R., 149
Bohr, Niels, 8
Bok, S., 115
Bonjean, C., 84
Boruch, R., 117, 283, 284, 285
Boundary examination in problem solving, 127
Bradburn, N., 161, 162, 163, 164, 165, 166, 168, 169
Bradshaw, G., 127
Braukmann, C., 146

Brenner, M., 168
British Medical Association, 101, 103
Broad, W., 12, 54
Broadbent, Donald, 30
Brockhuysen, G., 6
Bronowski, J., 70
Brown, Fredric, 119
Bruner, J., 221
Bryant, B., 146
Bryant, F., 260, 261
Bryson, M., 160
Burch, P., 105
Bustad, L., 301
Butler, Samuel, 97

C

Campbell, D., 91, 129, 198, 209, 283, 284, 285
Campbell, J., 72
Cancer
  coffee drinking and, 58, 60
  smoking and, 101–6, 190
Carlson, R., 245
Cartwright, Rosalind, 30
Case studies, 99, 100, 101–2, 174–77
  after-the-fact, 175–76
  complex analytic after-the-fact, 176–77
  single case experimental designs, 177
Casey, J., 149
Cassileth, B., 116
Caton, J., 104
Causation
  correlation and, 183, 191, 196–99
  effect and, 72–73
Cecil, J., 117
Cederlof, R., 103, 106
Cell biology, values and theoretical models of, 13
Chalmers, A., 8
Chalmers, T., 283
Chamberlin, T., 70
Chance, P., 30
Chance reliability (CR), formula for estimating, 148
Chapanis, A., 20, 145, 217
Chapman, J., 141
Chapman, L., 141
Checklists as observational data, 141–42
Child abuse prevention programs, evaluating, 273–75

Child Development Abstracts, 36
Chi square table, 403
Chomsky, Noam, 30
Cicchini, M., 323–24
Clark, D., 161
Clark, V., 159
Classification schemes, 80–81
Cliches, using, for problem solving, 130–31
Closed design, 254–55, 257–58
Clucas, Geralyn Rodriguez, article by, 374–80
Cluster sampling, 159, 167
Clutton-Brock, T., 310, 311
Cochrane, A., 14
Coefficient of determination, 187
Cohen, D., 30, 31, 32
Cohen, J., 255
Cohen, M., 303
Cohen, W., 319
Coile, D., 120
Colgrove, M., 127, 133
Coming of Age in Samoa (Mead), 149
Computer searches of scientific literature, 41, 357–73
Condition(s)
  of a desired end-point, 67–68
  determining cause/effect of, 72–73
  variance between and within, 210 (table)
Cone, J., 48, 49
Confidence limits, 331–33
  calculating, for proportions, 332
  calculating, for two independent groups, 332–33
  for correlations, 333
Confound variables, 46, 150
Constructs, 79
Construct validity, 89–90
  problems with, 45, 47, 56
Construct veracity for self-reports, 319
Contemporary Psychology, 41
Content analysis, 265–68
  data-analytic techniques, 267–68
  integrative complexity, 267
  reliability, 266–67
  sampling of sources in, 265
  scoring, 265–66
Content veracity for self-reports, 319
Control groups, 46–47, 104
  problems of, in prospective studies, 103–4

Control series design, 285–86
Conventional wisdom, testing, 72
Converse, P., 153, 158, 167
Cook, L., 235
Cook, T., 209, 285
Cooper, H., 349
Cooper, J., 117
Copernicus, 132
Correlation(s), illusory, 141
Correlational questions, 66–67
Correlational studies, 99, 102–4, 178–200
   best use of, 100
   calculating confidence limits for, 333
   causation and correlation relationships, 183, 191, 196–99
   contingency coefficient, 191
   correlation coefficients, 179–80, 186–87
   item analysis, 189
   partial correlation, 190–91
   Pearson product-moment correlation, 185–86
   phi coefficient (dichotomous measurement), 189–90
   point-biserial correlation, 186, 194
   potential problems in, 187–89
   quick method of computing statistical significance of correlations, 195
   regressions equations, 183–85
   Spearman rank-order correlation, 192–93
   test-retest reliability, 188
   when to use, 181–83
Correlation coefficients, 179–80, 186–87
   for variables measured with interval scales, 185–86
   z transformation to, 351, 401–2
Cost effectiveness analysis of social/medical programs, 280–81
Cournoyer, N., 160
Cowen, E., 278, 279, 281
Crandall, R., 110, 111, 115
Cravens, J., 145
Creative problem solving, 126–34
   considering alternatives, 133
   evaluating and selecting ideas, 131
   generating ideas, 129–31
   implementing ideas, 131
   positive moods and, 133
   pretending and role playing in, 133
   redefining and analyzing problems, 127–29
   relaxation and, 132–33
   tenacity at working on problems, 132
Crick, Francis, 35
Criteria for program outcomes, 277–79
   principles for choosing good, 279
   problems in choosing good, 278
Criterion-related validity, 87–89
Criterion veracity for self reports, 319
Critical incident method, 325
Cronbach, L., 49, 166, 217, 232, 264
Crook, J., 312, 313
Cross-lagged panel correlation, 198–99
Crossley, H., 163
Cross-sectional designs, 291–92, 293
   combined with longitudinal approaches, 294–95
Crowne, D., 170
Cullen, E., 311
Current Contents, 41
Cuvier, George, 201

d

Dabbs, J., 144
Daft, R., 72
Dalbey, W., 104
Dancer, D., 146
Darwin, Charles, 4, 7, 8, 10, 132, 138, 140
Data
   alternative interpretations for, 204, 218
   analysis of existing data (*see* Secondary analysis)
   collectors of, vs. scientists, 18–20
   influence of scientist's history on observational, 12
   influence of scientist's values on interpretation of, 13
   measurement (*see* Variables, selecting and measuring)
   observational (*see* Observational data)
   private sources of, 315–26
   problems in relationship between theory and, 54, 55 (box), 57
   sequential analysis of, 250–58

Data collection. *See also* Experiment(s); Observation; Survey(s)
   behavioral dimensions in, 50
   external validity issues in, 48–50
   methods/strategies for, 49, 50, 99–106
Data interpretation
   alternative, 204, 218
   eliminating alternative, through experimentation, 203, 204
Dauben, K., 133
Daughter of Time (Tey), 264, 268
Davidshofer, C., 181
Davies, N., 311
Dawis, R., 85
Daydreaming, 321
Death, statistics on time of, 262
Death penalty, 58
DeBono, E., 127
Deception
   ethics of, in research, 115, 149
   self-reports on, 321
Definitions, 11, 77
   operational, 78–79
DeLamater, J., 161, 168
DeMaio, T., 159
Demand characteristics, 50
Denzin, N., 112
Dependent variable (DV), 18
   correlations with behavior of interest as, 67
   nominal measurement of, and analysis of two-group experiments, 222–24
   self-reports as, 321–22
   validity considerations in selecting, 90
Depression and learned helplessness, 302
DerSimonian, R., 349
Descartes, Rene, 7
Descriptive questions, 65–66
Descriptive studies, 99
Design of social/medical programs, evaluating, 273–75
   criteria for determining if objectives are met, 275
   documentation, 273
   stated activities for meeting objectives, 274–75
   stated goals and objectives, 273–74
Dewsbury, D., 19, 310
Diener, E., 110, 111, 115

Dillman, D., 160
Dimensional sampling, 244
Disease, animal models for human, 301
Doll, R., 101, 102, 103
Domer, F., 140
Dontenwill, W., 104
Double blind experiments, 57
Down's syndrome, 71
Draw-a-Person test (DAP), 141
Drawing for problem solving, 131
Drug(s)
   creativity and, 17
   ethical issues in research using, 110, 111, 118–19
   habituation process and, 5
   problems in research on hunger and, 45–48
   problems in research on psychiatric problems and, 175
   screening, with animals, 88
   stress and use of, 197–98
   variable complexity when researching properties of, 219
Drug abuse prevention program, evaluating, 275
Druley, K., 283
Dukes, W., 176

e

Eating, causes of, 73 (table)
Ebbinghaus, H., 245
Ecological psychology, 221
Ecology, behaviors related to, 311–13
Edgerton, R., 313
Edgington, E., 244
Edison, Thomas, 133
Edwards, A., 161
Effect size
   combined, 352–53
   formula, 351
   measuring, 331
   similarities in, 351–52
Efficiency in research, 18
Einstein, Albert, 8, 10, 63, 71, 132
   on scientific research and religiousness, 4
Eisenberg, L., 114
Ember, C., 313
Ember, M., 313
Empiricism in science, 9, 64
Employment discrimination, 275

Environmental psychology, 221
Epidemiology, hypothesis development in, 138–39
Epstein, C., 71
Erlenmeyer-Kimling, L., 179
Eron, L., 292
Erosion measures, 92
Errors in research, 18
   cheating and distortions, 54–57
   construct validity, 45, 47
   data-theory relationship, 45, 54
   examples, 58–59
   external validity, 45, 48–50
   internal validity, 45, 46–47
   sampling errors (see Sampling error(s))
   statistical validity, 45, 51–54
   in voting surveys, 169–71
Esecover, H., 117
Ethics in research, 109–25, 217
   American Psychological Association's guidelines on, 112–13, 121
   example of an ethical dilemma, 118–19
   examples of unethical research, 110–11
   issues of, when using animal subjects, 120–23
   issues of, when using human subjects, 111–16
   deception, 115, 149
   methods for reducing ethical problems, 116–18
   suppression of research due to ethical concerns, 119–20
Evaluations of social and medical programs, 271–89
   policymaker's use of, 287–88
   political issues in, 272
   program design, 273–75
   program implementation, 276–77
   program outcomes, 277–86
Evans, C., 303
Evenson, R., 176
Evolution, animal studies related to, 307–13
Experiment(s), 99, 100, 104. See also Research
   checklist for problems in, 56–57 (box)

defining characteristics of, 104, 202
double blind, 57
generalizing results of, 220–21
reasons for conducting, 202–3
secondary analysis of results of, 264
sequential sampling (see Sequential analysis)
Experiment(s), two-group design, 201–48
   calculating confidence limits for, 332–33
   experimenter's judgement needed when evaluating results of, 218–22
   factorial designs, 232–41
   more statistics for analyzing results of two-group, 222–24
   randomly assigning subjects in, 208–9
   reasons for conducting, 202–3
   sample size in, 203–8
   studying an IV at more than two levels, 229–31
   variability in, 209–18
   within-subject designs, 242–49
Experimental groups, 46–47
Experimenter(s)
   bias in, 47, 144
   effect of, 47, 92
   need for judgement in, 218–22
   as subject of experiments, 245–49
   as a variable, 241
External validity, problems with, 45, 48–50, 56
Eysenck, H. J., 30, 78, 105, 243

f

Fact organization in science, 8–9
Factorial designs for experiments, 232–41
   2×2, 233–35, 237
      computational procedures for, 238–39 (box)
   3×5, 233
   blocking subjects in, 241
   example, 236–37
   experimenter as variable in, 241
   interpretation of interactions, 235–36, 237, 240
   subject variables in, 240

Faden, R., 111
Faillace, L., 303
Falk, J., 68–69, 102, 301, 303
Falsifiable theory, 9
   introspective data and, 320
Faraday, Michael, 71
Faust, D., 86
Federalist Papers, content analysis of, 266
Feekes, F., 6
Feldman, H., 149
Fellows, E., 235
Fenichel, O., 78
Fermi, Enrico, 35
Festinger, L., 30, 275
Feyerabend, P., 8, 22
Field experiments, 221–22
Fischer, C., 324
Fisher, S., 70
Fisher exact probablity test, 223
   critical values table, 385–99
Fisher z transformation, 351–52, 401–2
Flaherty, J., 321
Focused interview method, 323
Food dyes, effect of, on children, 58
Foulds, G., 283
Fox, M., 120
Frankel, J., 160
Franklin, J., 266
F ratio, 207–8
   distribution table, 381–84
   for fictitious 2 × 2 factorial studies, 240 (table)
Fraud, scientific, 54–57
Freeman, Derek, 149
Freeman, H., 273, 275, 277
Frerichs, R., 159
Freud, Anna, 10, 138
Freud, Sigmund, 7, 70
Friberg, L., 103
Fries, J., 116
Frogs
   body part regeneration in, 306
   Sapolsky's research on Rorschach imaging of, 55
Functional relationships, 333–35

g

Galileo, 7
Gallup Poll, 154, 155 (table), 168
Garcia, John, 71
Gardner, H., 79
Gardos, G., 110
Garfield, P., 246
Garfinkel, L., 102
Garmezy, N., 242
Gay, C., 114
Gender, S., 300
Generalizations
   from literature reviews, 354
   to other situations, 221–22
   to other subjects, 220–21
   from outcome evaluations, 287
   problems with, 48
Gerard, D., 279
Gergen, K., 216, 217
Geschwind, Norman, 67
Gilbert, J., 114–15, 282, 283, 284
Glaser, E., 13
Gleser, G., 49
Glickman, S., 311, 312
Goldstein, J., 282
Goodman, J., 293–94
Gordon, W., 127
Gottman, J., 143, 144
Goudy, W., 171
Graham, S., 241
Grant, E., 143
Gray, B., 116
Greenberg, R., 70
Greenwald, A., 67, 68
Gregg, Alan, 137
Griesemer, R., 104
Griffin, D., 307
Grunder, T., 116
Guerin, M., 104

h

Habituation process, drugs and, 5
Hadley, S., 243
Hallberg, S., 143
Hamilton, Alexander, 266
Hammer, E., 241
Hammond, E., 102
Hankins, W., 71–72
Hannum, K., 267
Hansel, C., 13
Hanson, N., 12
Hardy, G. H., 65
Harford, R., 320
Harlow, H., 295
Harlow, M., 295
Harm, research-created, 112–15
Harris, B., 42

Harris Poll, 168
Hartmann, E., 145
Hartshorne, H., 77, 82
Harvey, P., 310, 311
Hatchett, S., 168
Hauser, P., 261
Hayes, J., 320
Head Start programs, 278
Hefferline, R., 320
Heise, D., 191
Heisenberg, Werner, 64
Hempel, Carl, 34, 36
Henshel, R., 304
Herbert, J., 147
Hermann, J., 147, 148
Hilgard, E., 317–18
Hill, A., 101, 103
Hill, John, 101
Hill, R., 84
Hinde, R., 142
History, effects of, on data interpretation, 204
Hite, S., 171
Hobbes, Thomas, 7
Hodgkin, A., 306
Holsti, O., 260
Hopkins, B., 147, 148
Houghton, J., 6
Hrdy, S., 12
Hudson, Liam, 30
Huesman, L., 292
Hulin, C., 72
Hull, Cordell, 18
Humans and animal research
   animal models of human conditions, 301–3
   effect of independent variables on, 300–301
   learning about human systems with animal models, 303–5
Human subjects, 111–16
   deceiving, 115
   informed consent of, 116
   invading privacy of, 115, 278
   risk of harm to, 112–15
   using homogeneous, to reduce within-group variability, 213, 216
Humphreys, W., 29
Hutt, C., 143
Hutt, S., 143
Huxley, A., 306
Huxley, Thomas, 15–16
Hyman, H., 260, 261

i

Ideas in creative problem solving
  evaluating, 131
  generating, 129–31
  implementing with potential problem analysis, 131
Illusory correlations, 141
Implementation of social/medical programs, evaluating, 276–77
Independent variable (IV), 18
  correlations with behavior of interest as, 67
  generalizations about, 48
  studying, at more than two levels, 229–31
  testing effects of, on humans, with animal subjects, 300–301
  validity considerations in selecting, 90
Indexes, measuring with, 84–86
Index Medicus, 40, 344
Informants, data collection through, 50, 165
Informed consent, 116
Instrumentation, effects of, on data interpretation, 204
Integrative complexity, 267
Intellectual values and history, influence of scientist's, on research, 12–14
Intelligence, correlation of child's and parent's, 196
Intelligence tests, 79, 293–94
Internal validity, problems with, 45, 46–47, 56
Interrupted time-series design, 285
Interval data, sequential analysis of, 255–58
Interval Data Master Sheets, 404–12
Interval scales, 80
  correlation coefficient for, 185–86
Interviews, 99
  assuring confidentiality and truthfulness of, 117–18
  different interviewers and discrepancies in, 168
  focused, 323
  personal, 161
  telephone, 161–62
Introspective data, 315–26
  falsifiability of hypotheses generated from, 320
  methods of collecting, 323–25
  using self-report methods and, 320–23
  verifiable, 318–19
Isen, A., 133
Item(s)
  computing reliability with changing number of, 83 (box)
  correlational analysis of, 189 (box)
  data collection and, 49
  multiple-item scales, 84–86
Ittelson, W., 221
Iverson, D., 280

j

Jacobson, L., 196
Jaffe, J., 277
James, Bill, 262, 263
Janis, L., 117
Jarvik, L., 179
Johnson, J., 165
Johnson, Samuel, 33
Jones, J., 110, 217
Jones, R., 143, 147
Journal of Experimental Psychology, 344
Journals. *See* Scientific literature
Jouvet, Michel, 30
Jurs, S., 280
Just, E. E., 13

k

Kagan, J., 292
Kamin, L., 349
Kanowitz, J., 147
Karegianes, M., 349
Karpf, D., 317
Kasperson, C., 35
Kazdin, A., 148, 150, 242, 244
Keenan, B., 320
Keir, G., 182
Keith, P., 171
Kekule, August, 133
Kelman, H., 110, 111, 113, 115
Kelvin, W., 29, 132
Kent, R., 147
Kepler, Johann, 7, 29
Keppel, G., 229
Keynes, R., 306
Kiewiet, D., 167
Kimmel, M., 32
King, F., 120
Kinsey, Alfred, 171
Kinsey Report, methodological evaluation of, 171–72, 319
Kintz, B., 241
Kirigin, K., 146
Knowledge
  results of experiments interpreted within existing, 219
  writing survey questions about, 165
Koestler, A., 4, 29
Kohlberg, L., 77
Kolansky, H., 175, 176
Kolb, L., 279
Kortlandt, A., 145
Kramer, M., 84
Krebs, J., 311
Krippendorff, K., 266, 268
Kropotkin, Pyotr, 4
Kruuk, H., 6
Kunze, M., 102
Kurman, P., 64

l

Labouvie-Vier, G., 293
Laing, R. D., 31
Laird, N., 349
Lamb, D., 50
Lamberg, L., 30
Langley, P., 127
Lasagna, L., 117
Laudan, L., 14
Lazarsfeld, P., 80–81
  his approach to correlation studies, 197–98
Leacock, Stephen, 9
Lear, J., 110
Learned helplessness, models of, 302
Learning
  animal models of, 304
  paradoxical behavior in, 71–72
Leavitt, F., 82, 176
Le Bon, Gustave, 7
Lederman, L., 65
Lee, S., 283
Lefkowitz, M., 292
Lehner, P., 144
Leibovitz, B., 295
Leippe, M., 67
Lenneberg, E., 176
Leopold, A., 68
LeSage, J., 280
Levin, H., 280, 281

Levine, A., 142
Levine, M., 127, 129, 317
Liban, C., 165
Lick, 49
Lieberman, D., 317, 318, 319, 323
Liebson, I., 303
Light, R., 282, 349, 350, 354
Likert Scale, 85–86, 189
Linton, M., 246
Lipton, D., 287–88
Livingston, R., 48
Livson, N., 179
Lockard, R., 304
Loftus, E., 116
Loneliness, self-reports on, 322
Longitudinal designs, 292–94
   combined with cross-sectional, 294–95
Lowe, C., 305
Luborsky, L., 20, 283
Lumsdaine, A., 278
Lundman, T., 103
Lykken, D., 55, 91

m

McClearn, G., 303
McClelland, D., 31, 179
McConnell, D., 241
MacDonnell, H., 176
McGuigan, F., 241
McGuire, M., 143
McGuire, W., 232
McKeachie, W., 232
McKellar, P., 325
MacKenzie, B., 120
McKillip, J., 165
McKinley, John, 88
McLellan, A., 283
McLemore, D., 84
MacMahon, B., 58
McPeek, B., 114–15
McSweeney, R., 284
Madison, James, 266
Mahrer, A., 65
Maier, N., 127, 133
Maini, S., 35
Malitz, S., 117
Mandler, G., 181
Mann, L., 117
Margolis, J., 36
Margulis, S., 221
Marler, P., 295

Marlowe, D., 170
Marsupial development, 306
Matched-pair design, analysis of variance with, 213
Maturation, effects of, on data interpretation, 204
May, M., 77, 82
Maxims, using, in problem solving, 130–31
Maxwell, Clerk, 71
Mead, Margaret, 149
Measure(s), 79–93, 98
   multiple, 91, 277
   nonreactive, 92–93
   properties of good, 81–93
      reliability, 82–84
      sensitivity, 84–86
      validity, 86–93
   scales, 79–80, 84–86
   self-reports as, 322–23
Medawar, Peter, 8, 21, 64
Medical doctors, influence of values on readings of, 13
Medical programs, evaluating design, implementation, and outcomes of, 271–89
Medley, D., 141
Meehl, Paul, 51, 70, 86, 103, 330, 334
Mello, N., 303
Melzack, Ronald, 31
Mendelson, J., 303
Mental Health Book Review Index, 41
Mental Measurements Yearbook, 41
Merton, R., 68
Meta-analysis, 265, 349
Micheletti, P., 267
Miller, D., 140
Miller, N., 120
Miller, Neal, 31, 91
Milton, John, 7
Minnesota Multiphasic Personality Inventory (MMPI), 88–89
Mischel, W., 232, 317
Mitzel, H., 141
Models, creating animal, of human conditions, 301–3
Monod, J., 8
Mood, positive, in creative problem solving, 133
Moore, W., 175, 176
Morality development in children, 77
Morgan, T. H., 13
Morse, W., 305

Mortality, effects of, on data interpretation, 204, 218
Mosteller, F., 114–15, 266, 282
Motley, M., 70
Muller-Lyer illusion, Clucas article on, 374–80
Multiple-baseline design, 244
Multiple-item scales, 84–86
Multiple measures, 91, 277
Munsinger, H., 349
Murphy, K., 181, 182

n

Nanda, H., 49
Nardi, G., 58
Narratives as observational data, 140
National Academy of Sciences, 282
National Center for Health Statistics, 168, 170
National Organization of Women, content analysis project by, 265–66
Nature, descriptions of, through observation, 138
Nerve cells, squid studies on, 306
Nesselroade, J., 295
Nettesheim, P., 104
Netting, R., 313
Neurosis, animal models of, 301
Newicki, G., 133
Newton, Isaac, 7, 8
Nicolle, Charles-Jean-Henri, 97
Nietzel, M., 243
Nominal measurement scale, 79
Nonrandom assignment, alternatives to, 285–86
Nonreactive measures, 92–93
   using, in observation, 145
Nordbeck, B., 35
N-rays, 12
Nunnally, J., 331

o

Oakland Tribune, editorial on animals used in research, 122–23
Objective test, 99
O'Brien, C., 283
Observation, 28, 93, 99, 137–51
   obtrusive vs. unobtrusive, 49, 145
   participant, 99, 149–50
   problems in, 150–51

process/steps of good, 140–48
reasons for doing observational studies, 138–40
Observational data
bias in, 150
checklists as, 141–42
estimating reliability of, 148
illusory correlations in, 141
influence of intellectual history on, 12
measuring reliability of, 147–48
narrative form of, 140
ratings as, 141, 146–47
relative frequencies of, 150
Observational statements, 11
Obtrusive observation, 49
Ogborne, A., 287
Ogburn, W., 343
Olby, R., 35
O'Leary, K., 147
Olian, S., 48
Olson, R., 7
Open design, 252–54, 255–57
Operational definitions, 78–79
self-reports and, 322–23
Oppenheim, A., 167
Ordinal data, sequential analysis of, 252–54
Ordinal measurement scale, 79
Orians, Gordon, 66
Orlans, F., 120
Orne, M., 50, 221
Ortega y Gasset, Jose, 19
Osborne, A., 127
Outcomes of social/medical programs, evaluating, 277–86
cost effectiveness analysis, 280–81
criteria, 277–79
principles for choosing, 279
problems of choosing good, 278
nonrandom assignment, alternatives to, 285–86
random assignment, 281–86
decision-making facilitated by, 282–84
sometimes impossible, 281–82
usually possible, 284–85

p

Paradox, 71–72
Parry, H., 163
Partial correlation, 190–91

Participant observation, 99, 149–50
Pasteur, Louis, 4, 132
Pastore, N., 55
Patterson, R., 147
Pavlov, Ivan, 83, 301, 304
Payer, L., 13
Pearson product-moment correlation, 185–86
Personality and body type, 182
Petzel, T., 165
Pflaum, S., 349
Phenomenological method, 324–25
Phi coefficient, 189–90
Phillips, D., 262, 268
Philosophy
influence of science on, 7
of science, 8
Phobia, experimentally induced, 111
Physical evidence
data collection with, 99
measurement with, 92
Piaget, Jean, 77, 138
Piliavan, I., 114
Piliavin, J., 114
Pillemer, D., 349, 350, 354
Pinney, J., 101
Piotrowski, Z., 241
Platt, J., 70, 132
Platt, W., 132
Pleck, Joseph, 32
Point-biserial correlation, 194 (box), 331
critical values for significance in, 186 (table)
Policy-making and program evaluation, 272, 282–84, 287
Polsky, R., 143
Polya, G., 66, 127, 129
Popper, Karl, 8, 9, 70, 202
Positive correlation, defined, 179
Positive results lacking statistical validity, 53
Power of experiment samples, 205–8
Powers, E., 171
Practice in observation, 145–47
Pratkanis, A., 67
Pratt, D., 120
Preble, E., 149
Predictive validity, 87
Presidency, statistical analysis of American, 264
Presser, S., 154, 169

Pretesting
of questionnaires, 163, 167
of Solomon experimental groups, 218–19
Price, E., 304
Privacy
invasion of subject's, 115, 278
self-reports on, 324
Product-moment correlation, 185–86
critical values for significance in, 186 (table)
Prohansky, H., 221
Projective (indirect) tests, 99
Proportions, calculating confidence limits for, 332
Prospective studies, 102–4
Proverbs, using, in problem solving, 130–31
Psychological Abstracts, 36, *37–38,* 344
Psychology Today, 30
Psychopharmacology, 344
Psychopharmacology Abstracts, 36
Psychotherapy, value of, 78 (box), 243
Ptolemy, 5

q

Quasi-experiments, 285
Question(s), scientific, 62–75
answerability of, 17
features of good, 63–65
influence of values on, 12
range of phenomena addressed by, 17–18
types of
answerable by science, 10–11
leading directly to research projects, 65–68
for organizing research programs, 68–73
Questionnaires, 86, 162–67. *See also* Survey(s)
assembling, 167
assuring confidentiality and truthfulness of, 117–18
classes of questions on, 162
mailed, 160–61
ordering questions on, 166
pretesting questions on, 163, 167
writing questions about attitudes for, 165–66
writing questions about behavior for, 163–65
writing questions about knowledge for, 165

R

Rajaratnam, N., 49
Random assignment, 46, 104
   alternatives to nonrandom assignment, 285–86
   of experiment subjects, 208–9
   program outcome analysis and, 281–86
   of survey samples, 158
Rank-order correlation, 192–93
Rasher, S., 349
Ratings
   of descriptions of verbal statements, 146–47
   as observational data, 141
Ratio measurement scales, 80
Rats and mice, experiments using
   aging in, 304
   excessive drinking in rats, 68–69, 91
   generalizations based on, 48
   learning paradoxes, 71–72
   Norway rat's dietary needs, 301
   sexual behavior in rats, 87
   specialized species, 306
Raven Progressive Matrices Test, 182
Ray, R., 143
Rayner, R., 42, 111
Razran, G., 83
Reading about science, 33–43. *See also* Scientific literature
   computer searches as an aid for, 41, 357–74
   finding research problems by, 28, 29
   influence of values on, 13
   literature reviews, 348–55
   strategies for, 41, 42
Reese, H., 295
Regan, T., 120
Regression analysis, 183–85
   functional relationships of, 334–35
Regression discontinuity design, 286
Reid, J., 147
Reines, B., 120, 301
Reinforcement of behavior, 305
Relaxation in creative problem solving, 132–33
Reliability
   of content analysis, 266–67
   of measures, 81, 82–84
   of observations, 147–48
   of tests, 181–83

Research
   annual reviews of, 41
   checklist for experiments, 56–57 (box)
   data collection (*see* Data collection)
   efficiency and, 18
   errors in, 18, 44–61
   ethics in, 109–25
   evaluation (*see* Evaluations of social and medical programs)
   influence of values on, 14
   measures in (*see* Measure(s))
   methodology utilized by average persons, 15–17
   questions leading directly to, 65–68
   questions that organize programs of, 68–73
   suppression of, due to ethical concerns, 119–20
Research problems, finding interesting, 27–32
   becoming a careful observer, 28
   examples of, 30–32
   noticing routine and unusual occurrences, 28
   reading to stimulate ideas, 28–29
   seeking analogies, 29
   solving creatively, 126–34
Research reports, writing. *See* Writing research reports
Retrospective studies, 102
Reversible perspective figures, *12, 13*
Richardson, A., 319, 323
   methods of collecting self-report data by, 323–25
Richter, C., 301, 304
Rimington, J., 102
Rival hypotheses, 91
Rivers, D., 167
Rivlin, L., 221
Rogers, Carl, 20
Rogers, H., 320
Role playing in creative problem solving, 133
Rootman, I., 169
Rosenblum, L., 142, 145
Rosenthal, MaryLu C., 357
Rosenthal, R., 47, 54, 196, 221, 241, 349, 351
Rosnow, R., 221
Ross, H., 281

Rossi, P., 273, 275, 277
Roszak, T., 15
Rowan, A., 120
Rugg, D., 88
Rusinak, K., 71–72

S

Sabine, 57
Sackett, G., 143
Saenger, G., 279
Saint Simon, Comte de, 7
Samples, experiment, 203–8
   power of, 205–8
   sequential, 250–58
   types of errors, 203–5
Samples, survey, 154–60
   choosing individual subjects for, 157–59
      cluster sampling, 159
      simple random sampling, 158
      stratified sampling, 159, 265
      systematic sampling, 158
   errors in, 154–57, 167
   nonrespondents in, 159–60
   size of, 154–57
      calculating, 156 (box), 157 (table), Appendix E
Sampling error(s)
   experiments, 203–5
   surveys, 154, 155, 167
      calculating acceptable, 156 (box), 157 (table)
Sapolsky, A., 55
Sarason, S., 181
Saufley, W., 229
Scales, 79–80
   books with evaluations of, 85
   interval (*see* Interval scales)
   Likert, 85–86
   multiple-item, 84–86
   observational data gained through rating, 141
Schaie, K., 293
Schewe, C., 160
Schumaker, J., 146
Schuman, H., 154, 168, 169
Schuster, D., 58
Schuster, L., 58
Schwartz, R., 91, 169

Science, 3–23
    defined, 8–10, 340
    human nature and investigatory,
        15–17
    influence of, on modern thought, 7
    limitations of, 10–15
    objectives for this book related to,
        20–22
    philosophy of, 8
    readings in, 33–43
    rewards of, 5–7
Science Citation Index, 36
Scientific discipline, choice of,
        influenced by values, 12
Scientific literature, 6, 29
    Index Medicus index to, 40
    journals
        abstracts of, 36, *37–38*
        citation indexes of, 36, *39–40*
        conservatism of, 55
        current contents of, 41
        on-line databases of, for ten
            disciplines, 360–73 (table)
        referees for, 56–57
        misquotes in, 57
    on-line computer searches of, 41,
        357–73
    reviews of groups of studies, 348–55
        evaluating findings of, 354
        organizing a strategy for, 350–51
        quantitative expressions of,
            351–53
        using visual displays to
            illustration data from, 353
    strategy for reading, 42
    writing for, 342–46
Scientific research. *See* Research
Scientist(s)
    average person as, 15–17
    characteristics of good, 17–20
    vs. data collectors, 18–20
    influence of values/interests of, on
        research, 12–14
Scorer and scoring
    based on many items, 84–86
    data collection and, 49, 144
    drift in, 148
    reducing errors in, 216
    reliability of, 82, 189
    in secondary analysis, 265–66, 267

Scott, J., 266
Scudder, S., 145–46
Seasonal affective disorder (SAD), 302
Sechrest, L., 91
Secondary analysis, 99, 104, 259–70
    advantages of, 260–61
    baseball abstract example, 263
    integrative complexity in, 267
    locating appropriate data, 261
    types of, 261–69
        content analysis, 265–68
        direct use of statistical material,
            262
        documents used to develop/
            support positions, 262–64
        meta-analysis, 265
        reanalysis of experiments and
            surveys, 264
    weaknesses of, 268–69
Selection, effects of, on data
        interpretation, 204
Self-reports, data collection by, 50,
        315–26
Seligman, M., 32, 302
Selye, Hans, 35, 63, 64–65
Semmelweiss, Ignaz, 34, 36
Sensitivity of measures, 81, 84–86
Sequential analysis, 250–58
    with interval data, 255–58
        closed design, 257–58
        open design, 255–57
    with ordinal data, 252–54
        closed design, 254–55
        open design, 252–54
    steps in sequential analysis, 252
Seredipity, 176
Setting, data collection and, 49–50, 145
Sexual Behavior in the Human Female
        (Kinsey), 171
Sexual Behavior in the Human Male
        (Kinsey), 171
Sharman, G., 306
Sharp, L., 160
Sheldon, W., 182
Sidman, M., 91, 242, 243, 244
Siegel, B., 295
Siegel, S., 222–23
Sigelman, L., 169
Sign test, 207–8, 214–15 (box), 243
Silverman, I., 221

Simon, H., 127
Simple random sampling, 158
Singer, J., 321
Single case experimental designs,
        242–44
Single subject research. *See* Case
        studies
Sjoberg, G., 278
Sjoholm, N., 241
Skinner, B., 53, 175
Skinner, Burrhus, 32
Smart, R., 165, 169, 287
Smoking and lung cancer, 101–6, 190
Social programs, evaluating design,
        implementation, and
        outcomes of, 271–89
Social Science Citation Index, 36,
        39–40
Social Science Research Council, 162
Sociological Abstracts, 36
Soderstrum, E., 284
Solomon, R., 218
Solomon design for experimental
        groups, 218–19
Spearman Rank-Order correlation,
        192–93 (box)
Speed, G., 266
Speers, W., 303
Sperry, R., 306
Split-half reliability, 49, 181
Sroges, R., 311, 312
Standard deviation, 210
Stanford-Binet test, 182
Stanley, J., 284, 285
Statistical significance, 51–54
Statistical significance, testing for,
        327–36
    effects of, on theory, 329–31
    measuring effect size and, 331
    null hypothesis and, 328–29
Statistical validity, problems of, 45,
        51–54, 56
Steinberg, L., 354
Stewart, D., 261
Stone, Calvin, 86, 87
Storrs, E., 306
Stout, A., 102
Stratified sampling, 159, 265
Stratton, G., 245

Stress
    conception and, 58–59
    drug use and, 197–98
    operational definitions of, 77–78
    performance affected by, 229
Strupp, H., 20, 243
Stuart, A., 158
Subject(s) of experiments. *See also* Animal subjects; Human subjects
    blocking, 241
    effects of independent variables dependent on changing, 240
    experimenter as, 245–49
    random assignment of, 208–9
    selection of, to decrease within-group variability, 212
        uniform treatment of subjects, 216
        using homogeneous subjects, 213, 216
Subjects of surveys, choosing individual, 157–59
Sudman, S., 161, 162, 163, 164, 165, 166, 168, 169
Suedfeld, P., 50
Sum of squares, 210
Surgeon General's Report, 106
Survey(s), 86, 88 (box), 99, 101, 152–73
    best use of, 100
    case study (Kinsey Report), 171–72
    choosing data collection method, 160–62
    generalizing results of, 220
    nonresponse to, 159–60
    political, 153, 155, 167, 168, 169–71
    questionnaires used in, 86, 162–67
        assuring confidentiality and truthfulness of, 117–18
    reasons for discrepancies in, 167–69
    secondary analysis of data from existing, 264, 265
    selecting samples for, 154–60
    sources of inaccuracy in, 169–71
Sutter, A., 149
Sutton-Smith, K., 116
Systematic desensitization, 317
Systematic sampling, 158
Szuk, E., 6

t

Taylor, C., 182
Taylor, S., 149
Teaching methods, factorial study of effects of, on math test scores, 236–37, 240
Technology, use of, in observational data collection, 144
Telephone interviews, 161–62
Terman, L., 171–72
Test(s). *See also* Item(s)
    draw-a-Person test (DAP), 141
    number of items in, and reliability calculations, 83 (box)
    objective (direct), 99
    projective (indirect), 99
    reliability of, evaluated with correlational studies, 49, 82, 181–83, 188 (box)
Test Anxiety Questionnaire, 181
Test-retest reliability, 49, 82, 181, 188 (box)
Tetlock, P., 266, 267, 268
Tey, Josephine, 264
Theory, 19
    animal-human comparisons made to generate hypotheses, 310
    falsifiable, 9, 320
    generating/testing hypothesis with observation, 138–39
    influence of intellectual values on models of, 13
    problems in relationship between data and, 54, 55 (box), 57
    rival hypotheses, 91
    testing of, 70, 202–3
Thirst, evaluation of measures illustrated with, 81–84, 86, 91
Thompson, M., 281
Thompson, W., 48
Time, data collection and, 49, 144
Tims, F., 275
Tinbergen, Niko, 32, 139
    on gull nest-cleaning function, 6, 69
Tokuhata, G., 106
Toner, J., 235
Toulmin, S., 8
Traugott, M., 153, 158, 160, 167
Trichopoulos, D., 58

Trivers, Robert, 12
Trotter, R., 32
Trower, P., 146
Truth, absolute, unattainable by science, 11–14
Tuma, A., 242
Twins studies, 243
2 × 2 factorial study
    computation procedures for, 238–39 (box)
    eight possible outcomes of, 233–35
    fictitious data for, 233 (table)
    F ratios, 240 (table)
Type I and Type II errors in experiments, 203, 205 (table), 206

u

Underwood, B., 221, 222
Unger, 49
U.S. Department of Health and Human Services, 103
U.S. Public Health Service, 102
Unity from diversity concept, 70–71
Unobtrusive measures. *See* Nonreactive measures
Unobtrusive observation, 49
Urey, Harold, 63

v

Validity of measures, 81, 86–93
    construct, 89–90
    criterion-related, 87–89
    dependent variable selection and, 90
    independent variable selection and, 90
    using multiple measures for, 91
    using nonreactive measures for, 92
Values of scientists, influence of, on research, 12–14
Value statements, 11
VanGundy, A., 127, 129
Variability in experiments, 209–18
Variable(s). *See also* Dependent variable (DV); Independent variable (IV)
    confound, 46, 150
    correlation between, 66–67 (*see also* Correlational studies)

dichotomous measurement of, 189–90
experimenter as, 241
linear relationship between, 179, *180, 181*
purpose of isolating, 217
sensitivity to complexity of, 219
value of experiments for controlling, 203
Variables, selecting and measuring, 76–94
classification principles, 80–81
properties of good measures, 81–93
reliability, 82–84
sensitivity, 84–86
validity, 86–93
scales of measurement, 79–80, 84–86
Variance
analysis of (*see* Analysis of variance)
calculating, 187
standard deviation and, 210
table of, between and within conditions, 210 (table)
Vigderhous, G., 160
Viney, L., 267
Visual displays of data from literature reviews, 353
Vokopenic, P., 159
Voltaire, 126
Von Felsinger, J., 117
Von Frisch, Karl, 307, 308–9
Voting behavior, 155 (table), 169–71
Vutue, C., 102

W

Waddington, C. H., 13
Wade, N., 12, 54
Wagner, R., 182
Wahlke, J., 153
Walberg, M., 349
Wald, A., 251
Walder, L., 292
Wallace, A. R., 4–5
Wallace, D., 266

Walton, Dennis, editorial on use of animals in research, 122–23
War, statistical analysis of causes of, 262–64
Warga, C., 31
Warner, K., 101, 163
Warren, K., 58
Warwick, D., 113
Watson, J., 42, 111
Watson, John, 317
"Weapon, The" (short story), 119–20
Webb, E., 91, 92, 105
Webb, N., 264
Weinberg, Steven, 6
Weiss, B., 58
Weiss, S., 280
Weller, L., 48
Werth, L., 321
Weschsler Adult Intelligence Scale, 181
Whimbey, A., 320
White, N., 143
White, R., 260, 262, 264
Wicker, A., 221
Wiebecke, B., 104
Wile, R., 279
Wilkens, B., 117
Willems, E., 139
Williams, R., 16, 306
Williams, S., 55–56
Willick, D., 169
Willner, A., 146
Willner, P., 302
Windle, C., 278
Winget, C., 84
Winkel, G., 221
Within-group variability, 209–10
evaluation of reductions in, 216–18
techniques for decreasing, 212–16
Within-subject design, 242–49
experimenter as subject, 245–49
sequential analysis, 251
single case experimental design, 242–44

Wohlwill, J., 293
Wolf, M., 146
Wolf, S., 241
Wolfe, A. B., 34
Wolins, L., 264
Women on Words and Images, 265–66
Women Studies Abstracts, 36
Wonderlic Personnel Test, 182
Woody, G., 283
Worden, M., 275
Wortman, P., 260, 261
Wright, H., 138
Writing research reports, 339–47
characteristics of good writing, 340–42
example, 374–80
qualities unique to science writing, 342–44
questions to ask, 346 (table)
for scientific journals, 344–46

Y

Yen, S., 58
Yukawa, Hideki, 132

Z

Ziman, J., 35, 340
Zimbardo, P., 73
Zimmerman, D., 161
Zingg, Robert Mowry, 150
Zirkle, C., 83
Zucker, I., 67, 69, 102, 301, 302
Zupkis, R., 116
z values, 400
transformation to correlation coefficient, 351, 401–2
Zytkow, J., 127